Lecture Notes in Artificial Intellig

Subseries of Lecture Notes in Computer Scienc
Edited by J. Siekmann

Lecture Notes in Computer Science

Edited by G. Goos and J. Hartmanis

G. Comyn   N. E. Fuchs
M. J. Ratcliffe (Eds.)

Logic Programming in Action

Second International Logic Programming
Summer School, LPSS '92
Zurich, Switzerland, September 7-11, 1992
Proceedings

Springer-Verlag

Berlin Heidelberg New York
London Paris Tokyo
Hong Kong Barcelona
Budapest

Series Editor

Jörg Siekmann
University of Saarland
German Research Center for Artificial Intelligence (DFKI)
Stuhlsatzenhausweg 3, W-6600 Saarbrücken 11, FRG

Volume Editors

Gérard Comyn
Michael J. Ratcliffe
European Computer-Industry Research Centre (ECRC)
Arabellastr. 17, W-8000 Munich 81, FRG

Norbert E. Fuchs
University of Zurich-Irchel, Computer Science Institute
Winterthurerstr. 190, Ch-8057 Zurich, Switzerland

CR Subject Classification (1991): I.2, D.1.6, F.4.1, D.2.1, J.1, J.6

ISBN 3-540-55930-2 Springer-Verlag Berlin Heidelberg New York
ISBN 0-387-55930-2 Springer-Verlag New York Berlin Heidelberg

Typesetting: Camera ready by author/editor
Printing and binding: Druckhaus Beltz, Hemsbach/Bergstr.
45/3140-543210 - Printed on acid-free paper

Preface

While the First Logic Programming Summer School, *LPSS '90*, addressed the theoretical foundations of logic programming, the Second Logic Programming Summer School, *LPSS '92*, focuses on the relationship between theory and practice, and on practical applications.

Logic programming enjoys a privileged position. On the one side, it is firmly rooted in mathematical logic, on the other side it is immensely practical as a growing number of users in universities, research institutes and industry are realising. Logic programming languages, specifically Prolog, have turned out to be ideal as prototyping and application development languages. Often, one defines an application-specific language that can be translated into a logic language. In this case, logic programming not only helps to conveniently define the syntax of the application-specific language, but also to express its semantics in a direct and understandable way.

There is an interplay between the theory and practice of logic programming that has been essential for its progress. In the introduction to this volume Robert Kowalski - one of the pioneers of the field - addresses this interplay, and identifies a number of problems where further research will be necessary to improve the relation between theory and practice. Much of this research is being done in the framework of the Basic Research Project *Compulog*, and the Network of Excellence in Computational Logic *Compulog-Net* of the European Community's ESPRIT program.

The interplay between theory and practice is also reflected in the relationships between logic programming and other fields of computer science, e.g. deductive databases, knowledge-based systems, computational linguistics, and software engineering. On the one side, these fields have borrowed concepts and methods from logic programming, while on the other they have strongly influenced its research directions. This has led to a strong synergy. To name only two examples, Prolog was originally developed for writing natural language processing applications, while knowledge-based systems continue to profit from the powerful metaprogramming techniques provided by logic programming.

The contributions contained in this volume fall into two categories: tutorials and project presentations. Four tutorials provide an overview of the relation of logic programming to constraint logic programming, deductive databases, language processing and software engineering as well as some theoretical background. Each topic is expanded by project presentations which give detailed accounts of existing applications, some of which are in the prototype stage, while others are in daily use.

In their tutorial, *Constraint Logic Programming - An Informal Introduction*, ECRC's *CORE* team give an insight into constraint logic programming which is a relatively new but rapidly expanding subfield of logic programming. Constraint logic programming (CLP) combines the power of logic programming languages with efficient constraint solving methods. CLP has proved to be extremely useful for scheduling, planning, and optimisation problems. This is borne out by the following project presentations. Michel d'Andrea, in his contribution *Scheduling and Optimisation in the Automobile Industry*, describes a prototype for a job scheduling system developed by Bull for the Renault group. In *Factory Scheduling using Finite Domains,* Owen Evans shows the advantages that constraint logic programming offers for solving problems in a factory environment using the DecisionPower system sold by ICL. Finally, Pierre-Joseph Gailly and his colleagues report on the ESPRIT *Prince Project and Its Applications* in which a practical constraint solving system based on the logic programming language Prolog III is being developed.

Logic programming has always had strong relations with deductive databases and expert systems. In his tutorial, *A (Gentle) Introduction to Deductive Databases,* Shalom Tsur recalls how the limitations and weaknesses of relational databases, especially of relational query languages, led to the ideas of deductive databases, and points out the many interconnections to logic programming. Christoph Beierle presents *Knowledge Based PPS Applications in PROTOS-L* which shows how an enhanced Prolog developed in the context of the Protos Eureka project can be used for Knowledge-Based scheduling applications. Carlo Chiopris describes the development of the *SECReTS Banking Expert System from Phase 1 to Phase 2;* the application is being used by several Italian banks for the analysis of client data. In his contribution, *Logic Engineering and Clinical Dilemmas,* John Fox, who works at the Imperial Cancer Research Fund, focuses on the advantages of logic programming for clinical decision making, while Edward Freeman shows how *A Knowledge-Based Approach to Strategic Planning* helps corporations define their strategic directions based on models that relate critical business factors to business targets. In cooperation with the German mining industry, Lutz Plümer developed two *Expert Systems in Mining,* that are near practical applications: *Schikorre* helps to locate geological seams, while *BUT* solves the planning problem for underground illumination.

As mentioned above, the processing of natural language led to the development of Prolog, i.e. logic programming and language processing have been related from the very beginning. In his tutorial *Natural and Formal Language Processing,* Michael Hess identifies machine translation, interaction with computers in natural languages, and accessing information in natural language as three main goals of natural language processing, and shows how logic programming continues to contribute to achieve these goals. In her project presentation, Deborah Dahl introduces *Pundit - Natural Language Interfaces.* To be domain independent, Pundit consists of a number of modules that separately perform the tasks of syntactic, semantic and pragmatic analysis. The *ESTEAM-316 Dialogue Manager* presented by Thomas Grossi, Didier Bronisy and François Jean-Marie model a part of human dialogue, viz. advice giving in the domain of financial investments. Robert Kowalski points out the syntactic similarities of legal language and logic programming languages, and shows how the formalisation of *Legislation as Logic Programs* suggests ways in which logic programming could be extended. Knowledge representation is essential in natural language processing. Udo Pletat presents in *Knowledge Representation for Natural Language Processing* the knowledge representation formalism L$_{LILOG}$ which has the power of first-order predicate logic and offers a type system similar to the one in KL-ONE. Language processing is not restricted to natural language alone. Peter Reintjes has developed *A Set of Tools for VHDL Design* which convincingly demonstrates Prolog's strength as an implementation language for language-oriented work in general, and hardware description languages in particular.

Software engineering is another field that profits enormously from the power and conciseness of logic programming languages. Keywords that come immediately to mind are executable specifications, program synthesis and program transformations. Based on the great experience of his many years in the field, Alan Bundy shows how reasoning about logic programs helps to improve the efficiency and the reliability of programs. In his paper, *Tutorial Notes: Reasoning About Logic Programs,* he presents a unified view that encompasses the problems of verification, termination, synthesis, transformation, and abstraction. Formal specifications in logic programming languages are the topic of Abdel Ali Ed-Dbali and Pierre Deransart. In their contribution *Software Formal Specification by Logic Programming: The Example of Standard Prolog* they use as a concrete example the formal specification of the language Prolog itself. This work is part of the emerging international Prolog standard. One of the largest problems facing software engineering is the mass of existing programs, many of them badly or not at all documented. Peter Breuer presents a set of tools demonstrating *The Art of Computer Un-Programming: Reverse Engineering in Prolog* These tools were developed with the goal of improving the comprehensibility and maintainability of existing COBOL programs. A variety of methods for debugging have been suggested in the logic programming community. Though extremely powerful, these methods are not necessarily practical. Mireille Ducassé describes *OPIUM - An Advanced Debugging System* that is based on traces of program executions and combines the power of logic programming with great practicality.

Strangely enough, teaching is not normally considered as an application field though its importance cannot be underestimated. In the framework of the Swiss National Research Project NFP 23, Fabio Baj and Mike Rosner have developed *Automatic Theorem Proving within the Portable AI Lab.* This theorem proving tool helps to teach basic and advanced topics of logic and logic programming.

Producing the contributions for this volume has involved a lot of time and expense. Several organisations have generously supported us in this and we gratefully acknowledge their contributions:

    Bull
    Commission of the European Communities
    European Computer-Industry Research Centre (ECRC)
    ICL
    Industrade AG (Apple Computer Division)
    University of Zurich

The papers appearing in this volume demonstrate convincingly that logic programming fruitfully combines theory and practice. Realistic applications have already been successfully constructed using logic programming languages. We hope this volume will provide inspiration for others in the future.

Gérard Comyn
Norbert E. Fuchs
Michael Ratcliffe

July 1992

Contents

Introduction

Constraint Logic Programming

Deductive Databases and Expert Systems

Processing of Natural and Formal Languages

Software Engineering

Education

Theory and Practice in Logic Programming

Robert Kowalski

Department of Computing, Imperial College
London, England, U.K.
April 1992

Abstract. Logic Programming enjoys a relatively good relationship between its theory and its practice. Nonetheless, this relationship needs to be improved, and doing so is an important direction for research in the future. The European Community Basic Research Project, Compulog, and the more general "network of excellence", Compulog-net, are concerned with developing such improvements.

1 Procedural versus Declarative Interpretations

The procedural interpretations of Horn clauses and of negation as failure are the basis for both the theory and practice of logic programming. For many applications (e.g. databases and program specifications) the declarative view needs to dominate the procedural. For other applications, the procedural is more important. In many cases, both views are necessary and a smooth progression and interrelationship between the two is necessary. Achieving a harmonious balance is not always as easy in practice as it should be in theory.

Two areas where future research would be useful are improving the data structures and improving the link with object-orientation. Array-like data structures supporting destructive assignment are convenient in practice. At present the theory allows recursive data structures and various approximations of arrays. Sets of clauses, viewed as updatable databases, are a promising alternative.

Many suggestions have been proposed for combining logic programming with object-orientation. In some of these proposals objects are interpreted as terms; in others as predicates; in still others as "theories" or sets of clauses. All of these proposals and their relationships need to be investigated further.

2 Metaprogramming

Programs which manipulate other programs (or sets of clauses) are an important logic programming technique, used for such applications as providing metadata, implementing metainterpreters, programming in the large, and distributed intelligent systems. The Gödel logic programming language is currently under development in Compulog, motivated to a large extent by the goal of providing improved metaprogramming facilities. Additional work is necessary to reconcile the metalogical techniques which have proved useful in practice with the foundations that are needed in theory.

3 Negation as failure

Non-monotonic reasoning is necessary for many applications, including temporal reasoning in artificial intelligence and database systems. In recent years it has been recognised that negation as failure in logic programming provides a practically effective and theoretically sound technique for non-monotonic reasoning. Further research is necessary to understand better the relationships between different semantics for negation as failure and to develop appropriate extensions for disjunctive reasoning, integrity constraints, and the combination of explicit negation and implicit negation as failure.

4 Abduction

Very recently the extension of logic programming to include abductive (hypothetical) reasoning has begun to be investigated. This extension is related to other extensions such as constraint logic programming and conditional answers. It can also be used for non-monotonic reasoning and negation as failure. Further work is needed to relate better its semantics (viewed as a program specification) with its implementation.

5 Program optimisation

One of the main purposes of semantics is to provide a foundation for proving program equivalence and to justify program transformations and optimisations. Such transformation and optimisation can make a major contribution to improving programmer productivity. A number of powerful optimisation methods have been investigated. Much more can be done to put the theory into practice.

6 Wider implications

Logic programming, appropriately extended (e.g. with explicit negation, disjuction, abduction), begins to achieve the expressiveness of a complete, symbolic knowledge representation formalism. It has proved especially promising for formalising legal reasoning. This application is important, both because legal reasoning can be regarded as prototypical of practical reasoning in general, and because rule-based legal reasoning integrates naturally and comfortably with other kinds of reasoning, including case-based reasoning with open-textured concepts. The strong links between logic programming and legal reasoning provide evidence that logic programming may one day prove as useful for computing as legal reasoning is for human affairs. More importantly, it may help us better to achieve the goals of human logic itself: to reason more clearly and effectively as human beings, even without the use of computers.

Related Reading

1. C. Hogger, R. Kowalski: Logic Programming. In Encyclopedia of Artificial Intelligence (ed. S. Shapiro), (second edition, 1992) Vol. 1 (A-L), pp. 873-891

2. R. Kowalski: Problems and Promises of Computational Logic. In Proceedings Symposium Computational Logic (ed. J. Lloyd), Springer-Verlag 1990, pp. 1-36

Constraint Logic Programming
-
An Informal Introduction*

Thom Frühwirth, Alexander Herold, Volker Küchenhoff,
Thierry Le Provost, Pierre Lim, Eric Monfroy, Mark Wallace

ECRC
European Computer-Industry Research Centre
Arabellastr. 17, D-8000 Munich 81, Germany

email: {thom, herold, volker, thierry, pierre, eric, mark}@ecrc.de

Abstract. Constraint Logic Programming (CLP) is a new class of programming languages combining the declarativity of logic programming with the efficiency of constraint solving. New application areas, amongst them many different classes of combinatorial search problems such as scheduling, planning or resource allocation can now be solved, which were intractable for logic programming so far. The most important advantage that these languages offer is the short development time while exhibiting an efficiency comparable to imperative languages. This tutorial aims at presenting the principles and concepts underlying these languages and explaining them by examples. The objective of this paper is not to give a technical survey of the current state of art in research on CLP, but rather to give a tutorial introduction and to convey the basic philosophy that is behind the different ideas in CLP. It will discuss the currently most successful computation domains and provide an overview on the different consistency techniques used in CLP and its implementations.

1 Introduction

During the last decade a new programming paradigm called *"logic programming"* has emerged. The best known representative of this new class of programming languages is *Prolog*, originated from ideas of Colmerauer in Marseille and Kowalski in Edinburgh. Programming in Prolog differs from conventional programming both stylistically and computationally, as it uses logic to declaratively state problems and deduction to solve them.

It has been argued in the literature [Kow79, Ste80] that a program is best divided into two components called *competence* and *performance* or *logic* and *control*. The competence component describes factual information - statements of relationships - which must be manipulated and combined to compute the desired result. The performance component deals with the strategy and control of the manipulations and combinations. The competence part is responsible for the correctness of the program; the performance part is responsible for the efficiency. An ideal programming

---

* This work is partially funded by the ESPRIT project CHIC, Nr. 5291

methodology would first be concerned with the competence (*"what"*), and only then, if at all, worry about the performance (*"how"*). Logic programming provides a means for separation of these concerns. It is based on *first order predicate logic*, and the performance component is mostly automatic by relying on a built-in computation mechanism called *SLD-resolution*.

In this way, logic programming has the unique property that its *semantics*, operational and declarative, are both simple and elegant and coincide in a natural way. These semantics, however, have their limitations. Firstly the objects manipulated by a logic program are uninterpreted structures - the set of all possible terms that can be formed from the functions and constants in a given program. Equality only holds between those objects which are syntactically identical. Every semantic object has to be *explicitly* coded into a term; this enforces reasoning at a primitive level. *Constraints* on the other hand are used to *implicitly* describe the relationship between such semantic objects. These objects are often ranging over such rich computation domains, as integers, rationals or reals.

The second problem related to logic programming stems from its uniform but simple computation rule, a depth-first search procedure, resulting in a *generate and test* procedure with its well-known performance problems for large search applications. Constraint manipulation and propagation have been studied in the Artificial Intelligence community in the late 1970s and early 1980s [Mon74, Ste80, Mac86] to make search procedures more intelligent. Techniques like local value propagation, data driven computation, forward checking (to prune the search space) and look ahead have been developed for solving constraints. These techniques can be summarised under the heading *"Consistency Techniques"*.

Constraint Logic Programming (CLP) is an attempt to overcome the difficulties of logic programming by enhancing a Prolog-like language with constraint solving mechanisms. Curiously both of these limitations of logic programming can be lifted using "constraints". However, each limitation is treated by a quite different notion of constraint. CLP has hence two complementary lines of descent.

Firstly it descended from work that aimed at introducing richer data structures to a logic programming system thus allowing semantic objects, e.g. arithmetic expressions, directly to be expressed and manipulated. The core idea here is to replace the computational heart of a logic programming system, unification, by constraint handling in a constraint domain. This scheme, called CLP(X), has been laid out in the seminal paper of Jaffar & Lassez [JL87]. X has been instantiated with several so called computation domains, e.g. reals in CLP($\mathcal{R}$), rationals in CLP($\mathcal{Q}$), and integers in CLP($\mathcal{Z}$).

Secondly CLP has been strongly influenced by the work on consistency techniques. With the objective of improving the search behaviour of a logic programming system Gallaire [Gal85] advocated the use of these techniques in logic programming. He proposed the active use of constraints, pruning the search tree in an a priori way rather than using constraints as passive tests leading to a "generate and test" or "standard backtracking" behaviour. Subsequently the different inference mechanisms underlying the finite domain part of the CLP system CHIP [DVS+88] were developed. The key aspect is the tight integration between a deterministic process, constraint evaluation, and a nondeterministic process, search. It is this active view of constraints which is exploited in CHIP to overcome the well-known performance problems of "generate and test". This new paradigm exhibits a data-driven computation and can be characterised as *"constrain and generate"*.

Constraint solving has been used in many different application areas such as engineering, planning or graphics. Problems like scheduling, allocation, layout, fault diagnosis and hardware design are typical examples of constrained search problems. The most common approach for solving constrained search problems consists in writing a specialised program in a procedural language. This approach requires substantial effort for program development, and the resulting programs are hard to maintain, modify and extend. With CLP systems a large number of constrained search problems have been solved, some of them were previously solved with conventional languages. CLP languages dramatically reduce the development time while achieving a similar efficiency. The resulting programs are shorter and more declarative and hence easier to maintain, modify or extend. The wealth of applications shows the flexibility of CLP to adapt to different problem areas. Many Operations Research problems have been solved with the CLP system CHIP [DVS+88, Van88, DSV90]. Another very promising application domain is circuit design [Sim92, FSTW91]. Extensive work has also been devoted to financial applications [Ber89, LMY87]. More recently applications in user interfaces [HHLM91] and in databases [KKR90] have been studied. As the subsequent tutorial in this summer school focusses on industrial applications of CLP, we will not further discuss them in this article.

The aim of this informal tutorial is to present the most prominent ideas and concepts underlying CLP languages. It is not intended to present the underlying theory of this new class of programming languages or to give an overview on the current state of art in CLP research. There are already technical surveys in the literature, giving more details on those aspects. In particular the article of [Van91] is worth reading. A restricted view is presented in [Coh90, Frü90] discussing work around the CLP scheme. For the usage of "consistency techniques" in CLP, [Van89] is a valuable source going from theory to application with a large number of programming example.

This tutorial is organised as follows: In the next section we will introduce the CLP scheme and review the most important computation domains that have been developed so far, linear and non-linear arithmetic and boolean constraints. Then we will introduce the concept of finite domains, consistency techniques and their extension to arbitrary domains. Next we will explore ways of extending and tuning constraint systems. Then the work on search and optimisation in CLP will be presented. Finally current CLP implementations will be reviewed, amongst them the most well-known systems: CHIP [DVS+88], CLP($\mathcal{R}$) [JMSY90] and Prolog III [Col90].

2 The CLP Scheme

In this section we will introduce in an informal way the basics of the Constraint Logic Programming Scheme (called CLP(X)), as developed by Jaffar and Lassez [JL87]. The key aspect in the CLP scheme is to provide the user with more expressiveness and flexibility concerning the primitive objects the language can manipulate. Clearly the user wants to design his application using concepts that are as close as possible to his domain of discourse, e.g. he wants to use sets, boolean expressions, integers, rationals or reals, instead of coding everything as uninterpreted structures, i.e. finite trees, as is advocated in logic programming. Associated with each computation domain are the usual algebraic operations, including set intersection, conjunction of boolean expressions or multiplication of arithmetic expressions. These computation domains also have certain relations defined on them, such as set equality, equality between boolean expressions or equality, disequality and inequality between arithmetic expressions.

The constraint logic programming scheme admits computation directly over these domains. Special function and predicate symbols are introduced into logic programming, whose interpretation in the domain of computation is fixed. The relations over the domain of discourse are termed "constraints". Formulae involving the special function and predicate symbols are called "constraint formulae". Informally the word "constraint" is used also for constraint formulae.

When constraints are introduced into logic programming, a mechanism to solve them must also be introduced. In traditional logic programming the only constraint is equality between terms, and the unification algorithm is used to solve such constraints. There are two aspects related to unification. Firstly it tells us if the equation $t_1 = t_2$ has a solution. Secondly in case there exists a solution, it gives us a most general solution, which is logically equivalent to the original equation. The important aspect of unification is the first one deciding whether a constraint (or a set of constraints) has a solution or not. In other computation domains, where such a most general solution may not exist, the system can continue manipulating the original set of constraints. Therefore in order to accommodate constraints in logic programming the unification algorithm needs to be replaced by a decision procedure telling us whether a constraint or a set of constraints is satisfiable. In the following we will call such a decision procedure a *constraint solver*.

One reason for the success of CLP in recent applications has been the choice of constraint systems integrated into the different implementations. The selection of new constraint domains needs to satisfy both technical and practical criteria [DVS+88, JL87, SA89]. Most important are

- the expressive power of the computation domain,
- the existence of a complete and efficient constraint solver,
- its relevance in applications.

The constraint solver is complete if it is able to decide the satisfiability of any set of constraints of the computation domain. To achieve efficiency the constraint solver needs to be incremental, i.e. when adding a new constraint C to an already solved set of constraints S, the constraint solver should not start solving the new set $S \cup \{C\}$ from scratch.

In the following we will illustrate the operational behaviour of a CLP(X) system and the two most successful constraint domains, arithmetic and Boolean constraints. A description of other interesting domains may be found in section 6 where specific constraint languages are described.

2.1   The Arithmetic Domain

Linear Constraints Providing arithmetic was one of the motivations behind the research in combining logic programming with constraints. Although Prolog has built-in facilities for evaluating arithmetic expressions the behaviour is not what one would ideally expect. Prolog cannot handle equations like $X - 3 = Y + 5$. In Prolog the term $X - 3$ is not equal to the term $Y + 5$ as Prolog knows only about uninterpreted structures. The programmer needs to resort to the built-in arithmetic. And here the problems are the same as in any other programming language. Indeed the programmer needs to know which of the variables will be instantiated first and then he can use assignment (is) to instantiate the other. CLP($\mathcal{R}$) [JL87] was the first constraint programming language to introduce arithmetic constraints. There is a caveat. The decision procedure is only complete for linear arithmetic constraints. Nonlinear constraints are suspended until they become linear. Linear constraint handling turned out to be sufficient in many applications such as simulation of circuits and devices, decision-support systems and geometrical problems.

Linear arithmetic expressions are terms composed from numbers, variables and the usual arithmetic operators: negation $(-)$, addition $(+)$, subtraction $(-)$, multiplication $(*)$ and division $(/)$. For the condition of *linearity* to be satisfied it is required that in a multiplication at most one of the components is a variable and that in a division the denominator is a number. An arithmetic constraint is an expression of the form $t_1 \; R \; t_2$ where R is one of the following predicates $\{>, \geq, =, \leq, <, \neq\}$.

There are several decision procedures for deciding a system of linear arithmetic constraints. Usually a combination of Gaussian elimination and a modified *Simplex* algorithm is employed. The Simplex algorithm is required as soon as inequality constraints need to be solved. The Simplex algorithm is used because it has quite a good behaviour on average, it is well-understood, and it can be made incremental.

We now present the execution mechanism for CLP languages informally through a small example. Consider the following problem from [Col90].

> *Given the definition of a meal as consisting of an appetiser, a main meal and a dessert and a database of foods and their calorific values we wish to construct light meals i.e. meals whose sum of calorific values does not exceed 10.*

A CLP program (in an arithmetic domain) for solving this problem is given below.

```
lightmeal(A,M,D) :-
    I > 0, J > 0, K > 0,
    I + J + K <= 10,
    appetiser(A,I),
    main(M,J),
    dessert(D,K).

main(M,I) :-
    meat(M,I).
main(M,I) :-
    fish(M,I).
```

```
appetiser(radishes,1).
appetiser(pasta,6).

meat(beef,5).
meat(pork,7).

fish(sole,2).
fish(tuna,4).

dessert(fruit,2).
dessert(icecream,6).
```

A CLP program is syntactically a collection of *clauses* which are either *rules* or *facts*. Rules are as in Prolog, with the addition that they may contain constraints in their premises. Rules describe the conclusions that can be reached given certain premises. For our example we read "The meal consisting of foods A, M and D is a light meal if A is an appetiser (with a positive calorific value I), M is a main meal (with positive calorific value J), D is a dessert (with positive calorific value K) and I + J + K is less than or equal to 10". The premise of a rule is a conjunction of *constraints*, e.g. I + J + K <= 10 and *atoms* e.g. appetiser(A,I). Facts express known relationships. In our case, the calorific value of beef (which is a meat) is 5.

We shall describe the intermediate results of an execution of a CLP program as *computation states*. A computation state consists of two components, a *constraint store* and the remaining goals. We shall separate the constraint store from the remaining goals by the symbol ◇. The constraint store consists of the set constraints collected during the computation so far. CLP programs are executed by reducing the goals in the computation state using the facts and rules. In each intermediate computation state the constraint store must be consistent. Consider the general query ?- lightmeal(A ,M ,D) asking for all light meal plans. This corresponds to the initial computation state

```
◇ lightmeal(A, M, D).
```

For our first reduction step we first have to choose an atomic goal to reduce. There is only one possibility i.e. lightmeal(A, M, D). Next we need to choose an applicable rule. Again there is only one possibility i.e. the rule with the consequent lightmeal(A, M, D). The next step is to form equations between variables in the consequent of the rule and the selected atom. The constraint store of the new computation state consists of the current constraint store, this equation set and the set of constraints in the premise of the rule. The atom set of the new computation state is the current atom set where the selected goal is replaced by the atoms of the premise of the rule (as in the case of Prolog). Thus our first reduction step produces the following computation state:

```
I + J + K <= 10, I > 0, J > 0, K > 0 ◇
appetiser(A,I), main(M,J), dessert(D,K).²
```

---

² In the examples trivial equations are omitted

A CLP system *searches* for all solution by systematically trying all possible rules (and facts) for the reduction of all the atoms in the atom set. Therefore any one possible alternative is in fact a sequence of reduction steps called a *derivation*. A derivation terminates when there are no more atoms to be reduced and the final constraint store is consistent. For the first example a *successful* derivation is the following:

```
A=radishes, I=1, 1+J+K <= 10, 1>0, J>0, K>0
◇ main(M, J), dessert(D, K)

A=radishes, I=1, M=M1, J=I1, 1+J+K <= 10, 1>0, J>0,K>0
◇ meat(M1, I1), dessert(D, K)

A=radishes, I=1, M=beef, J=5, M1=beef, I1=5, 1+5+K <= 10, 1>0, 5>0, K>0
◇ dessert(D, K)

A=radishes, I=1, M=beef, J=5, M1=beef, I1=5, D=fruit, K=2, 1+5+2 <= 10,
1>0, 5>0, 2>0 ◇.
```

Note that the answer to this query is given by the constraint store. A simplified answer in terms of the input variables is `A=radishes, M=beef, D=fruit`.

If the constraint store becomes inconsistent, the derivation *fails*. An example of a failed derivation is now presented. We begin with the same initial computation state as above but make some different choices in the rules and facts to apply.

```
A=pasta, I=6, 6+J+K <= 10, 6>0, J>0, K>0
◇ main(M,J), dessert(D,K)

A=pasta, I=6, M=M1, J=I1, 6+J+K <= 10, 6>0, J>0, K>0
◇ meat(M1,I1), dessert(D,K)
          .
A=pasta, I=6, M=beef, J=5, M1=beef, I1=5, 6+5+K <= 10, 5>0, 6>0, K>0
◇ dessert(D,K) (inconsistency)
```

If the last computation state for this derivation is examined it can be seen that the constraint store containing `6+5+K <= 10` and `K > 0` is not satisfiable.

The answer `A=radishes, M=beef, D=fruit` is *definite* in the sense that a constant is equated with each variable in the query. However, in general answers can also be *indefinite*, i.e. the answer consists of a set of constraints representing a possibly infinite set of solutions. An example of this kind will be presented a little later when nonlinear constraints are discussed. How to extract an understandable answer from the constraints in the constraint store is an active field of research [JMSY92].

Nonlinear constraints To introduce nonlinear arithmetic constraints we shall use a program multiplying two complex numbers R1 + I*I1, R2 + I*I2 taken from [JL87]:

```
zmul(R1, I1, R2, I2, R3, I3):-
     R3 = R1*R2 - I1*I2,
     I3 = R1*I2 + R2*I1.
```

If the query `zmul(1,2,3,4,R3,I3)` is given, then the nonlinear equations become linear at run time, and the answer produced by e.g. CLP($\mathcal{R}$) is:

```
R3 = -5
I3 = 10
```

*** Yes

If we ask the query `zmul(1,2,R2,I2,R3,I3)`, the solution is a conjunction of two linear equalities:

```
I2 = 0.2*I3 - 0.4*R3
R2 = 0.4*I3 + 0.2*R3
```

*** Yes

This answer is an example for an indefinite solution. The solution is an infinite set of points that is represented by a minimal set of constraints stating relations between the variables of the query. To obtain precise values for I2 and R2 (i.e. to obtain I2 equal to a constant and R2 equal to a constant), the user has to further instantiate I3 and R3.

For the two previous queries, there is no need for a nonlinear solver. But for the query `zmul(R1,2,R2,4,-5,10)`, R2 < 3 nonlinear constraints appear in the solution. CLP($\mathcal{R}$) gives the answer:

```
R1 = -0.5*R2 + 2.5
3 = R1 * R2
R2 < 3
*** Maybe
```

This is due to the property of CLP($\mathcal{R}$), whose decision procedure can only solve linear arithmetic. When a nonlinear constraint is encountered during computation, then it is delayed until it becomes linear. For the previous query, two nonlinear equations are encountered during computation. They are delayed, but no instantiation makes them linear. So at the end of the computation CLP($\mathcal{R}$) gives back the delayed constraints without knowing if there are some solutions or not (***** Maybe**).

This introduces the need for nonlinear arithmetic solvers in constraint logic programming. Nonlinear constraints arise for instance in computational geometry [PS85], and financial applications. Several algorithms can be used to solve nonlinear constraints. Their capacities and complexities are quite different (see [Mon92a] for a comparison of different solvers). For example Gröbner bases [Buc85] treat only equations whereas quantifier elimination [Col75] can handle all (well formed) formulae over the reals at, sometimes, considerable extra cost.

For the first two queries of the previous example (multiplication of complex numbers) the answer given by nonlinear solvers is the same as the one from CLP($\mathcal{R}$). But the last query `zmul(R1, 2, R2, 4, -5, 10)`, R2 < 3 is completely solved, and the answer is definite:

```
R1 = 1.5
R2 = 2
```

Gröbner Bases are used in CAL [AH92], and in the system of [Mon92b]; and an improved version of quantifier elimination [Hon90] is used in RISC-CLP [Hon92].

2.2 The Boolean Domain

The most prominent applications of boolean constraints are in the area of circuit design [Sim92], here in particular hardware verification [FSTW91], and in *theorem proving* in the domain of propositional calculus [SD90, Col90]. Such applications motivated the incorporation of boolean constraint solvers into constraint logic programming languages.

Boolean terms are built from the *truth values* (*false* and *true*, represented sometimes also by 0 and 1), from *variables* and from *logical connectives* (e.g. $\vee$, $\oplus^3$, $\wedge$, *neg*). The only constraint between boolean terms is the equality (=). In some implementations (e.g. CHIP) additional constants can be used in the construction of terms. This is particularly important in hardware verification as these constants can be used to represent symbolic names for input arguments of circuits.

Each of the systems mentioned above employs quite different ways of handling boolean constraints. A *Boolean unification* algorithm [BS87] is used in the case of CHIP. In the literature a number of different unification algorithms for Boolean constraints are reported [MN90, Bue88]. Another possibility is to implement boolean constraint solving as a special case of numerical constraint solving. A modified version of the Gröbner bases algorithm [ASS+88] is used in CAL. Prolog III uses a saturation method to solve boolean constraints [Col90]. This method does not compute a most general solution and is hence not easily applicable to circuit verification. Since boolean constraint solving provides a decision procedure for propositional calculus and is therefore NP-complete, any algorithm for boolean constraints has an exponential worst case complexity. It is thus very important to use a compact description of boolean terms to achieve efficiency. CHIP [DVS+88], for example, represents boolean terms as directed acyclic graphs, which are manipulated by special purpose graph algorithms [Bry86].

The following classic example coming from hardware verification illustrates how boolean constraints can be solved by boolean unification.

% Full-adder circuit example
$add(I1, I2, I3, O1, O2) :-$
  $X1 = I1 \oplus I2,$
  $A1 = I1 \wedge I2,$
  $O1 = X1 \oplus I3,$
  $A2 = I3 \wedge X1,$
  $O2 = A1 \vee A2.$

---

[3] $\oplus$ is the exclusive or

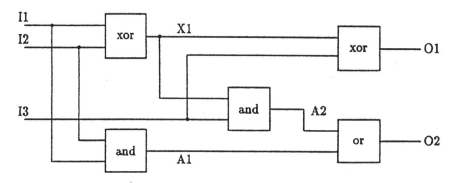

Figure 1: Full Adder Circuit

The computation of an answer to the query $add(a, b, c, O1, O2)$ gives the following set of intermediary constraints:

$$X1 = a \oplus b$$
$$A1 = b \wedge a$$
$$O1 = a \oplus b \oplus c$$
$$A2 = c \wedge (a \oplus b)$$
$$O2 = a \wedge b \oplus a \wedge c \oplus b \wedge c.$$

The boolean solver hence produces the answer:

$$O1 = a \oplus b \oplus c, \quad O2 = a \wedge b \oplus a \wedge c \oplus b \wedge c$$

which describes the logical function of the piece of hardware. The output parameters are expressed as boolean expressions constructed from the input parameters. These boolean expressions can now be compared with the specification of the circuit, which is also expressed in terms of boolean expressions.

In case of hardware verification the full power of boolean unification is needed. But obviously boolean unification is a very costly method. For simulation tasks for instance, where the input parameters are not symbolic constants but the ground values 0 or 1, this power is not needed and other methods are more efficient. In section 3.2 and 4.2 we will describe such other techniques.

3 Consistency Techniques

3.1 Finite Domains

Consistency techniques were first introduced for improving the efficiency of picture recognition programs, by researchers in artificial intelligence [Wal72]. Picture recognition involves labelling all the lines in a picture in a consistent way. The number of potential labellings can be huge, while only very few are consistent.

Consistency techniques effectively rule out many inconsistent labellings at a very early stage, and thus cut short the search for consistent labellings. These techniques have since proved to be effective on a wide variety of hard search problems, made even wider since their integration into a logic programming framework in CHIP and subsequent CLP implementations.

The handling of constraints using consistency techniques is unlike constraint solving in the CLP Scheme, as described earlier, in that it does not guarantee to detect inconsistency of the (global) constraint store until the labelling of the problem variables is complete. Instead consistency techniques provide an efficient way to extract from the constraint store new information about the problem variables.

A Scheduling Example To illustrate one such consistency technique let us take a very simple scheduling problem, with six tasks to be scheduled into a five-hour day, where each task takes an hour. The following diagram shows tasks on the left which must precede other tasks on the right:

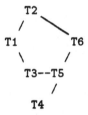

In addition we impose the constraint that tasks $T2$ and $T3$ cannot be scheduled at the same time.

To express this as a constraint satisfaction problem, we associate a variable Ti with the start time of each task, whose domain of possible values is $\{1, 2, 3, 4, 5\}$. We then impose the constraints

```
before(T1,T2)
before(T1,T3)
before(T2,T6)
before(T3,T5)
before(T4,T5)
before(T5,T6)
notequal(T2,T3)
```

Consistency techniques work by *propagating* information about the variables via the constraints between them. For example given that $T1 \in \{1, 2, 3, 4, 5\}$ and that $T2 \in \{1, 2, 3, 4, 5\}$, then based on the constraint **before(T1,T2)** our consistency technique deduces the information that $T1 \in \{1, 2, 3, 4\}$ and $T2 \in \{2, 3, 4, 5\}$. The value 5 is removed from the domain of $T1$ because there is no value in the domain of $T2$ which is consistent with it - that satisfies the constraint **before(T1,T2)**. The value 1 is removed from the domain of $T2$ for the same reason. (This consistency technique, which removes values inconsistent with a single constraint between two variables, is termed *arc consistency* [Mac77].)

Propagation continues until no further new domain reductions can be extracted from the constraints. The effect of applying arc consistency in our example is to reduce the domains associated with the tasks' start times as follows:

$$T1 \in \{1,2\}, T2 \in \{2,3,4\}, T3 \in \{2,3\}, T4 \in \{1,2,3\}, T5 \in \{3,4\}, T6 \in \{4,5\}$$

Consistency techniques alone can rarely be used to solve a problem, since in general there remain combinations of values in the resulting domains which are inconsistent. For example the constraint before(T1,T2) has been used during propagation to reduce the domains of $T1$ and $T2$, but it is still not satisfied by all the values of the resulting domains of $T1$ and $T2$. (Although $T1 = 2$ is consistent with some values in the domain of $T2$, it is not consistent with the value $T2 = 2$.)

To find a solution to this scheduling problem the system therefore performs some search, by labelling a variable with some value in its domain (search is discussed in detail in section 5 below). This choice (which may prove later to have been erroneous), allows further propagation to be attempted. For example suppose $T1$ is labelled with the value $T1 = 2$. Propagation yields $T2 \in \{3,4\}$ and $T3 \in \{3\}$. At this point the constraint notequal(T2,T3) is used actively for the first time to produce the information $T2 \in \{4\}$. Propagation continues until the following information has been extracted: $T1 \in \{2\}, T2 \in \{4\}, T3 \in \{3\}, T4 \in \{1,2,3\}, T5 \in \{4\}, T6 \in \{5\}$.

Propagation versus Solving The treatment of the notequal constraint with arc consistency is a typical example of how and why consistency techniques differ from constraint solving. If variables X and Y each have domains with more than one value, then the constraint notequal(X,Y) will not yield any new information. The reason is that every value in the domain of Y will be consistent with at least one value in the domain of X, and vice versa. Propagation on the constraint notequal can be implemented very efficiently. The constraint yields no information until one of the variables has a domain with only one remaining value. This value is immediately removed from the domain of the other variable, and the constraint is satisfied. It can never again yield new information.

However if the constraint notequal is handled by a constraint *solver* it can yield more information than propagation. For example suppose variable X, Y and Z all have two-value domains: $X \in \{1,2\}, Y \in \{1,2\}, Z \in \{1,2\}$. The constraints

notequal(X,Y) notequal(Y,Z) notequal(Z,X)

are not satisfiable. Although this is detected by a solver for the notequal constraint, arc consistency yields no information.

For simple examples, such as this, the solver can detect the inconsistency at little cost. However non-trivial problems involve a reasonably large number of constraints and domains containing a reasonably large number of values; and in this case the cost of solving the constraints increases very quickly (exponentially) with the number of variables involved. For such problems it is often too expensive to attempt constraint solving on the notequal constraints, and constraint propagation proves to be a more effective technique.

Constraint Driven Computation Consistency techniques extend the notion of data driven program execution. The arrival of "data" no longer means the arrival of a specific value for a variable, but rather any reduction of the domain associated with the variable. We call it *constraint driven*. In this framework new "data" may arrive many times on a single variable - each time its domain is reduced. Much research has been published on constraint propagation and its complexity, and we list some important references [MH86, Mon74, Fre78, HE80, Mac77, MF85].

For handling constraints defined extensionally as relations, there is a range of standard consistency techniques. However for particular constraints, specialised consistency techniques can be applied which take advantage of their particular semantics. The specialised techniques can support more efficient constraint propagation than the standard techniques [DV91].

For problems modelled using integers (like the scheduling example above), the constraints most often required are equations and inequations between mathematical expressions (involving the predicates $=$, $> \geq$, $<$ and $\leq$). These can be efficiently handled by reasoning on maxima and minima. For example suppose X, Y and Z each have domain $\{1, \ldots, 10\}$. Reasoning on the constraint 2*X + 3*Y + 2 < Z we use maxima and minima to remove inconsistent values from the domains of all three variables. Since 10 is the maximum possible value for Z, we can deduce that $2 * X + 3 * Y < 8$. Since the minimum value for Y is 1, it follows that $2 * X < 5$. Consequently the domain of X can be reduced to $X \in \{1, 2\}$. Similarly $Y \in \{1\}$. Finally by reasoning on the minima of X and Y we conclude that $Z \in \{8, 9, 10\}$.

Of particular importance for current day computing systems is that constraint propagation can be performed in parallel. Propagation on the different constraints can occur concurrently and asynchronously, and as long as it continues until no more domain reductions are possible the result is independent of the precise behaviour.

Embedding in CLP We now illustrate the embedding of consistency techniques in a logic programming system, by expressing a couple of problems in the CHIP language (see section 6 for information about CLP languages).

The above example can be encoded in CHIP as follows:

```
?- [X,Y,Z]::1..10,
2*X + 3*Y + 2 #< Z,
indomain(X), indomain(Y), indomain(Z).
```

First the finite domain variables X, Y and Z over the subrange 1..10 are declared. Then the constraint that must hold between X, Y and Z are stated as a goal[4]. This goal is recognised by its syntax to be a constraint that will be handled by propagation. Finally the search for admissible values of X, Y and Z is expressed using the goal **indomain**. This goal instantiates its argument to a value in its current domain. This instantiation will cause constraint propagation to take place, which may reduce the domains of the remaining variables, or even cause a failure. If this choice proves later to have been wrong, and the system backtracks, another value in the domain will be chosen, until all the alternatives have been exhausted.

---

[4] The symbol #< stands for < on finite domains.

Because the domains are pruned by propagation, the two admissible combinations of values are found without any wrong guesses. For this simple example, it is an interesting exercise to write a logic program without constraints that avoids unnecessary search. For real-life problems, such an exercise is no longer interesting, and it can easily lead to unmaintainable and even incorrect logic programs. Using CLP however, we use a simple standard program structure and rely on consistency techniques for efficiency. The structure is as follows:

- Declare problem variables and their finite domains
- Set up the constraints
- Search for a solution

Notice that consistency techniques are deterministic, as opposed to the search which is non-deterministic (and usually entails backtracking). This standard structure ensures that deterministic computation during propagation is performed as soon as possible and non-deterministic computation during search is used only when there is no more propagation to be done. The importance of prioritising deterministic computation has been recognised as an important principle in the logic programming community.

It is also possible to specify user-defined predicates as constraints for propagation, by a declaration such as lookahead.[5]

Thus in the following program goals for the predicate less will be treated using consistency techniques, whilst goals for the predicate gteq will be treated by choice and backtracking in the normal fashion of logic programming.[6]

```
lookahead less(d,d).
less(1,1).
less(1,2).
less(2,2).
less(2,3).

gteq(2,1).
gteq(3,2).
```

The query

```
?- [X,Y,Z]:: [1,2,3,4,5], less(X,Y), less(Y,Z), gteq(X,Z)
```

is evaluated as follows. As soon as the constraints (less(X,Y), less(Y,Z)) are set up, the domains of the variables are reduced by consistency techniques to $\{1, 2, 3\}$. Now the goal gteq(X,Z) is invoked; the system selects the first clause defining gteq and attempts to add the constraints $Z = 1, X = 2$ to the constraint store. Arc consistency on less(X,Y) reduces the domain of Y to $Y \in \{2, 3\}$, then propagation on less(Y,Z) reveals an inconsistency. Thus the attempt to match the first clause for gteq fails, and the second clause is tried. This fails similarly, and so the whole query fails.

As usual for consistency techniques, the evaluation of the constraint goals less are constraint driven, and there is no backtracking on these goals.

---

[5] "Looking ahead" is another name used for consistency techniques [HE80]

[6] The d in lookahead less(d,d) signifies that this argument of less is a domain variable.

3.2 Generalised Propagation

The study of constraint propagation has been recently extended to remove the requirement for finite domains associated with the variables. One step in this direction is to admit intervals instead of finite domains (eg $1 < X < 10$ for real X) [Dav87]. However, more radically, it is possible to perform propagation without requiring either domains or intervals to be associated with the problem variables. This technique has been named *generalised propagation* [LW92a]. Generalised propagation integrates the CLP scheme, described in section 2 above, and constraint satisfaction techniques, described in this section.

In the CLP scheme an answer to a goal is a (consistent) set of constraints on the problem variables. Standard logic programming is a particular instance of the CLP scheme, where answers are expressed using equations on terms. Thus if predicate p is defined by

```
p(1,1).
p(2,2).
```

the query ?- p(X,Y) has two answers $X = 1, Y = 1$ and $X = 2, Y = 2$. The idea of generalised propagation is to enable p(X,Y) to be used as a constraint, even though there are no domains or intervals associated with its arguments. Instead of extracting information in the form of reduced domains for X and Y, the information extracted is in the form of constraints in the current computation domain - i.e. equations between terms.

As with finite domain propagation, the information extracted must not exclude any answers to p(X,Y). Thus generalised propagation only extracts information common to *all* the answers to p(X,Y). Over this computation domain, the information extracted from a goal is technically the "most specific generalisation" of all the answers to the goal. In this case the most specific generalisation is $X = Y$.

If p is handled as an ordinary predicate in the query ?- p(X,Y), p(V,W), Y=V, notequal(X,W), the system will backtrack four times before failing. To use p(X,Y) and p(Y,Z) as constraints for generalised propagation, it is merely necessary to annotate the query as follows:

```
?- propagate p(X,Y), propagate p(V,W), Y=V, notequal(X,W)
```

The annotation **propagate** Goal tells the system to perform generalised propagation on Goal, instead of treating it as an ordinary logic programming goal. Generalised propagation will immediately deduce that $X = Y$ and $V = W$. Consequently when the goals Y=V, notequal(X,W) are executed, the inconsistency will be detected without any backtracking.

Another example of generalised propagation is its application to the predicate and, defined as follows:

```
and(0,0,0).
and(0,1,0).
and(1,0,0).
and(1,1,1).
```

Consider the query ?- **propagate and(X,Y,Z), Rest** where **Rest** is some goal that performs search, eventually yielding further information about the variables X, Y and Z. Initially no information can be extracted from the constraint **and(X,Y,Z)**. However as further information is added to the constraint store, during evaluation of **Rest**, interesting propagations on **and(X,Y,Z)** may become possible. For example if the constraint $X = 0$ is added to the constraint store, generalised propagation on **and(X,Y,Z)** immediately yields the new equation $Z = 0$. Alternatively if $X = 1$ is added to the constraint store, generalised propagation yields $Z = Y$.

Like propagation, generalised propagation is a form of constraint driven computation. As more information about the problem variables becomes available, via the constraint store, further information is extracted from the constraints. All the extracted information is added to the constraint store, which enables further propagation to take place. Propagation is repeatedly attempted on all constraints until there is no more information to be extracted.

In section 4.2 below, it is described how the user can explicitly program the handling of constraints, so as to achieve a similar constraint driven behaviour for the constraint **and(X,Y,Z)**. The advantage of generalised propagation is that such constraint driven behaviour is achieved by a single annotation, and without risk of incorrectness or potential omission of possible propagation steps.

Generalised propagation yielding equality constraints, as in the above examples, has been implemented in a system called Propia [LW92b]. Programming in Propia has shown three advantages of generalised propagation.

- It is relatively simple to encode the constraints of real problems in Propia, and there is no need to explicitly add finite domains. (In fact current systems only admit finite domains of integers which implies an extra encoding step).

- It is very natural to encode a problem using a logic program without regard for efficiency. To turn such a program into a Propia program utilising generalised propagation, it is merely necessary to add annotations as in the above example. Consequently it is easy to experiment with different ways of executing the program by changing the annotations. The final program still has the same structure as the original logic programming "specification" and is therefore easy to maintain.

- Even some problems which involve finite domains prove to be solved more efficiently when encoded in Propia, than is achieved with finite domain propagation. For propositional logic problems, which can be encoded using finite domains with two values, generalised propagation produces more information than arc consistency. In fact Propia turns out to be broadly as efficient as specialised programs on a current benchmark of such problems. For problems which involve large finite domains, on the other hand, generalised propagation scores again by its simplicity: it extracts less information but it avoids wasting storage and execution time doing so. Consequently Propia can solve problems which are too big and too slow to run on existing CLP systems with finite domains.

4 Extending and Specialising the Constraint System

A given constraint system supports certain computation domains, and certain consistency techniques, enabling it to solve a range of problems efficiently. However, specialised problems may require specialised constraints, with specialised solving and consistency techniques. Two different approaches have been developed to tackle this problem. The first approach consists in identifying frequently occurring constraints and offering them via a system library. The second consists in offering the user a language to define his own constraints and the necessary propagation.

4.1 Specialised Constraints

In the past a variety of frequently occurring constraints, have been identified, which caused problems if they are encoded using the standard built-in constraints. For these, specialised constraint solving algorithms have been developed. We shortly mention some of those, developed within the CHIP system. Note that the user of these constructs need not be concerned about the implementational aspects, as they all have a declarative reading.

The Element Constraint Many constraint problems use the notion of a cost function associated with a choice. This can describe the cost which we want to optimise, or it can be just an internal figure that has to be kept within certain limits. For example in a production unit switching a job from one machine to another involves a certain setup time. Now the overall time needed is restricted by some constraints. These constraints provide a pruning on the possible choices of jobs. An efficient implementation of arc-consistency for the functional constraint between choices (jobs in this example) and their costs (here the setup times) is supported via a special *element* constraint. It has the following structure: element(N, List, Value), with the reading: *Value* is the *N*-th value of the list *List*.

```
[M1,M2,M3] :: 1..5,        % 3 Machines, 5  jobs
alldistinct([M1,M2,M3]), % no job is done twice
element(M1,[3,2,6,8,9],C1),
element(M2,[4,6,2,3,2],C2),
element(M3,[6,3,2,5,2],C3),
C1+C2+C3 #=  Cost,
Cost #<= 9.
```

Running this query will give the result that $M1$ does Job 1 or 2, $M2$ can do all jobs except Job 2, and $M3$ all except Job 1. The cost is guaranteed to be between 6 and 9.

Note that this constraint works in all directions, e.g. restrictions of the possible values also prune the associated index.

A variety of special constraints on lists of choices have been developed. They express e.g. that all the elements have to be different (alldistinct); certain values may not occur more than a certain number of times (atmost, as exampled below);

that only one variable may take a certain value, etc. A special constraint - *cumulative* - developed for scheduling and loading problem has been recently presented in [AB92].

```
[chipc 7]: [A,B,C] :: 0..5, atmost(1,[A,B,C],5),B=5.

A = $_267  [0..4]
C = $_287  [0..4]
B = 5
yes.
```

4.2 User Defined Constraints

The implementation of special constraints can only be done by the system designer. But as it is useful to have special constraint solving mechanisms available the trend now is to develop tools to allow the constraint solving behaviour necessary for the specific application to be defined by the application programmer. Given these tools provide means for a simple declarative specification, they once again support one key concept behind logic programming: the programming time is reduced, different possibilities can be tested easily, and support of the software becomes easier.

In this section we discuss facilities for the user to control the evaluation of constraints, to specify constraint-driven computation, and to define constraint solvers for new constraints.

Delay Declarations We saw above that for certain constraints (like non-linear constraints in $CLP(\mathcal{R})$) it is necessary to delay their handling until certain variables have a specific value. In some systems the delaying of the appropriate constraints is built into the system. Often, however, the user needs to be able to control the delaying of goals and constraints. An example of a declaration to delay the handling of a goal till a certain condition is satisfied is

```
delay employee(Nr,Sal) until ground(Nr)
```

Such a declaration will prevent the system from trying to look up salaries for employees until a specific employee number is known. The declaration would also postpone the application of consistency techniques to this goal, in case **employee** was a constraint.

Declarations are annotations applied to a program which refer to the program text. As such they are termed "meta-commands" to distinguish them from commands within the program which manipulate the data.

Guards There is another approach to providing user control based on the concept of a *guard* The guard defines a logical condition, and is part of the program itself rather than a meta-command. An example of a guarded clause is

```
and(X,Y,Z) <=> X=0 | Z=0
```

The guard is $X = 0$. When the current set of goals include an atomic goal of the form ?- and(A,B,C) the guard is used to control when, or if, the clause can be applied. Specifically it can be applied as soon as the constraint store contains, or implies that, $X = 0$. As soon as this is true, the atomic goal can be rewritten into its body (in this case $Z = 0$). Hence the query ?- and(X, Y, Z),X = 0 will result in a constraint store X = 0, Z = 0.

A special feature of definitions by guarded clauses is that when a guard is satisfied, the system *commits* to the clause and there is never any backtracking to alternative clauses. This means that guarded clauses define a computation with "don't care" nondeterminism, rather than the "don't know" nondeterminism of logic programming which involves backtracking to check the other alternatives. The declarative semantics of logic programming is sacrificed with the move to don't care nondeterminism, unless strict conditions are met by the guarded clauses as given in [Mah87]. An advantage is that the guards can be evaluated concurrently, which is why guarded clauses are interesting for concurrent CLP, discussed later in this section.

The control offered by the guards is precisely constraint driven computation, without backtracking, as needed to explicitly encode constraint propagation

Example We shall take as an example the **and** constraint used earlier in our discussion of generalised propagation.

Declaratively **and** is defined as follows:

```
and(0,0,0).
and(0,1,0).
and(1,0,0).
and(1,1,1).
```

We can specify a propagation behaviour for handling **and** goals using the following guarded clauses:

```
and(X,Y,Z) ⇔ X=0 | Z=0.
and(X,Y,Z) ⇔ Y=0 | Z=0.
and(X,Y,Z) ⇔ Z=1 | X=1,Y=1.
and(X,Y,Z) ⇔ X=1 | Y=Z.
and(X,Y,Z) ⇔ Y=1 | X=Z.
and(X,Y,Z) ⇔ X=Y | Z=X.
```

Notice that the information $Z = 0$ is not sufficient to allow any further consequences to be extracted from the **and** constraint. Thus if the constraint store only contains $Z = 0$ none of the guards are satisfied. In this case more information on X or Y will be needed before any of the clauses can fire.

Consider the full-adder circuit

```
add(I1,I2,I3,O1,O2) :-
    xor(I1,I2,X1),
    and(I1,I2,A1),
    xor(X1,I3,O1),
    and(I3,X1,A2),
    or(A1,A2,O2).
```

together with rules for the logical gates (as was exemplified by the rules for the and-gate).

The query add(I1,I2,0,O1,1) will produce I1=1,I2=1,O1=0. The computation proceeds as follows: Because I3=0, the result of the and-gate with input I3, the output A2, must be 0. As O2=1 and A2=0, the other input A1 of the xor-gate must be 1. Because A1 is also the output of an and-gate, its inputs I1 and I2 must be both 1. Hence the output X1 of the first xor-gate must be 0, and therefore also the output O1 of the second xor-gate must be 0.

In this particular case the same behaviour is obtained by applying generalised propagation to the declarative specification of **and**. However the facility to define explicitly what propagation is to take place on a given goal means that tailored propagation behaviour can be obtained for particular applications.

Embedding in CLP CHIP was the first constraint logic programming language to introduce constructs to specify user-defined constraint propagation. Their need was realised in applications for diagnosis and test pattern generation of digital circuits [SD87, Sim89]. They have been called "demon constructs" [DVS+88] because of their event-driven activation. CHIP introduces in addition conditional propagation with the **if-then-else** construct. A framework for using guarded rules for constraint handling is given in [Smo91].

Constraint Solving To express constraint *solving* it is necessary to be able to handle the interaction of multiple constraints. Consequently a multi-headed guarded rule is introduced. A unified approach encompassing single- and multi-headed guarded clauses has been developed under the name Simplification Rules [Frü92]. Two rules encoding a solver for the **greater** constraint are as follows:

```
greater(X,Y) <=> X=Y | fail                          % irreflexivity
greater(X,Y), greater(W,Z) => Y=W | greater(X,Z)     %transitivity
```

(If the second clause is executed it does not *replace* the goal with the body, it merely *augments* the current set of remaining goals with the clause body.)

The above rules capture the transitivity and irreflexivity of **greater** but not its semantics: "less" is also transitive and irreflexive! We now add one further guarded rule to check that **greater** is indeed the same as the built-in comparator ">":[7]

```
greater(X,Y) <=> ground(X), ground(Y) | X>Y
```

Concurrent Constraints User-defined constraint propagation and simplification is a very active area of research in constraint logic programming. A framework including a powerful set of constraint constructors is described in [VD91]. The concept of constraint agents, and their transformational semantics underlies much ongoing work, e.g. [Sar92, Van91]. The idea behind all these approaches is to express

---

[7] Since *groundness* is a meta-concept, some people prefer to use the delay declaration instead of a guard for this control. The framework of simplification rules supports control by both guards and delays

constraint evaluation in terms of concurrent computations. The first such concurrent constraint logic programming language has been suggested in [Mah87]. In [Sar92] a general framework for these languages has been developed based on the notion of ask & tell. The basic operation in these languages, besides telling (adding) a constraint to the constraint store and deciding its consistency, is to ask for a constraint, i.e. to decide if this constraint is entailed (implied) by the constraint store. Algorithms for constraint entailment are extensions of constraint solving algorithms. In case of demons above this simplifies to deciding whether the variables in the guard have certain values or not.

5 Search and Optimization in CLP

As outlined above the key idea behind constraint reasoning systems is to tackle complex tasks by incrementally inferring properties of the problem solutions and using this information to enforce consistency [Van89]. This deterministic knowledge is acquired in an explicit form. It is therefore possible to prune the space of possible alternatives, i.e. excluding certain cases (choices) that need not be considered in the future.

As in general the solution cannot be inferred right away after the deterministic reasoning steps some assumptions about the problem solution have to be made. Those assumptions are fed back into the constraint reasoning scheme, thus yielding more information about the solution. This process continues until a solution is obtained.

If an inconsistent solution description is obtained the assumptions have to be withdrawn. In this case the process has to be continued with alternative assumptions. This process is usually referred to as backtracking. The nature of this is another reason why constraint propagation fits well in the Prolog language, which supports a backtracking mechanism.

Note that the *generate and test* approach uses the same schema, but the inference engine is only used when complete solutions are obtained, i.e. only a test is done, if a complete solution candidate has been produced.

The constraints reasoning schema depends crucially on two aspects:

– The inference power of the reasoning engine.
– The strategy to make the assumptions.

In this section we will concentrate on the second aspect. In general this is referred to as the *search*. We will concentrate on the *finite domain* case, where this process is also called *labelling*.

5.1 Aspects of Search

In AI, problem solving is classically seen as a state space search: solving a problem is to find a path from an initial state to the goal state - representing the solution. Within that framework search is the general mechanism that is used when no other, better method is known.

Similarly in constraint reasoning we refer to search, if the constraint handler cannot provide us with more information. But note that we deal here with partial solutions: e.g. in each state of the search we know some variable values but not all. In a traveling salesman problem (TSP), for example, the instantiated variables represent known parts of the route. Once we do a search step we assume that a certain city should be visited best at a certain point of the trip.

Taking a search step within a constraint reasoning framework involves two decisions:

1. On which aspect of the problem do we want to make an assumption ?
2. What should that assumption should be ?

The Right Granularity In general it is important that the granularity and the strategy of the search process fits well with the constraint handler. The right choices here are crucial for the performance of the overall system. The assumptions made during search perform two roles. First they are queries about the solution. Secondly, and even more important, they provide input to the constraint handler which performs reasoning on the constraints and their impact on other problem variables. With the right input the solver will be able to prune large parts of the search space, thus yielding a good problem solving performance.

Declarativeness Within the approaches discussed below the strategy for selecting variables/values can be defined declaratively. This means that the complexity of the program used to define the strategy is independent of the complexity of the strategy itself. This has the important consequence that certain real world problems can still be tackled with this declarative technology, while specialised procedural constructs are hard to build. In fact it is has been our experience that CLP solves problems that are new in the sense that they have not been solved systematically by software so far - despite the fact that specialised algorithms have been known.

5.2 Labelling Strategies

Within the CHIP system the user is free to program his own search strategy. This can be done easily with the support of the underlying Prolog system. As some general approaches have given good results they are already incorporated into the system. They make labelling based

– on individual problem variables and
– on single values for those variables.

The problem variables in the TSP example are the stops on the tour, the values are the location of those stops.

In many cases it is most effective to use the variable with the smallest remaining domain for labelling. This principle is often referred to as *first fail principle* as with fewer choices possible we will find out earlier if those were right or wrong. Alternatively the variable which occurs in most constraints can be chosen. Several combinations of these principles are possible [Van89].

```
% label(Problem_variables)
label([]).
label(Problem_variables) :-
        deleteff(Var, Problem_variables, Rest_vars),
            % choose var with minimal choices
        indomain(Var),
            % choose a value from its domain
        label(Rest_vars).
```

Which value then to give that chosen variable is harder to answer in general. For some problems it is possible to define a metric, with the 'smallest' values being most promising.

For the map colouring problem good results have been obtained by rotating the colours used for labelling. I.e. for the country A use the first colour, for country B the second and so on. This approach has the effect of the intuitively appealing idea of using different colours whenever possible, as connected countries have to have different colours.

```
% special labelling routine for map colouring example
% label_colour(Countries, Available_Colours)

label_colour([],_).
label_colour([First|Rest],Colours):-
        member(First,Colours),
        rotate(Colours,Colours1),
        label_colour(Rest,Colours1).

rotate([A,B,C,D],[B,C,D,A]).
```

Within the generalised propagation schema it can be very natural to use an entire tuple of values that satisfies a constraint, as the tuples satisfying / defining a constraint are usually available. E.g. if we want to solve a crossword puzzle, it makes sense to put (assume) a word in a certain position, which means labelling a set of variables with characters at the same time.

Labelling with several values In some cases selecting a specific value for a variable can be a very strong assumption. It can therefore be better to make an assumption on the set of possible values of that variable. The classical approach here is to make a binary chop of the domain. This means that we cut the domain in two halves and then assume that the value is in one half. This can be done by stating an additional constraint which excludes the other half. This technique has been used successfully for the cutting stock application [DSV88].

```
% binary chop labelling routine

label_chop([]).
label_chop([X|Vars]):-
        mindomain(X,Min),
        maxdomain(X,Max),
        Mid is (Min + Max)//2,
        above_or_below(X,Mid),
        label_chop(Vars).

above_or_below(X,Mid) :-
        X #<= Mid.  % set up additional constraint
above_or_below(X,Mid) :-
        X #> Mid.
```

5.3 Branch and Bound

Due to the incremental approach of constraint solving branch and bound strategies
fit well with it. For a constraint problem with minimization the current minimal
value of the target function is maintained. As soon as a choice / search step is done
that would increase that value again, this is rejected. Thus parts of the search tree
need not be considered. If a new minimum value has been obtained a new branch
and bound run with that value can be invoked. Note that the previously considered
combinations need not be considered again, as the current minimum is known to be
optimal with regard to the search space already considered.

Given the classical setup of a CHIP program:

```
solve(Vars) :-
        define_vars(Vars),
        setup_constraints(Vars),
        label(Vars).
```

the program for the minimal solution can be written easily: a labelling routine that
produces the cost values is combined with the minimise declaration.

```
% 2-dimensional cutting stock example
% Vertical and horizontal cuts, Waste Produced

label_min(Vert_Hor_Cuts) :-

        minimise(
           label_waste(Vert_Hor_Cuts, Waste),
                   % labelling routine that also computes the waste
           Waste).
                   % minimise Waste value
```

As seen in the example in CHIP the declaration to use the branch and bound minimization schema is very simple to be added to a program. For some problems this
approach gives quite good performance results.

5.4 Optimization and Advanced Search

For optimization problems it is not always easy to infer deterministic information about the optimal solution. If fewer inferences can be made the proper choice of assumptions will become more important.

Local Search One approach is to improve the current assumptions by local search. The idea is here that an initial solution - satisfying the constraints - is improved in terms of the cost function to be minimised. An operator is defined that maps one solution to others that are similar (in the sense that most of the variables retain the same value). The operator must, of course, ensure the constraints are still satisfied.

Search in this framework means applying the operator to the current solution. If the new solution has lower cost it becomes the current solution, and search continues until a solution is reached which cannot be improved upon by a single application of the operator. The final solution is better than its immediate neighbours, but there may be still better solutions in another part of the search space. In other words the final solution may only be a "local" optimum. This approach works well for the unconstrained traveling salesman problem [LLKS85], where a typical operator is one that exchanges two edges of a tour. To apply this approach to constrained problems, it is necessary to impose constraints on the operator that maps solutions to new solutions. Currently available systems do not offer this feature.

Novel Search Techniques The so called 'novel search techniques' suggest different ways of moving through the search space, while more or less implicitly information about the solution is acquired and used to further guide the search. Currently they offer the best approaches to solving many important classes of optimization problems, as e.g. the TSP. For an overview of these techniques see [Küc92b].

The main disadvantage of these techniques is their missing completeness and correctness properties. There is no guarantee that a certain mechanism will ever find an optimal or even constraint satisfying solution. Therefore these approaches are often ruled out for real world applications where certain requirements - hard constraints - definitely have to be met. On the other hand it is not necessary to always obtain *the* optimal solution with regard to the cost function - which may be very hard to compute - but rather a good solution can be sufficient [HT85]. Many real world problems are a mix of constraint satisfaction problems and optimization problems. A classical example is the vehicle scheduling problem. A fleet of vehicles has to deliver goods to customers with minimal effort. The problem is related to the TSP-optimization problem, but additional constraints also have to be met. Those are for example the capacity constraints of the vehicles. It may thus be permitted to offer a solution that is not optimal with regard to the length of the proposed tour, but in any case none of the vehicles may be overloaded.

In an ideal system for constraint optimization an advanced search mechanism is combined with a constraint solver. It is a current research topic to consider such a combination in detail.

6 CLP systems

This section reviews some constraint programming systems and discusses briefly their most important features. This list cannot be complete and is not intended to be. The objective of the single descriptions is not to be exhaustive, but rather to give a rough idea of the presented system. The interested reader is referred to the cited literature for each of the systems.

6.1 CHIP

The Constraint Logic Programming language CHIP [DVS+88] has been developed at the European Computer-Industry Research Centre (ECRC). The most important feature of the CHIP system is the introduction of arithmetic constraints over finite domains solved by consistency techniques. In addition CHIP provides a rich set of symbolic constraints. Minimization is done by a branch and bound technique.

Beside constraints over finite domains CHIP provides the following constraint solvers:

- Boolean constraints are solved with a Boolean unification algorithm
- Linear rational constraints are handled by an extended Simplex algorithm.

Finally as already mentioned CHIP gives the user the possibility to define his own constraints and control their execution. The demon rules are most prominent. Conditional propagation based on an if-then-else construct is another way to control the evaluation of constraints.

Based on the CHIP technology there are currently four different commercial products available or under development. Bull is offering the finite domain technology within its CHARME system, ICL has a product called DECISION POWER based on the CHIP/SEPIA compiler [MAC+89, AB91]. Siemens-Nixdorf Informationssysteme are currently developing their new version of SNI-Prolog, which will incorporate the whole CHIP technology. Finally the CHIP interpreter has been productised by the French company COSYTEC.

CHIP's finite domain constraints, and generalised propagation, have been integrated into the OR-parallel logic programming platform ElipSys [DSVX91]. Currently a successor to CHIP is under development at ECRC. It will provide integration of new constraint solvers [Mon92b]; generalised propagation [LW92b] working on various computation domains; constraint simplification rules [Frü92]; and novel search techniques [Küc92a].

6.2 CLP($\mathcal{R}$)

The Constraint Logic Programming language CLP($\mathcal{R}$) [JMSY90] has been developed as a demonstrator for the CLP(X) scheme at Monash University, IBM Yorktown Heights and Carnegie Mellon University. The constraint domain of CLP($\mathcal{R}$) is real linear arithmetic. As already mentioned non-linear constraints are delayed. The underlying constraint solver is an extended Simplex algorithm. Currently there are two implementations available from IBM / Carnegie Mellon University, an interpreter and since recently a compiler-based version.

6.3 Prolog-III

PROLOG III [Col90] is the CLP language developed at the University of Marseille and at Prologia in France. It includes three new constraint domains: linear rational arithmetic, boolean terms and finite strings (or finite lists).

- Linear rational arithmetic is handled via an extended Simplex algorithm.
- The boolean constraint-solver is based on a saturation method.
- The facilities of PROLOG III for finite string (lists) processing is explained below. The constraint solver is based on a restricted string unification algorithm.

For finite strings there exists a single function to concatenate two strings, denoted by ".". and the only constraint is the equality constraint. To illustrate how these finite strings may be used consider the following problem (from [Col90]).

Find the string(s) Z such that `<1,2,3>.Z = Z.<2,3,1>`

There are in fact an infinite number of solutions. Hence Prolog III delays the evaluation of such constraints until their length is known. Let us consider the string length 10 (the length operator is infix in Prolog-III and denoted by the operator ::).

`{ Z :: 10, <1,2,3>.Z = Z.<2,3,1> } ;`

The system comes back with the single solution:

`{ Z = <1,2,3,1,2,3,1,2,3,1> }`

PROLOG III is a commercial product of Prologia, Marseille.

6.4 Trilogy

Trilogy [Vod88] is a constraint programming language developed at Complete Logic Systems in Vancouver. The constraint domain of Trilogy is integer arithmetic, i.e. it allows linear equations, inequations, and disequations over integers and integer variables to be expressed. The solver is based on a decision procedure for Presburger Arithmetic. Unlike other CLP systems TRILOGY is not integrated into a Prolog environment, but it is based on an own "theory of pairs" [Vod88]. Trilogy is compiled into native code for PCs as target machines. It can be acquired via Complete Logic Systems in Vancouver.

6.5 CAL and GDCC

CAL [ASS+88] (Constrainte Avec Logique), developed at ICOT, Tokyo, was the first CLP language to provide non-linear constraints. During the last few years a parallel version of CAL has been developed at ICOT, called GDCC [AH92]. The system can handle constraints in the following domains:

- Non-linear real equations are solved with a Gröbner Base algorithm.
- The constraint solver for boolean constraints is based on a modified Gröbner Base algorithm.
- Linear rational arithmetics are again solved with a Simplex algorithm. A branch and bound method has been implemented on top of this constraint solver to solve integer optimization problems.

Both CAL and GDCC are available from ICOT, Tokyo.

6.6 BNR-Prolog

BNR-Prolog [Bel88] has been developed at Bell-Northern Research, Ottawa. It has been specifically designed for Apple Macintosh. The interesting feature of BNR-Prolog from a CLP point of view is the introduction of the so called relational arithmetic. This new constraint domain is based on a new interval variable representing a real number lying between lower and upper bound of this interval. The constraint handler is based on interval arithmetic [OV90]. The system can be acquired from Bell-Northern Research, Ottawa.

6.7 RISC-CLP

RISC-CLP [Hon92] is a prototype system in the domain of real arithmetic terms. It has been developed at the RISC, Linz. It can handle any arithmetic constraints over the reals. The constraint solver behind is an improved version of Tarski's quantifier elimination method [Hon90].

7 Conclusion

This paper aimed at giving an informal introduction into the different concepts of CLP. It tried to explain the philosophy behind the main ideas in CLP and illustrate them by examples. Emphasis has been put on the practically relevant parts.

CLP is successfully employed in a large variety of applications, in particular ones that can be expressed as constrained search problems. While keeping the main features of logic programming, i.e. declarativeness and flexibility, CLP brings into these languages

- the efficiency of special purpose algorithms written in imperative languages and
- the expressiveness of the different constraint domains it embodies.

The main advantage of CLP compared to other approaches is that it drastically reduces development time and provides more flexibility offered by the solution while showing an efficiency comparable to solutions written in procedural languages.

Constraint Logic Programming is moving out of the research labs into the commercial world. A number of products based on this technology are offered today. These products have been applied in a large range of very different application which are in use. Amongst them:

- At Hongkong International Terminal and in the Harbour of Singapore the resource planning and scheduling system controls ships, cranes, containers and stacks.
- At Cathay Pacific the Movement Control Systems supports the planning and scheduling of their entire fleet.
- At the French national railway SNCF movements of empty waggons are optimised.
- Within Siemens CLP is supporting the circuit designers with their Circuit Verification Environment.

- In the ESPRIT projects APPLAUSE, CHIC and PRINCE a large number of applications is currently under development.

CLP is still very much under development. The main practical systems are implemented to run on a single processor. Many researchers are studying concurrent constraint handling and parallel implementations of CLP. Secondly constraints in existing systems need to be well-understood by the end user if he is to obtain maximum benefit of them in his programs. The development of cleaner and simpler ways to specify constraint behaviour will be essential for its future industrial acceptance. Thirdly the practical requirement to integrate constraint handling with other software techniques and systems is becoming pressing. The current work on integrating CLP with data base technology is an important step in this direction.

Acknowledgements

The authors would like to thank the following people for the discussions they had with them and for their encouragement and support: Abder Aggoun, Nicolas Beldiceanu, Françoise Berthier, Gérard Comyn, Mehmet Dincbas, Hervé Gallaire, Thomas Graf, Micha Meier, Joachim Schimpf, Helmut Simonis, Pascal Van Hentenryck, André Veron. Many thanks to Norbert Eisinger for reading a draft of this paper and suggesting improvements.

References

[AB91]   A. Aggoun and N. Beldiceanu. Overview of the CHIP Compiler System. In
           K. Furukuwa and P. Deransart, editors, *Proceedings of the 8th International
           Conference on Logic Programming*, pages 775–789, June 1991.

[AB92]   A. Aggoun and N. Beldiceanau. Extending CHIP in order to solve complex
           scheduling problems. Technical report, COSYTEC, 1992.

[AH92]   A. Aiba and R. Hasegawa. Constraint Logic Programming Systems - CAL,
           GDCC and Their Constraint Solvers. In *Proceedings of FGCS 92*, pages 113–
           131, 1992.

[ASS+88]   A. Aiba, K. Sakai, Y. Sato, D. J. Hawley, and R. Hasegawa. Constraint Logic
           Programming Language CAL. In *Proceedings of the International Conference
           on Fifth Generation Computer Systems (FGCS-88), ICOT, Tokyo*, pages 263–
           276, december 1988.

[Bel88]   Bell-Northern Research Ltd BNR. BNR-Prolog User Guide. Technical report,
           Bell-Northern Research Ltd., 1988.

[Ber89]   F. Berthier. A financial model using qualitative and quantitative knowledge.
           In F. Gardin, editor, *Proceedings of the International Symposium on Compu-
           tational Intelligence 89*, Milano, Italy, September 1989.

[Bry86]   R. Bryant. Graph based algorithms for boolean function manipulation. *IEEE
           Transactions on Computers*, 35(8):677–691, 1986.

[BS87]   W. Buettner and H. Simonis. Embedding Boolean Expressions into Logic Pro-
           gramming. *Journal of Symbolic Computation*, 4:191–205, October 1987.

[Buc85]   B. Buchberger. Gröbner Bases: an Algorithmic Method in Polynomial Ideal
           Theory. In N. K. Bose Ed., editor, *Multidimensional Systems theory*, pages
           184–232. D. Reidel Publishing Company, Dordrecht - Boston - Lancaster, 1985.

[Bue88]   W. Buettner. Unification in finite algerbas is unitary (?). In *Proceedings CADE-
           9*. LNCS 310, Springer-Verlag, 1988.

[Coh90]   J. Cohen. Constraint logic programming languages. *Communications of the
           ACM*, 33(7):52–68, July 1990.

[Col75]   G. E. Collins. Quantifier Elimination for Real Closed Fields by Cylindrical
           Algebraic Decomposition. In *Proceedings of the Second GI Conference on Au-
           tomata Theory and Formal Languages*, pages 515–532. Springer Lecture Notes
           in Computer Science 33, 1975.

[Col90]   Alain Colmerauer. An introduction to prolog-III. *Communications of the ACM*,
           33(7):69–90, July 1990.

[Dav87]   E. Davis. Constraint propagation with interval labels. *Artificial Intelligence*,
           32:281–331, 1987.

[DSV88]   M. Dincbas, H. Simonis, and P. Van Hentenryck. Solving a Cutting-Stock Prob-
           lem in Constraint Logic Programming. In *Fifth International Conference on
           Logic Programming*, Seattle, WA, August 1988.

[DSV90]   M. Dincbas, H. Simonis, and P. Van Hentenryck. Solving Large Combinatorial
           Problems in Logic Programming. *Journal of Logic Programming*, 8(1-2):74–94,
           1990.

[DSVX91]   M. Dorochevsky, K. Schuerman, A. Véron, and J. Xu. Constraints Handling,
           Garbage Collection and Execution Model Issues in ElipSys. In Springer Verlag,
           editor, *Parallel Execution of Logic Programs, ICLP'91 Pre-Conference Work-
           shop Proceedings*, pages 17–28, Paris, June 1991.

[DV91]   Y. Deville and P. Van Hentenryck. An efficient arc consistency algorithm for a
           class of csp problems. In *Proc. of the 13th IJCAI*, Sydney, Australia, August
           1991.

[DVS+88] M. Dincbas, P. Van Hentenryck, H. Simonis, A. Aggoun, T. Graf, and F. Berthier. The Constraint Logic Programming Language CHIP. In *Proceedings on the International Conference on Fifth Generation Computer Systems FGCS-88*, Tokyo, Japan, December 1988.

[Fre78] E.C. Freuder. Synthesizing constraint expressions. *Communications of the ACM*, 21(11):958–966, November 1978.

[Frü90] Thom Frühwirth. Constraint logic programming - an overview. Technical Report Technical Report E181-2, Christian Doppler Laboratory For Expert Systems, August 1990.

[Frü92] Thom Frühwirth. Simplification rules. Technical report, ECRC, Munich, Germany, 1992.

[FSTW91] T. Filkorn, R. Schmid, E. Tiden, and P. Warkentin. Experiences from a large industrial circuit design application. In *ILPS*, San Diego, California, October 1991.

[Gal85] H. Gallaire. Logic programming: Further developments. In *IEEE Symposium on Logic Programming*, pages 88–99. IEEE, Boston, July 1985.

[HE80] R.M. Haralick and G.L. Elliot. Increasing tree search efficiency for constraint satisfaction problems. *Artificial Intelligence*, 14:263–314, October 1980.

[HHLM91] R. Helm, T. Huynh, C. Lassez, and K. Mariott. A linear constraint technology for user interfaces. Technical Report RC 16913, IBM Yorktown Heights, 1991.

[Hon90] Hoon Hong. *Improvements in CAD-Based Quantifier Elimination*. PhD thesis, Ohio State University, Computer and Information Science Research Center, Colombus, Ohio, USA, 1990.

[Hon92] H. Hong. Non-linear Constraints Solving over Real numbers in Constraint Logic Programming (Introducing RISC-CLP). Technical report, RISC, Linz, 1992.

[HT85] J. Hopfield and D. Tank. 'Neural' computation of decisions in optimization problems. *Biological Cybernetics*, 52:141–152, 1985.

[JL87] Joxan Jaffar and Jean-Louis Lassez. Constraint logic programming. In *Proceedings of the 14th ACM Symposium on Principles of Programming Languages, Munich, Germany*, pages 111–119. ACM, January 1987.

[JMSY90] Joxan Jaffar, Spiro Michaylov, Peter Stuckey, and Roland Yap. The CLP($\mathcal{R}$) language and system. Technical Report RC 16292 (#72336) 11/15/90, IBM Research Division, November 1990.

[JMSY92] J. Jaffar, M. Maher, P. Stuckey, and R. Yap. Output in CLP($\mathcal{R}$). In *Proceedings the FGCS'92*, Tokyo, 1992.

[KKR90] P. Kanellakis, G. Kuper, and P. Revesz. Constraint query languages. In *Proceedings of PODS 90*, pages 299–313W, 1990.

[Kow79] R Kowalsi. *Logic for Problem Solving*. North-Holland, New York, Amsterdam, Oxford, 1979.

[Küc92a] Volker Küchenhoff. Clp and novel search techniques: an integration. Technical report, ECRC, Munich, Germany, 1992.

[Küc92b] Volker Küchenhoff. Novel search techniques - an overview. Technical report, ECRC, Munich, Germany, January 1992.

[LLKS85] E. Lawler, J. Lenstra, R. Kan, and D. Shmoys. *The Traveling Salesman Problem*. John Wiley and Sons, 1985.

[LMY87] Catherine Lassez, Ken McAloon, and Roland Yap. Constraint logic programming and options trading. *IEEE Expert, Special Issue on Financial Software*, (3):42–50, August 1987.

[LW92a] T. Le Provost and M. Wallace. Constraint Satisfaction Over the CLP Scheme. Technical Report ECRC-92-1, ECRC, 1992.

[LW92b]   T. Le Provost and M. Wallace. Domain Independent Propagation. In *Proceedings on the International Conference on Fifth Generation Computer Systems 1992 FGCS-92*, pages 1004–1012, Tokyo, Japan, June 1992.

[Mac77]   A.K. Mackworth. Consistency in networks of relations. *Artificial Intelligence*, 8(1):99–118, 1977.

[Mac86]   A.K. Mackworth. Constraint satisfaction. In *Encyclopedia of Artifical Intelligence*, 1986.

[MAC+89]   M. Meier, A. Aggoun, D. Chan, P. Dufresne, R. Enders, D. Henry de Villeneuve, A. Herold, P. Kay, B. Perez, E. van Rossum, and J. Schimpf. SEPIA - An Extendible Prolog System. In *Proceedings of the 11th World Computer Congress IFIP'89*, San Francisco, August 1989.

[Mah87]   M. J. Maher. Logic semantics for a class of committed-choice programs. In *Proc. 4th International Conference on Logic Programming*, pages 858–876, Melbourne, Australia, May 1987.

[MF85]   A.K. Mackworth and E.C. Freuder. The complexity of some polynomial network consistency algorithms for constraint satisfaction problems. *Artificial Intelligence*, 25:65–74, 1985.

[MH86]   R. Mohr and T.C. Henderson. Arc and path consistency revisited. *Artificial Intelligence*, 28:225–233, 1986.

[MN90]   U. Martin and T. Nipkov. Boolean unification - the story so far. In C. Kirchner, editor, *Unification*. Academic Press, 1990.

[Mon74]   U. Montanari. Networks of constraints : Fundamental properties and applications to picture processing. *Information Science*, 7(2):95–132, 1974.

[Mon92a]   E. Monfroy. A Survey of Non-Linear Solvers. Technical Report 91-15i, ECRC, Munich, Germany, January 1992.

[Mon92b]   E. Monfroy. Non Linear Constraints: a Language and a Solver. Technical Report ECRC-92, ECRC, Munich, Germany, 1992. to appear.

[OV90]   W. Older and A. Vellino. Extending prolog with constraint arithmetics on ral intervals. In *Canadian Conference on Computer and Electrical Engineering*, Ottawa, Canada, 1990.

[PS85]   F. P. Preparata and M. I. Shamos. *Computational Geometry: An Introduction*. Springer-Verlag, New York, 1985.

[SA89]   K. Sakai and A. Aiba. A Theoretical background of Constraint Logic Programming and its Applications. *Journal of Symbolic Computation*, 8(6):589–603, December 1989.

[Sar92]   V. A. Saraswat. *Concurrent Constraint Programming Languages*. MIT Press, 1992.

[SD87]   H. Simonis and M. Dincbas. Using Logic Programming for Fault Diagnosis in Digital Circuits. In *German Workshop on Artificial Intelligence (GWAI-87)*, pages 139–148, Geseke, W.Germany, September 1987.

[SD90]   H. Simonis and M. Dincbas. Propositional calculus problems in chip. In H. Kirchner, editor, *Proceedings of the 2nd International Conf on Algebraic and Logic Programming*, Nancy, France, October 1990. CRIN and INRIA-Lorraine, Springer Verlag.

[Sim89]   H. Simonis. Test Generation Using the Constraint Logic Programming Language CHIP. In *Proceedings of the 6th International Conference on Logic Programming*, Lisbon, Portugal, June 1989.

[Sim92]   H. Simonis. *Constraint Logic Programming as a Digital Circuit Design Tool*. PhD thesis, 1992. (submitted).

[Smo91]   Gerd Smolka. Residuation and guarded rules for constraint logic programming. Technical report, Digital Equipment Paris Research Laboratory Research Report, June 1991.

[Ste80]   G. L. Steele. The definition and implementation of a computer programming language based on constraints. Technical Report MIT-AI TR 595, Dept. of Electrical Engineering and Computer Science, M.I.T., August 1980.

[Van88]   P. Van Hentenryck. A Constraint Approach to Mastermind in Logic Programming. *ACM Sigart*, (103), January 1988.

[Van89]   P. Van Hentenryck. *Constraint Satisfaction in Logic Programming*. Logic Programming Series. MIT Press, Cambridge, MA, 1989.

[Van91]   Pascal Van Hentenryck. Constraint logic programming. *The Knowledge Engineering Review*, 6(3):151–194, 1991.

[VD91]   Pascal Van Hentenryck and Yves Deville. The cardinality operator: A new logical connective for constraint logic programming. In *Proc. of the 8th Int. Conf. on Logic Programming*, pages 745–759, Paris, France, 1991. MIT Press.

[Vod88]   P. Voda. The constraint language trilogy: Semantics and computations. Technical report, Complete Logic Systems, North Vancouver, BC, Canada, 1988.

[Wal72]   D. Waltz. Generating semantic descriptions from drawings of scenes with shadows. Technical Report AI271, MIT, Massachusetts, November 1972.

Scheduling and Optimisation
in the Automobile Industry

Michel d'Andrea

Département S.E.R.P. - Bull CEDIAG
Louveciennes - France

Abstract. This article sets out a scheduling spare parts arrival prototype realized in CHARME language for MCA (Maubeuge Construction Automobile).

We point out the approach feasibility relative to a flow-job scheduling problem solved in its real complexity (combinatorial aspects and heterogeneity of the constraints).

The use of CHARME lead us to write efficient programs in terms of constraints and data definition.

The real efficiency of the prototype got MCA to ask us a proposal for a final product.

1 Introduction

Scheduling problems are often described and generally treated with classical Operations Research methods. It gives good results for many restricted cases of scheduling problems (single resource scheduling, job-shop, flow-shop scheduling, ...), but has a major drawback when faced with real problems where many different constraints have to be taken into account simultaneously.

The following scheduling problem we are going to develop belongs to the class of NP-hard problems.

Constraint logic programming, particularly CHARME is a good approach for solving scheduling problems with structural complexity, heterogeneity of constraints.

The prototype we present was realized in CHARME First.

First we'll describe the shop and the constraints used on the shop. Then when the required type of approach is caracterised we'll give a brief description of CHARME.

The prototype will be described in itself.

2 The MCA scheduling problem

2.1 Complexity of the problem

MCA (Maubeuge Construction Automobile), a branch of the RENAULT Group, assembles cars using spares parts. These spare parts are received in trucks which arrival hours are appointed in accordance with the carriers, thus allowing the establishment of a schedule for the operators to process the arriving boxes. Until now, this scheduling was hand-made on a grid, proceeding by trial and error.

The problem could be caracterized as follows:

This is a flow-shop scheduling with heterogeneous constraints. Scheduling the daily work of the different operator jobs (unloading, administrative and warehouse) should be done while respecting the legal hours of work. All through the day, the necessary resources (loading platforms, banks) should be allocated. This scheduling should bring on about 80 trucks.

2.2 The different tasks

Unloading: the arrival of a truck on a loading platform, is followed by its unloading through operators. They will lay off the spare parts on banks and then let the truck leave the platform.

Administrative: this concerns the spare parts counting and labeling.

Storage: during this task the spare parts are arranged in warehouse. Either these spare parts are heavy and should be drawn by tractors (*Retract*) or they are simply carried by hands.

Several breaks are included in the day-time.

2.3 The operators

7 operators are setted out as follows:

Unloading: 2 operators (*DECH*). The second one steps in only when the first one has reached saturation point.

Administrative: 2 operators (*ADM*).

Warehouse: 3 operators

- First, 2 of them (*RET*) use retracts to take the spare parts from the banks. They also store the heavy boxes. The second one steps in only when the first one has reached saturation point;
- The last one (*MAG*) stores the light boxes.

2.4 The flow-job

For obvious reasons, some constraints must be taken into account:
- The truck can't leave the loading platforms before the unloading is over;
- The administrative tasks start when the unloading one is 75% realized;
- Then when the admnistrative is 75% realized, the retract tasks start.
- The last task (MAG) will begin when the retract one is 25% realized.

These constraints are part of the problem definition.

2.5 Scheduling

Two types of trucks come to be discharged: *free "aheurage"* and *hard "aheurage"*.

Hard "aheurage" trucks are said planned ones. That means their (set) arrival time is known by MCA.

To return, MCA undertakes hisself to unload trucks within 30 minutes according to their arrival time.

Free "aheurage" trucks arrival should be planned within a 2 hours bracquet (Min and Max). They should be unloaded at the latest 30 minutes after the Max time planned.

The prototype aim: scheduling all the trucks arriving on the loading platform with respecting all resources and time constraints.

The global scheduling aim: MCA wishes to get 100% of the trucks planned. That means to secure each trademan of an immediate unloading (relatively to the arrival time and the available resources).

The prototype developed with CHARME gave them this answer.

3 The CHARME language

3.1 Presentation

Problems involving constraints are typical of the Operational Research field. However, although conventional solutions, using Operational Research libraries, are satisfactory in terms of performance, they usually require a considerable development time. These solutions result in heavy applications, which are difficult to modify or maintain.

An expert system has considerable problems handling the number of combinations, especially in these fields in which the expertise is difficult to clearly define. The expert system is the transcription of the expert's reasoning into a set of rules, representing the method in wich the solution is reached. However, it frequently occurs that the path of reasoning, leading to the solution, is unknown. Conversely, the specifications of the required solutions are clearly defined: the constraints which must be verified by the problem solution can be stated, even though the solution may not be known.

CHARME is a constraint-based programming industrial environment, which enables the constraints linking various objects to be expressed. Using the declarativity concept, as in Artificial Intelligence, together with powerful algorithms, as in Operational Research, this language provide the flexibility and efficiency required to solve constraint problems.

CHARME is a declarative language. Thus, the user does not need to express the problem solution in the form of an algorithm. The actual problem is described as a set of constraints, concerning specified objects.

This language has powerful predefined constraints including, for example, economics function optimisation constraints. Constraint processing is based upon the propagation mechanism, which is specialized for finite domains.

Its dynamic nature enables real simulation applications to be produced, which makes CHARME both a problem solving tool and a decision aid tool. Its declarativity and its conciseness are all significant advantages for the quick creation of short and efficient prototypes.

CHARME is written in C, its syntax is close to that of C and it is very open to C. It thus favors the perfect integration of the final application and guarantees efficient solution.

3.2 General description

Modelling a problem involves defining variables, which are then given a set of assumed values.

The variables represent the unknowns of the problem, while the constraint express restrictions on the possible values of the variables, within a solution.

Possible values associated with a variable constitute the domain of this variable. All acceptable variable values must be within the domain of the variable. The initial search area is thus the cartesian product of all the variable domains.

Finding a solution to the problem means defining a set of values which respect all the variable constraints. The instantiation of a variable means the attribution of a value to that variable.

This search is based upon the fundamental reduction mechanism. The constraints are used to reduce the initial search area, by permanently eliminating variable values which cannot occur in one of the solutions to the problem. This leads to the choices: a value is assigned to a variable and the constraints are then again used dynamically to reduce the current area. This process continues until all the variables have a value, and these values then constitute a solution.

3.3 CHARME Toolkit

CHARME is fully extendable and, using toolkit, modules written in C can be executed and the results can be recovered.
Toolkit enables the set of primitives of the language itself to be extended. A CHARME procedure can also be specifically modified, to improve program efficiency, by writing a program in C.
The graphics library enables user friendly interfaces to be produced and is the first example of the use of the CHARME Toolkit.

4 The Problem Solving

In this chapter, we'll try to explain how we have solved this problem with CHARME.
Programming in CHARME consists in : first describe the problem as a set of constraints on variables, describe the solutions as values of the variables respecting the set of constraints.
Thanks to the language declarativity, a CHARME program will state the problem to be solved and not the way to solve it. This is why a program is generally divided into 4 different parts:
- structures definition
- constraint setting
- generation
- interface
Here we have added a pre-processing phase in order to obtain simplified data when setting the constraints.

4.1 Data structuring

It's the step in wich are defined the data structures of the problem.

CHARME language objects are constants, variables, arrays and expressions. The most commonly used stucture is the array; it enables to set constraints easily on a set of data.
For instance, one can represent variables with arrays of integers, of symbols, of arrays.
Here we have defined 4 main arrays: tasks (*Taches*), material resources (*RessQuai* and *RessTravees*), operators (*RessHum*).

The time unit is the minute. So, all the times have been transformed in minutes.
The limits of the legal hours of work (6h - 22h) are contained in the constants *HMin* and *HMax*.

Arrays concerning trucks are defined as:

> array CamDuree :: [Camions,[adc,lib,adm,ret,mag]]
> array CamAutres::[Camions,[hmin,hmax,typah,typcam,cfj,nbtrav]]

The tasks to be executed are represented by the following array:

> array Taches :: [Camions,*[adc,lib,adm,ret,mag]*,[hdeb,hfin,duree]]

Loading platforms and banks correspond the following structures:

> array RessQuai :: [Camions,[hdeb,hfin,quai]]
> array RessTravees :: [Camions,Trav,[occupe,hdeb,hfin]]

MCA has many operators working on 4 different posts.
To achieve a given task on a truck, we'll use one of the operators (resources) assigned to the corresponding post.
These operators could be caracterized by the following structure:

> array RessHum :: [Camions,Postes,[hdeb,hfin,ress]]

4.2 Constraints setting

Setting the constraints is a main step in solving the problem as constraints are a translation of the problem statement.
Let's have a look to the constraints we used in the prototype definition.

4.2.1 Constraints relative to the Tasks
Let's have a truck I and a task J.
The following constraints should always be satisfied:

> Taches[I,J,hdeb] >= HMin
> Taches[I,J,hfin] <= HMax
> Taches[I,J,duree] = CamDuree[I,J]* CoeffTasks[J]

The CoeffTasks array contains some coefficients which are proportional to tasks occupation rate.
For each task I the precedence constraints are expressed as:

> Taches[I,adc,hdeb] >= CamAutres[I,hmin]
> Taches[I,lib,hdeb] >= Taches[I,adc,hfin]
> Taches[I,adm,hdeb] >= Taches[I,adc,hdeb]
> Taches[I,ret,hdeb] >= Taches[I,adm,hdeb]
> Taches[I,mag,hdeb] >= Taches[I,adm,hdeb]

More, every task has to begin and end in the legal hours of work. Then, for a given truck I and a given task J, we obtain the following constraints:

    Taches[I,J,hdeb] >= TrancheJ[Tr,hdeb]
    Taches[I,J,hdeb] <= TrancheJ[Tr,hfin]
    Taches[I,J,hfin] >= TrancheJ[Tr + T,hdeb]
    Taches[I,J,hfin] <= TrancheJ[Tr + T,hfin]
    Tr in 1..NbTranches
    T in 0..1

We also know that every ending time of a task is linked as a static mode to the corresponding beginning time. Nevertheless, it's possible that our task "cross" a break; in this case, we must take the breaking time into account.
So we obtain the following constraint:

    Taches[I,J,hfin] = Taches[I,J,hdeb] + Taches[I,J,duree] + Tpause

The two parameters Tpause and T are then linked by the constraint: if T = 0 then
Tpause = 0 else
    Tpause = TrancheJ[Tr + 1,hdeb] - TrancheJ[Tr,hfin]

4.2.2 Constraints relative to Loading platforms and Banks

The single constraint relative to Loading platforms is associated to the fact that a given platform cannot be used by several trucks in the same time. Once the platform *J* chosen (Cf generation), we'll have the following constraints:

    RessQuai[C,quai] = J
    domain_min(QuaiF[J],Min)
    RessQuai[C,hdeb] >= Min

where *QuaiF* is an indicator of the platforms occupation rate.

Concerning the banks, we'll further see that if our truck needs N banks, the prototype allocates to it the N banks that are available the soonest.
Supposing the 2nd bank necessary to the truck C has been assigned to the bank number 7, we obtain the constraint:

    RessTravees[C,2,occupe] = 1
    domain_min(Travees[7],Min)
    RessTravees[C,2,hdeb] >= Min

4.2.4 Constraints relatives to the Operators

We'll see in the Generation paragraph how the program assigns one of the two operators to the corresponding task. Anyway, the operation cannot begin before the operator is free. Supposing that for the administrative operation the prototype has chosen the 2nd operator, we'll obtain the following constraints:

```
RessHum[C,adm,ress] = 2
domain_min(Admin[2],Min)
RessHum[C,adm,hdeb] >= Min
```

The single warehouse operator MAG must not be used by several trucks in the same time. So we have the disjunctive constraint:

```
for C1 in Camions do {
        for C2 in Camions do {
                opdisjonc(C1,C2,mag)
        }
}
```

The *opdisjonc* procedure forces the warehouse operator to store C1 before C2, or inversely (no "lapping over").

4.3 Generation strategy

Setting constraints, as we saw in the last paragraph, and their propagation have reduced the initial search area.

Nevertheless it is still large, which led us to write a more efficient generation strategy than CHARME's standard one.

This strategy is the following:

- the basic tasks being sorted following the departures ascending order, operations will be instanciated in the same way, thus for a given operation, the truck will have its resources allocated the soonest possible. For example, unloading a truck can start at the end of its predecessor's unloading, if there is a platform, banks and an unloading operator available.

- when several resources are candidates for the truck, the chosen strategy generally consists in allocating to the truck, the soonest available resource. This choice is made through the indicators (*QuaiF*, *Travees*, *Admin*, ...). Nevertheless, for Unloading and Retract operations, the second operator will be chosen only if choosing the first one implies a delay for the operation (therefore, the first one is prefered).

- the banks case is an interesting one: indeed for trucks which unloading requires several banks, 3 for example, we have applied the following strategy:

  - To unload we choose (using the *Travees* indicator) among the banks set, the 3 soonest available banks. Then, the unloading is done progressively : the first bank will be occupied from the very start, while the third will only be allocated when the operators will have finished to load the two first ones.

- the desallocation is also done progressively: the first allocated bank will be the first desallocated. On the other hand, the last chosen will follow the whole retract operation.

5 Conclusion

This prototype, programmed in CHARME within 3 months comes up to the objectives setted out.

Today, this prototype is used every day by the Logistic Department of MCA. It does simulations which give a more global view of the workload. Acting in cooperation with workshop responsibles, it allows allocation of available resources with more finesse. The arrival hours can then be renegociated with the carriers.

The prototype has significantly increased productivity (about 15%).

It was developed on a Bull workstation (DPX 1000). Now it runs on a workstation which belongs to the Logistic Department (HP 9000/400). Its running takes about 2 minutes.

References

1. A. Oplobedu - Bull Cediag (Louveciennes, F), J. Marcovitch & Y. Tourbier - Renault (Rueil-Malmaison, F): CHARME, an industrial constraint programming language. Its application to plan design at Renault. (Avignon 89)
2. Bull Cediag (Louveciennes, F): CHARME V1 Reference manual. (jan 1990)

Factory scheduling using Finite Domains

Owen V. Evans

International Computers Ltd., Bracknell Berks RG12 4SN UK

1. Introduction

Factory scheduling can be regarded as the fitting of a set of tasks to a set of available resources - the latter normally being industrial plant of some form - but is equally applicable to other kinds of resource, including human resources. To be schedulable, a job must give rise to a sequence of predefined tasks, sometimes known as a *routing*, although this can sometimes reduce to a single task. The duration of the constituent tasks is determined by the batch size of the job, machine characteristics, setup time and shift characteristics. For each machine, these characteristics will be known in advance as a result of some previous planning exercise (not covered in this presentation). If these characteristics, and the repertoire of tasks defined for a factory never changed, it would reasonable, albeit expensive, to design a scheduler application dedicated to a particular factory using a conventional procedural language. Traditionally this is how schedulers have been, and continue to be designed in cases where this level of investment can be justified. By their nature such designs are inflexible and expensive to redesign to accommodate change. They do, however have the merit of being able to accommodate very large (12,000 plus) task loads.

One of the merits of the design approach to be described below is that with it, schedulers are sufficiently quick to build that not only can changes be made very rapidly, but it is reasonable to use them as a planning tool for investigating the behaviour of alternative factory layouts and routings - in effect running 'what if' scenarios. Another merit is that the design approach is completely declarative. All the conditions and constraints that have to hold are first defined and then values of variables tried until all of them are met. A very practical approach is adopted, using just the basic constraint propagation mechanisms available in DECISIONPOWER©, based on CHIP[1], and a small subset of the built-in constraint handling predicates.

2. Constraints and the Prolog computation model

The computation model of Prolog is based on the unification algorithm, a transformation mechanism which by computing the most general unifier of two terms is symmetric. For example:

$$X+2 = 3+Y. \qquad \text{\% will give X=3,Y=2} \qquad (1)$$

Prolog also has a less generalised arithmetical evaluator which enables programs to use the underlying arithmetic capabilities of the support computer. For example:

$$X \text{ is } Y+3. \qquad \text{\% will make X ground if Y is ground} \qquad (2)$$

In addition, DECISIONPOWER Prolog has a delay mechanism which will hold up the execution of *(2)* if Y is not ground. Some familiarity with these three conventional Prolog mechanisms is necessary to understand how constraints are propagated, since the propagation is both *symmetric* and may occur over *linear terms* of the form $A+k*Y$. Also, the delay mechanism will apply to arithmetic operations involving constrained variables.

Constrained variables must be explicitly defined, both to distinguish them from normal variables and to assign them a domain. Three forms of constrained variable can be defined:

`X::Min..Max`	defines a consecutive integer range
`X::Min:Max`	defines an interval (bound pair)
`X::List`	defines an enumerate set of integers

Only the first variant is used in the scheduler design.

Constraints are propagated using the special operators **#=, #<=, #<, #>=, #>, ##** to represent equality, the magnitude relations and inequality respectively. For example:

$$X::3..6, Y::2..5, X\#=Y \qquad \text{gives } X=\{3..5\}, Y=\{3..5\}$$

It is a charactesistic of current implementations of constraint progagation systems that constraints can be tightened but not be relaxed except in the course of backtracking. The only other details that need to be known of the constraint handling system are the behaviour of some of the built-in predicates. These will be defined as they occur in the description of the scheduler.

3. The generalised scheduler model

In order to use constraints in the scheduler model, it is useful to review the degrees of freedom that a schedule imposes on its constituent tasks. These are illustrated in Figure 1 where the use of discrete resources (the machines) is plotted against a continuous time axis. Normally all the routings of the jobs in a schedule would be shown in such a diagram, commonly known as a Gantt chart. These degrees of freedom are important. Without them the problem becomes completely deterministic and an AI approach to its solution ceases to be useful and no opportunities for optimisation arise.

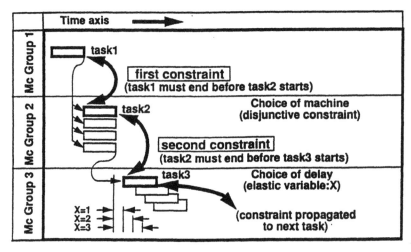

Fig. 1. Degrees of freedom for the tasks of one job

In a routing consisting of a sequence of tasks, clearly the first constraint must be that the start of a subsequent task must occur after the end of its predecessor. Such constraints are conjunctive, which means that they must be true for all routings for a valid schedule to be possible. The second constraint is that each task duration must correspond with an interval during which a machine that can run it is idle. Normally there will be a set of alternative members of a *machine group* on which any particular task can run. This set of alternatives is one of the degrees of freedom that the scheduler can exploit in order to produce a schedule. Because all alternatives need to be tried until a task can be fitted, they are termed *disjunctive constraints* and do not benefit from the type of propagation mechanisms that can be applied to conjunctive constraints. Even with several alternative machines on which each task can run, the competition for free resources may dictate that some tasks cannot be scheduled to start the moment their predecessors finish. Under these circumstances it is necessary to relax some of the sequencing constraints to achieve a solution. This relaxation can be achieved by adding the value of an *elastic variable* to the finish time of a task. Such a variable can be tested with a sequence of increments of ascending value to simulate degrees of relaxation of a constraint until a solution is achieved.

With these degrees of freedom, a selection process can be started whereby a succession of values can be given to all the variables representing members of a machine group and to all the elastic variables used to relax the hard due date or intertask constraints until a combination is found that gives a solution. This successive trial of different values is known as *labeling*. The merit of the constraint propagation mechanism is that it eliminates the need to label individual task start and finish values since the propagation mechanism will automatically derive the possible ranges of all task timings by propagating a job start time forwards through the routing or backwards from the due date.

4. External data formats

Two varieties of external data are needed as input to the scheduler. One is load-related and the other is factory-related. The output of the scheduler is a file — the *schedule*, also displayed as a chart.

Clearly, there must be a *job list* specifying the set of jobs to be loaded onto the factory - with each element in it specifying the batch size (or quantity of material) to be produuced, the due date (or start time), and some parameter specifying the routing that the job is to take through the factory. The latter can be specified as a predefined set *of task profiles*. The elements of the job list have the following format:

```
[job_ID,batch_size,due_date,task_profile_ID]
```

The second class of data that needs to exist, the *setup data*, defines the permanent characteristics of the factory. In it the the number and identity of the machines in each machine group is specified together with, for each individual machine, the identities of the tasks that can be run on it and information that enables the set up and run time to be calculated once the batch size of the task is known. The setup data can exist as a set of (Prolog) facts in a variety of forms depending on how one chooses to sort it. In the model described this data has been ordered on task identity. Each task is associated with all the machines it can run on and their characteristics. These facts have the form:

```
task(task_ID,mgs,
     1  -mc_IDₐ-[ ..properties of mca ..],
     2  -mc_ID_b-[ ..properties of mc_b ..],
     ....../
     mgs-mc_IDₓ-[ ..properties of mcₓ ..]).
```

The enumerable keys (1, 2, .. mgs) correspond to one of the degrees of freedom open to the scheduler — they are the way in which this class of disjunctive constraint is represented in the model and are different from the machine identifiers (mc_ID) by which the machines are known in the factory. Machine properties are numerical constants such as unit production time (e.g., time to machine one item, or rate in units/hour), various classes of setup time and any other machine-specific data needed to calculate the task run time for a particular job.

Finally the setup data must specify what sequence of tasks corresponds to each permissible routing. These routings or task profiles are represented as a set of facts:

```
profile(task_profile_1, [t_11, t_12, t_13, ... t_1n]).
profile(task_profile_2, [t_21, t_22, t_23, ... t_2n]).
....
profile(task_profile_m, [t_m1, t_m2, t_m3, ... t_mn]).
```

The format of the file output by the scheduler, the schedule, is a list of all the tasks successfully scheduled. Each element will include job identity, task identity, machine identity, task start time, task setup time, task run time and a measure of any delay applied to the task to fit it into the schedule.

5. Internal data formats

To schedule one routing, a particularised version of the setup data first needs to be generated that incorporates all possible alternative batch-specific setup and run times for each task that a job expands into, plus the start time (start times rather than due dates will henceforth be assumed). This is then repeated for all jobs in the job list. The elements of the resulting list, the *task list* take the form:

$$[job\_ID, task\_ID, m, \qquad\qquad (3)$$
```
      [1-mc_ID_a-t_setup_1-t_run_1,
       2-mc_ID_b-t_setup_2-t_run_2,
       ...,
       m-mc_ID_x-t_setup_m-t_run_m],
      Ts]
```

where m is the enumerable range of machines with ordinal keys 1 to m which can run the task and Ts is the task start time. Ts will be ground only for the first task. In all other cases it is defined as a finite domain variable.

To exploit the degrees of freedom available in the task list, a set of finite domain variables must be defined that can take on the successive values corresponding to the ranges present in each element of the task list. Likewise an elastic variable must be defined for each task that will be made subject to a discretionary delay. For any particular job they are defined as follows:

```
M_1 :: 1..m_1,
M_2 :: 1..m_2,
...
M_1 :: 1..m_1,
[Td_1, Td_2, Td_3, ...Td_n] :: 0..DL, ...
```

where each M assumes the enumerable range of machines for a task, and each Td assumes an elasticity that can stretch to DL hours (say) of delay.

One other set of finite domain variables remains to be defined before the scheduler proper can be run — these are the variables that hold the mappings of busy and idle time for each machine used by the scheduler. Since the variables are defined as continuous ranges of integers, it is possible to represent busy intervals as a voids in the variable. A variable is created for each machine and carried round in a list in the form:

$$[\ldots, \ mc\_ID_1-X_1, \ mc\_ID_2-X_2, mc\_ID_3-X_3, \ \ldots \ ]$$

The range of the finite domain variables X reflects the time period over which the schedule will run at the requisite granularity. Because of the way these variables are used it is notable that they never become ground.

6. Matching tasks to resources - the scheduler proper

The input to the scheduler, the task list and the list of (initially idle) machines, is a complete declaration of all the options that can be combined in order to create a schedule. In common with other programs based on CHIP, the way the scheduling algorithm works is first to create all the constaints, then test all the conditions and finally to generate values for the constrained variables until a solution is found. Normally the propagation of constraints in the second phase will reduce the range of the values that need to be generated in the final stage. In the model being described the only labeled terms that take part in the constraint propagation process are the elastic variables, which are given a fixed (and small) domain so that it is immaterial whether the labeling is done before or after the testing. It would have been equally possible to label the task start times and not use elastic variables at all, but this would have made delayed tasks difficult to distinguish from non-delayed ones. Since, as will be seen, the scheduler has a facility for manual intervention by the user, there is some merit in being able to highlight delayed tasks.

In order to understand what the scheduling algorithm is doing, the generator will be described first. The generator can be run on a per-job basis or on a per-task basis. To schedule the entire task list, all the tasks in a job can be labeled in one predicate, or a single task taken at a time. The determining factor is whether optimisation is being considered over all the tasks in a job just within a task. In labeling a job of say, n tasks, the generator would take the form of a call:

$$generate([M_1, M_2, M_3, \ \ldots \ M_n, Td_1, Td_2, Td_3, \ \ldots Td_n]).$$

Where the variables M select among the machine options for each task and Td are the elastic variables. `generate/1` uses a built-in predicate `indomain/1` to assign successive values to each variable in the list from the domain of of that variable.

The successive values of M are used to select a machine for each task from the task list *(3)* by matching the current value to one of the ordinal keys to extract the setup and run time of a particular machine. A simple membership test is used, one for each task, as:

`member(M-Mc_ID-T_setup-T_run, Mc_options_sublist),`

$Ttot$, the sum of setup and run times is used in the constraint propagation process, and `Mc_ID` to select the machine occupancy map. Again for all the tasks in a job, the constraints $Ttot$ and Td can be combined in the following constraint propagation sequence:

$$
\begin{aligned}
Tj_f \ &\#= \ Ts_n \ &+ \ Ttot_n \ &+ \ Td_n, \\
Tj_{n-1} \ &\#= \ Ts_{n-1} \ &+ \ Ttot_{n-1} \ &+ \ Td_{n-1}, \\
&\ldots \\
Tj_3 \ &\#= \ Ts_2 \ &+ \ Ttot_2 \ &+ \ Td_2, \\
Tj_2 \ &\#= \ Tj_s \ &+ \ Ttot_1 \ &+ \ Td_1,
\end{aligned}
$$

$\mathbf{Tj_s}$ is the job start time that gets propagated through to $\mathbf{Tj_f}$, the finish time in the process of which values are assigned to all intermediate task start and finish times.

With the selection of a set of machines and the calculation of the task start times for a job, the only remaining action of the scheduler is whether the resulting total task times coincide with points where the corresponding machine resources are unoccupied. First the relevant machine maps are picked from the list by matching on $\mathbf{Mc\_ID}$ to give the domain $\mathbf{X}$ with:

$$\mathbf{member(Mc\_ID-X,\ Mc\_list)},$$

The task start times and durations are combined to give absolute ranges in the units of resolution of the system (say, hours) using the following sequence (for each task):

$$
\begin{array}{llll}
Y_n   & :: Ts_n   & .. \ Tj_f, dom(Y_n,   & R_n \ ), \\
Y_{n-1} & :: Ts_{n-1} & .. \ Ts_n, dom(Y_{n-1}, & R_{n-1}), \\
\dots \\
Y_2   & :: Ts_2   & .. \ Ts_3, dom(Y_2,   & R_2 \ ), \\
Y_1   & :: TJ_s   & .. \ Ts_2, dom(Y_1,   & R_1 \ ), \ \dots
\end{array}
$$

(where $\mathbf{dom/2}$ is a built-in predicate used to turn the domain of variables $\mathbf{Y_r}$ into lists of integers $\mathbf{R_r}$) followed by a test of every element $\mathbf{E_i}$ of the lists $\mathbf{R_r}$ for coincidence with the corresponding values in the domains of $\mathbf{Y_r}$ with the built-in predicate $\mathbf{isindom(Y_r,E_i)}$. The merit if $\mathbf{isindom/2}$ is that the test does not alter the domains of $\mathbf{Y_r}$.

If all these tests succeed for the tasks of a job, the corresponding busy intervals can be created in the domains of the machines and the sublist of job identifier, task identifier, task start time, task setup time, task run time and degree of delay applied, appended to the schedule list. The scheduler can then proceed to deal with the next job. As implemented the scheduler will also represent the part schedule graphically as a set of bars on a Gantt chart.

The algorithm can equally well be adapted to scheduling tasks simultaneously over more than one set of resources. An example that has been succesfully tried is that of a textile mill scheduler, where two separately available kinds of resources are available — looms and beams. The jobs in this case are all one task long and can be fitted to looms with no idle time, but they must also be matched to beam availability as an extra condition. The extra condition is merely added to the scheduler — but it must be preceded by a task list generated from suitably modified setup data. For two different resources the lists of resource options in the setup data are arranged hierarchically with nested enumeration. All the beams that can be used for a task are listed in the setup data as the set of primary options, and for each beam so listed, the looms taht go with the beams as the second level option.

7. Prioritisation & Optimisation

The scheduler as described so far, if successful, will deliver a first solution. Normally the elements of the job list would be pre-sorted in ascending order of start time (or due date), which can make the delay of tasks of late starting jobs unavoidable. Where this is undesirable for particular classes of job, these can be given a priority number in the job list, and the list sorted on these priority values instead, with the most important jobs being ranked first.

Local optimisation is also possible on one or other of the terms influenced by the labeling process. Where there is a difference in performance between members of the same machine group due, say, to age or capacity, it is possible to aim to minimise task setup time or run time or both - or to minimise the cumulative run time for an entire job. Such a policy however may prove to be only partially successful since the faster machines will be favoured over other plant and for late starting tasks this minimisation of run time may be achievable only at the expense of unacceptable task delays. There may therefore be greater merit in minimising task delay by spreading the task load evenly over the members of each machine group. This latter optimisation avoids any accumulation of tasks on particular machines. In each case the goal over which the minimisation is run is the same. It is the basic job-level (optionally task-level) scheduling predicate described in the previous section but excluding the commitment stage, but in one case the objective function is ΣTtot and in the other ΣTd. DECISIONPOWER has a set of four higher level built-in predicates providing optimisation. The following sequence uses minimize/2 to optimise on the cumulative task delay for one job:

```
...
td_cost(Td_list,Td_sum),
minimize(sched_a_job(Task_list_jobset),Td_sum),
...
```

where td_cost/2 sums all the task delay values in one job (which must be domain variables), sched_a_job/2 is the job-level scheduler and Task_list_jobset is a list representing all the elements of the task list pertinent to one job.

8. Graphics and user intervention

The graphics facilities included in DECISIONPOWER are based on the object oriented principles originally developed for PCE[2] and have the merit that a mechanism exists for passing control of specified user-initiated screen events back to nominated predicates in the application program. Since the scheduler described can accommodate jobs that give rise to a sequence of tasks governed by precedence constraints, it is inevitable that idle time will occur. A measure of manual optimisation (or *schedule repair*) is therefore open to the user whereby jobs with delayed tasks can be have their start times moved to fill gaps in the schedule. Such a move, here achieved by "grabbing" the start bar of the job with the cursor and physically moving it over the screen, is made to cause the scheduler to rerun the job list from the earliest time point affected by the move. All jobs scheduled after this point including the moved job are considered to be invalidated and are moved to a new job list for scheduling. Before this can occur the busy intervals of the invalidated tasks need to be removed from the finite domain variables representing the machine use mappings. This involves a relaxation of the constraints imposed so far. Since this is not allowed in the current implementation, these finite domain variables are first converted into lists and the relaxation applied by refilling the voids originally created by the invalidated tasks. The information for this comes from the invalidated part of the old schedule. The domains for the machines are then defined anew and the still valid voids reinserted before the schedule is rerun. This procedure can be run as many times as necessary. An important cause of schedule repair is machine outage. Unforseen withdrawals from service of machines can be transferred to the screen by "clicking" the start & finish times of the unavailable interval with the middle button on the mouse, an action which will create a corresponding void in the domain of the affected machine and force a reschedule from the start of the interval.

The complete Gantt chart is currently created on the fly on a graphics canvas over which the portion viewable through the window can be scrolled. There is however a limit to the number of objects that it is sensible to maintain on the canvas as too many can start to introduce viewing difficulties and performance penalties. An alternative strategy is selectively to display a partial chart from the completed schedule. This is sensible if changes can be kept within the scope of the time interval being displayed.

9. Results and performance

The generalised scheduling problem comes under the complexity class of NP-hard problems. In fact, fitting tasks to machines without time, sequence or machine preference constraints reduces to a *Knapsack* problem. In practice, by exploiting the problem structure and aiming for a first solution that can be subsequently modified, the complexity can be reduced to near linear. Four factors have been found to influence the performance available from the scheduler. The most significant factor is the use of finite domain variables to map the busy intervals of machines. The original alternative of representing busy times as lists of intervals was three times slower and very demanding in memory resources. Using finite domain variables and no optimisation, solutions have been achieved for jobs producing over 5000 tasks. The other three factors influencing performance are optimisation, use of graphics and finite domain variable conversion costs. Local optimisation at task level will reduce the performance significantly but without affecting the linearity of the solution. Since the graphics package is run as a separate process, it will not affect the performance of the scheduler directly, but it does have an effect on elapsed time proportional to the number of objects that need to be displayed. Finally, the conversion of the finite domain variables to and from lists to store and relax them was found to contribute a small overhead. Figure 2 illustrates typical results achieved.

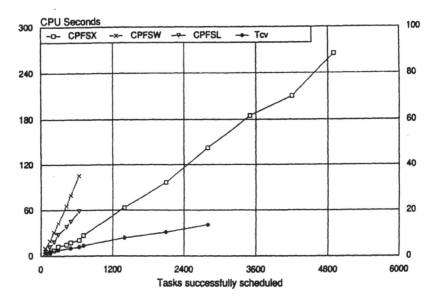

Fig2. Scheduler performance with and without the use of FDVs for mapping machine use

The two poorest performing traces are for cases where finite domain variables were not used for mapping machine use. The lowest line (using the right hand axis) shows how the conversion costs increase with task count. The input data for all the runs was generated by a program that produced start times, task profiles and batch sizes randomly between given limits. Versions of the algorithm have been successfully applied to the textile mill scheduling problem and have achieved 25% reductions in total setup time by preferentially mounting jobs on looms that retain some of the settings from the previous job that are also applicable to the next.

In practice, it is unusual for the output of a scheduler to be used as an exact timetable. Local conditions in factories including availability of staff and material would make this practically impossible. Where the schedulers can contribute significantly is as a means of confirming that the available resources can support a given job load. Also, although some local optimisations can be input via the screen, others, such as the switching of machines between machine groups would involve changes to the setup data and would require an interface other than a Gantt chart if they were to be performed on-line.

10. References

The first two references [1,2] provide an introduction to the technology that went into the design of DECISIONPOWER. In the former the examples should be favoured over the theory. Clearly anyone planning to use the latter will have access to the relevant reference manuals. A useful introduction of the sorts of constraints that can be applied to scheduling problems is given by Fox [3] but not in a logic language and without finite domains. Finally [4] is a useful introduction to the factors that go to influence the design of the setup data needed to run any scheduler.

[1]   Van Hentenryck, P.: *Constraint Satisfaction in Logic Programming*. MIT Press, 1989.

[2]   Anjewierden, A.: An Overview of PCE-Prolog: U. of Amsterdam, 1986.

[3]   Fox, M.S.: *Constraint-Directed Search: A Case Study of Job-Shop Scheduling*. Pitman,London1987.

[4]   Wright, P.K., Bourne, D.A.: *Manufacturing Intelligence*. Addison Wesley, 1988.

®   DECISIONPOWER is a trademark of International Computers Limited.

The work reported in this paper has been done as part of the Esprit CHIC Project (No. 5291) which receives support from the European Commission.

The Prince Project and its Applications

Pierre-Joseph Gailly[1], Wolfgang Krautter[2], Christophe Bisière[3], Sylvie Bescos[1]

[1] BIM, Everberg, Belgium
[2] FAW, Ulm, Germany
[3] CEFI, Les Milles, France

Abstract. The Esprit project Prince aims at development of an industrial Constraint Logic Programing environment based on the Prolog III language. In parallel, the current technology is being validated within the project on representative real-world examples. This paper describes the current status of the three application domains which were selected to demonstrate the applicability and usefulness of CLP. These applications deal with industrial systems engineering, medium term banking planning and jobshop scheduling as well as multiple plants global planning in the chemical industry.

1 General Goals of the Prince Project

The main goals of the Esprit II project Prince (P5246) are to develop a Constraint Logic Programing (CLP) system based on the Prolog III technology and to bring it to high industrial standards as well as to validate the technology by demonstrating its applicability and usefulness on real life applications. Prolog III [6] was developed by the group of Alain Colmerauer and the company PrologIA within the framework of the Esprit I project P1106. This previous project has demonstrated that CLP is an attractive solution for many advanced problems [13]. This paper will focus on application aspects but before that an outline will be given of the CLP system, called Prince Prolog, developed in this project.

Four essential aspects can be seen. First, in terms of expressiveness, it is intended not only to include the constraint domains already existing in Prolog III (i.e. infinite trees, linear algebra (over both floating point and infinite precision rational numbers), boolean algebra and lists; this last one being unique to Prolog III) but also to introduce new domains based on end-users' requirements (finite domains and interval arithmetic are being currently considered). Soundness and effectiveness of constraint solving algorithms receive careful attention. As was already the case with Prolog III, full integration of constraints into the kernel of the language and clear semantics of the interaction of constraints with the rest of the system are considered very important. Second is the efficiency issue: a completely new compiler-based system is being developed, drawing on the experience of two major Prolog manufacturers PrologIA and BIM and in particular the ProLog *by* BIM and Prolog II+ technologies. Besides a new kernel and improved constraint solvers is the third important aspect: investigation of global analysis and precompilation of CLP programs and incorporation of the results into the system. This longer term research is being performed by academic partners from the Universities of Bristol, Leuven and Madrid. Finally, the software engineering environment should provide easy, rapid and reliable development facilities as well as communication capabilities with the external world (Windowing, Databases, Network, other programming languages). In particular, constraints debugging is a new field which still requires much exploration.

Designing tools is one thing, putting them to practical use is another issue. This is even more true in software engineering. Industrial partners are validating the CLP technology by its "mise en œuvre" to solve real life problems. The manufacturers Bosch and MBB companies as well as the research institute FAW are involved in engineering applications for technical systems quality assurance. Two examples are under investigation: satellite attitude control and Failure Mode and Effect Analysis, the latter one being discussed in more detail below. The CEFI research center and La Hénin bank are developing a Decision Support System in the field of medium term banking planning. BIM is applying CLP technology to jobshop scheduling and multiple plants global planning tools in the chemical field. Besides showing the usability of the language, the application partners have provided important feedback to the implementors of Prince Prolog on the language itself, the selection of constraint domains, methodology and tools.

The different applications do not put forward the properties of CLP languages in the same way; however, the presentation will emphasize following characteristics:

- CLP languages are high level, making development and prototyping easier.
- Prolog enables the elaboration of reversible *symbolic* computer based model of real-life applications. CLP adds the domains dimension to the scene; in particular, Prolog III's *linear numerical and boolean* domains will be illustrated.
- Besides the built-in constraints, Prolog III's delayed goals can provide users with some facilities to break the linearity limitations.
- Combinatiorial search can be pruned more efficiently with the help of constraints.

2 The FMEA Application

2.1 Objectives

Increasing the quality of technical systems, in particular, identifying weak points, evaluating the effects of such weaknesses and the associated possible risks as well as determining their causes and investigating alternatives and improvements are very important. Failure Mode and Effect Analysis (FMEA) is a technique aiming at that.

In the Promotex project [13], FMEA of single components in stationary states has been studied. The project has shown that Prolog III is well suited to serve as a formal representation language to describe both the function of the components in the correct and faulty states and the structure of the system. In the Promotex system, component functions have been expressed in the form of systems of numerical constraints. In this project, the type of analysis has been extended at two levels: performing FMEA at the system rather than the component level and tackling dynamic aspects. Essentially two tasks are to be performed:

- investigate the consequences of known component failure modes,
- determine whether some faulty (possibly hazardous) system state can be reached as the consequence of failure modes of the components.

The first point can be addressed by classical simulation tools; the second requires more. The "multidirectionality" of Prolog III constraints enable reasoning on dynamic systems. Possible techniques are: numerical solution of differential equations, Petri Nets, methods of qualitative physics, simulation and process theory.

Figure 1 shows an example system: a closed loop speed controlled motor. The motor drives a sensor tooth wheel that produces 4 interrupts per revolution. The interrupt routine measures the time elapsed between two interrupts and estimates the motor speed. Each time a pulse is generated, the proportional controller changes the voltage with respect to the nominal speed. This system contains continuous elements (motor) as well as discrete events (pulse, interrupt). Possible failures are, for example, sensor wheel slip or tooth breakage; these are represented by parameters in the model. Possible failure effects are speed oscillations or deviations from the nominal speed.

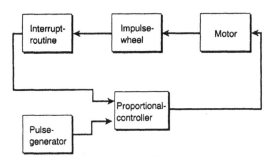

Fig. 1. Closed loop speed-controlled motor

2.2   Continuous Simulation

In continuous simulation, differential equations are solved numerically. If the solving methods are implemented using numerical constraints, the simulation may "run forward or backward", choosing to fix the value of some boundary conditions or system constants, leaving the others unknown and to be found by the simulation. Values cannot be fixed or left unknown arbitrarily as, at runtime, contraints must be linear. In addition to the common simulation run, the Prolog III mechanism allows the following queries:

- given the solution at certain time steps or intervals, infer single parameters of the differential equations. For this purpose, failure modes were modelled as parameters in the system.
- representation of parameters in input and output values as numerical intervals[4].

Example: the speed $n(t)$ of the motor can be represented with the differential equation

$$\frac{dn(t)}{dt} = \frac{-n(t) + k * v(t)}{Tc} \stackrel{\text{def}}{=} f(t, n(t))$$

where n is the speed and t is time; v, the voltage, k, an amplification factor, and Tc a time constant are model parameters. The trapezoidal method for solving a differential equation numerically is:

$$n(t_{i+1}) = n(t_i) + \frac{t_{i+1} - t_i}{2} * (f(t_i, n(t_i)) + f(t_{i+1}, n(t_{i+1})))$$

This can be implemented using Prolog III as follows. The **step** predicate has three arguments: the first represents the initial state in the form of a tuple <Time, Speed>; the second contains the system parameters (with $Dt = (t_{i+1} - t_i)$) and the third is a list of the successive states starting from the initial one[5].

---

[4] This was done using inequality constraints and **maximum** and **minimum** built-in predicates.

[5] The number of simulation steps being specified via constraints on the size of this list.

```
step(V, C, <>).
step(<Ti, Ni>, <Tc,K,V,Dt>, <<Tip1, Nip1>>.L) :-
  step(<Tip1, Nip1>, <Tc,K,V,Dt>, 1),
  {Ti  = Tip1 + Dt,
   Nip1 = Ni + Dt/2 * ((- Ni + K*V)) + (- Nip1 + K*V))/Tc)}.
```

Assume that the motor is initially stopped and that it should reach a speed between 5990 and 6000 during the last five steps of a 500 step simulation. The following query determines the range of possible values for the unknown amplification factor under the above constraints.

```
>step(<0,0>, <0.05, K, 6, 0.001>, L.<<T1,N1>,<T2,N2>,<T3,N3>,<T4,N4>,<T5,N5>>)
  minimum(K,M1) maximum(K,M2) out(<M1,M2>) line fail,
  {L::495,
    5990<N1<6000, 5990<N2<6000, 5990<N3<6000, 5990<N4<6000, 5990<N5<6000}.
```

Prolog III yields the answer: K should be in the range <998.387, 1000.5>. When dealing with more complex systems (e.g. the example system), the computation time gets critical because of the large number of choices induced by greater model complexity.

2.3 Petri Nets

Petri Nets are a method for performing discrete-event simulation, that has also been investigated for system safety analysis [10]. In particular, the ability to reverse the simulation process is of interest. To represent a system as a Petri Net, the standard Place-Token nets (that can be analyzed and reversed using linear algebra) are not sufficient. They have to be extended in two ways: addition of delays to the transitions to express the duration of the actions, individual firing conditions, data types and functional instructions in the transitions.

If the firing conditions and functional instructions are restricted to linear expressions, transition firing can be reversed. This is a limitation, but nevertheless an important enhancement over Place-Token Nets. Figure 2 shows the representation of the example system. The petri net is described with the help of the **transition** predicate

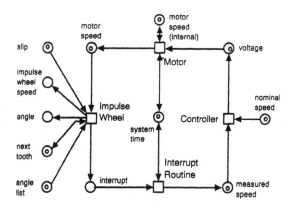

Fig. 2. The example system represented as Petri Net

in Prolog III. For example, the controller is defined by the following clause. It indicates that the measured and nominal speed are related to the voltage via the linear expression below, where R is retrieved from the model via the **const** predicate. Furthermore, this transition does not add any delay.

```
transition("Controller",
          input_places(<<"measured_speed", Ms>, <"nominal_speed", Ns>>),
          output_places(<<"voltage",V>>),
          delay(<0,0>)) :-
  const("R",R),
  { V = R * (2 * Ns  - Ms)}.
```

This transition description can be used both forward (computing output places from input ones) and backward. One of the drawbacks of Petri Nets is their emphasis on the event-oriented aspects of a system, while continuous processes are rather difficult to integrate. Furthermore, describing an industrial system in form of places and transitions requires a rather high level of abstraction.

2.4 Qualitative Methods

In addition to traditional simulation methods, using qualitative reasoning for system analysis purposes is being investigated. Qualitative reasoning describes a system and predicts its behaviour in qualitative terms, e.g. "raising", "oscillating" or "constant". This requires a higher level of abstraction than for a numerical simulation. Various qualitative methods have been proposed; for an overview see [14]. The possible use of Prolog III will be described on two examples:

- Kuiper's QSIM system [7] describes a system in terms of qualitative states, consisting of a qualitative value and a qualitative direction (one of decreasing, increasing and steady) at a time instance or during an interval. The behaviour of the system is then simulated by examining possible state transitions. Allowed transitions are held in a table. Between the functions the relations addition, multiplication, minus, proportionality and derivation may hold. QSIM then gives rules constraining qualitative values and direction of changes. Expressing these rules using Prolog III equality and disequality constraints makes both the programming easy and efficiently filters out forbidden states or transitions combinations.
- Allen's Calculus of Time [1] proposes a model based on time intervals. The relationship between two intervals can be expressed using 7 base relations and their inverses. The goal is to find the transitive relation between more than two intervals. In Prolog III, time intervals can be represented as tuples of two variables representing the start and end points. Constraints are used to express relations between intervals. The modelling of a relation will require to describe explicitly when the relation holds and when it does not hold. For example, the predicate relation(Name, Truth_val, Int_1, Int_2) will succeed whenever the relation called Name between intervals Int_1 and Int_2 has truth value Truth_val. The following clauses describe the during relationship which is satisfied if and only if the first interval is strictly enclosed in the second (where 1' and 0' represent the booleans true and false).

```
relation(during, 1', <S1,E1>, <S2,E2>) :- { S1 > S2,  E1 < E2 }.
relation(during, 0', <S1,E1>, <S2,E2>) :- { S1 <= S2 }.
relation(during, 0', <S1,E1>, <S2,E2>) :- { E1 >= E2 }.
```

With boolean constraints, all propositional logic operators can be used. For example, assume three intervals i1, i2, i3; also assume that one of the relations equal, during, overlapped_by, start, started_by, finish holds between i1 and i2. The following query determines which relations may hold between i1 and i3?

```
>   positive(<I1,I2,I3>),              % intervals must have a positive length
    relation( equal,        B1, I1, I2),
    relation( during,       B2, I1, I2),
    relation( overlapped_by, B3, I1, I2),
    relation( start,        B4, I1, I2),
    relation( started_by,   B5, I1, I2),
    relation( finish,       B6, I1, I2),
    relation( meet, 1', I2, I3),
    findall(R, relation(R, 1', I1, I3), L),
{ B1 | B2 | B3 | B4 | B5 | B6 = 1'}.
```

The answer is: L = [finished_by, contains, overlaps, before, meet]

Prolog III is well suited to implement the methods of qualitative physics. Yet, the loss of information as a consequence of abstraction is a problem inherent to qualitative methods. It is not yet clear if they can be used for FMEA purposes.

3   The Banking Application

3.1   General Overview

The central point in the development of a Decision Support System (DSS) in the field of medium term banking planning is reversibility, i.e. to bypass the existing frontier between *simulation* and *decision* systems in the financial domain. The DSS has been specified and a Prolog III prototype is being developed.

The application relies on a medium term banking model, mainly oriented towards interest rate risk management. Interest rate risk comes from a possible mismatch between asset and liability structure of the bank's balance sheet: if a loan is refunded by a borrowing of a different nature (i.e. short/long maturity, fixed/floating interest rate, etc.), the future net income of this production will be affected by fluctuations in interest rates. Because such a mismatch usually increases expected earnings and risk at the same time, funding decisions will be stated according to the bank's global risk preference. The model has been built in collaboration with the La Hénin bank. Basically, it can be viewed as a set of interconnected modules, each of them dealing in detail with a particular aspect of banking activity: outstanding loans evolution, funding policy, evaluation of potential risks, expected earnings, environmental constraints, etc. When used in simulation, this model is fed with a set of hypotheses and decisions. It then works out a medium term forecast for the main banking aggregates, under the form of a balance sheet, as well as an evaluation of the risk-return position that has been reached.

Of course, analysis of the consequences induced by alternative decisions (i.e. "what if" analysis) is an important step of the banking planning process. Nonetheless, this procedure remains limited, because the ultimate goal of decision makers is to choose an action which allows them to reach a given objective (i.e. "what for" analysis). Using conventional languages and approaches, two different methods can be used to perform this goal-oriented search. The first one is to repeatedly run the simulation algorithm with slight variations on the decision variables values (in particular, the funding policy), in order to get closer to the expected risk-return position. This method is time consuming and, more important, has little chance to "smoothly converge" to the desired point, because of the complicated form of a real-life banking model equations.

The second approach is to build a "reverse" version of the simulation model, which can generate the funding decision according to a given and specific risk-return position. The model has then to be rewritten for this particular purpose, and will have to be rewritten each time the head management wishes to ask another type of query. Besides development and maintenance costs, this solution also implies knowledge duplication.

In this context, CLP languages are of high interest. Their original features seem able to facilitate the development of flexible and powerful DSS in the financial domain [4, 5, 9]. Their declarative aspect and reversibility, extended to the numerical domain, leads to faster and more straightforward implementation of the model. The resulting "knowledge base" remains unique, but can be used for simulations as well as for goal seeking queries ([3] for a qualitative reasoning approach). The search space exploration algorithm, combined with pruning facilities, helps to efficiently solve complicated cases of goal seeking queries (involving piecewise linear, or non-monotonous functions). In addition, the goal delaying mechanism provided by Prolog III can be used to improve the treatment of non-linear equations; [8] provides a formal treatment of this problem.

How the linear part of the model is tackled will be explained first; then the treatment of nonlinearities will be explained. Because of the application domain involved, focus will be on numerical constraints.

3.2 Using CLP in the Linear Case

Prolog III cannot deal directly with non-linear constraints and uses a delayed constraint mechanism to handle multiplicative non-linearities[6]. Assume, for the moment, that the banking model is just a set of linear equations and inequalities. Then, this set can be directly implemented as a set of numerical constraints (S), which therefore represents the structure of the bank's behaviour. In this context, to "run" the model means to add to S another set (Q), which represents the user's query. This last set is mainly made of simple equations of the form $variable = value$. The resulting set $A = S \cup Q$ then represents the "answer" to the user's query. This very simple approach allows to integrate simulation and decision model, because of reversibility. The user is therefore free to ask any question that make sense to him.

Furthermore, this approach can be combined with the backtracking mechanism of CLP languages in the following way. When the banking application is started, the set S is created. From this unique "root", the software then develops small branches, each of them corresponding to a user's query. At each step, the system follows a branch in order to solve a query (i.e. to add a set Q and to display the result), and then backtracks to the root, ready for the next step. By doing so, it is possible to avoid an important part of the computational cost overhead induced by the non-specialisation of the decision support system. All this is summarised in Fig. 3.

3.3 Dealing with Hard Constraints

The actual model also contains non-linear equations. Two classes of non-linear equations can be distinguished: those corresponding to disjunctive constraints which can therefore induce a whole search process during goal seeking analysis (e.g. non-strictly monotonous functions, piecewise linear functions,...) and the others (e.g. involving logarithmic or exponential function or quadratic functions over positive numbers,...).

---

[6] I.e. when, at runtime, the added constraint involves the product of two unknown variables.

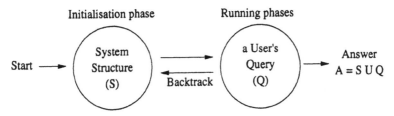

Fig. 3. Functioning of the application in the linear case

Consider a model made of linear equations and non-linear equations of the second class. During the initialisation phase of the application, only linear equations can be introduced in the set S application. The others have to be delayed, hoping that each set Q will bring enough information to introduce them in the whole set A. Consider the following non-linear equation:

$$P = \frac{C}{R} + \frac{NR - C}{R(1+R)^n}$$

Depending on the known variables, the following methods can be used:

- if R and n are known, the equation becomes linear,
- if all variables except n are known, n can be determined by calculus,
- if all variables except R are known, it can be iteratively approximated.

This simple strategy has been programmed in a generic way, using the **freeze** mechanism of Prolog III. A set of predicates has been designed, which facilitate the implementation of these non-linear equations. Using this toolbox, it is only necessary to "declare" the methods which can be used to incorporate an equation under the form of a linear constraint in A, and the conditions under which each method can be activated. Of course, this approach does not transform a CLP language into a general purpose mathematical solver, counter-examples (leading to "deadlocks") can be found easily. However, because of the particular semantics of the models which must be dealt with, these deadlocks can be precisely identified, and then solved according to the previous strategy.

Non-linear equations of the first class are handled in nearly the same way, taking care of the virtual combinatorial explosion they could induce in goal seeking mode. In the case of a piecewise linear function, this danger is clearly shown by the corresponding disjunctive writing. For a classical non-monotonous function, the problem arises from the fact that its inverse function is not univoque. Obviously, these equations cannot be activated during the initialisation phase of the application (see Fig. 3). The corresponding goals are therefore delayed, waiting for more information. Then, a strategy should be stated to organise for their best handling. In the application, a set of heuristics has been designed to achieve this task. This additional feature leads to the general structure drawn in Fig. 4.

The presentation only showed some particular aspects of the financial application. Other aspects include temporal shift detection or sensitivity analysis. In the future, one of the most ambitious features which will be addressed is the design an interface module, aimed at assisting users in their quest for a good decision. To do so, this module will have to "interpret" sets of constraints returned by the system.

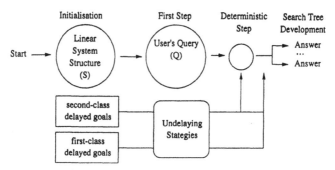

Fig. 4. Functioning of the application in the general case

4 Scheduling Applications

Job-shop scheduling of a chemical plant is a real-life example taken from the background of the EUREKA project PROTOS [2, 11]. A set of products has to be produced on different apparatus. A given apparatus can be used in the production of several products (multipurpose apparatus environment). Each of the products is produced in a single process where each process may have several (1 to 3) production variants (i.e. recipes). Each production variant is decomposed into a finite number (from 5 to 20) of steps. For each production step, one apparatus will be chosen from a set of alternatives. The production steps have to be performed in a specific continuous sequence (production cycle). When a larger quantity than the one delivered in one production cycle is required, the cycle is repeated. The problem consists in finding a time interval within the earliest possible starting date (all inputs have to be available) and the latest possible due date (management requirement) for the production process of each product.

The second example deals with the global planning of several plants. The whole production of a large Swiss pharmaceutical company is split over several plants. The aim is to compute a global production plan for all these plants. Up to now, no such global plan existed and all the coordination and production process adjustments between the different plants are achieved through phone calls between plant managers; there is no global control. This scheme works because of the plant managers' experience, but there is a high risk of the result not being optimal. If a good global plan could be provided, ensuring that no major coordination problem should occur, then each plant could make local optimisations as long as the constraints imposed by the global plan are respected; also the resulting production process would become much closer to optimality. As a side effect, this global plan would also reduce the need for the phone call based coordination, although it is not expected to suppress it totally.

As it is far too complex to take into account all details of the local data of each individual plant, the considered global planning tool is based on an approximation of the local reality: the abstraction of individual machines in machine groups (A machine group is a set of machines located physically close to each other, and each order can be completely executed using only machines within one machine group). Thus, the output of this tool is only a "rough" global plan, that will then be further refined at each plant, by the local scheduling tool.

Both examples were implemented; essentially, two types of constraints were used: precedence constraints, straightforwardly expressed by numerical inequalities, and non-overlapping tasks sharing the same resource, which can only be expressed by choices.

5 Conclusions

The previous examples have shown the interest of Constraint Logic Programming in several real life application fields. The experience of the end-users who developed these examples is that CLP and Prolog III are good software-engineering tools for dealing with such problems, leading to reasonably concise and elegant solutions. In particular, the ease to implement, and hence compare, variations of algorithms combining heuristic and symbolic computation with constraints, has been appreciated.

Both the technical systems and banking applications have shown CLP's adequacy to define models which encode both structure and functionnality. The possibility of avoiding as much operational aspects in the model's definition is very important and enables to use the same model in several ways: direct execution or simulation and backward execution or goal seeking analysis. This form of reversibility extends the knowledge representation paradigms and increases their declarativity for non-symbolic domains; linear numerical and boolean examples were given. Non linearities raise difficulties but some practical, even if not general, methods to deal with them were outlined.

Applications have also shown the current limitations of the Prolog III interpreter. Efficiency, a common concern, is being improved by the development of the Prince compiler. Having other or more specialised constraint domains (such as being able to use constraints as choices in scheduling) would increase the expressive power, and hence the ease of use of the language. Such topics are currently being intensively studied in the framework of the Prince project, but this falls outside the scope of this paper.

The authors would like to thank the members of the Prince teams and in particular Paul A. Massey, for fruitful discussions and comments on earlier drafts of this paper.

References

1. Allen, J.: Towards a General Theory of Action and Time. AI V.23, pp.123-154, 1984.
2. Appelrath, H.-J.: PROTOS: Prolog Tools for Building Expert Systems - a Project Overview. Procs. of the 1st. PROTOS Workshop, September 1989.
3. Berndsen, R., Berthier, F.: Goal Seeking in Qualitative Reasoning: an Implementation in CHIP. in Procs. of IMACS international symposium, March 1991.
4. Berthier, F.: Solving Financial Decision Problems in CHIP. Procs 2nd Conf. on Economics and Artificial Intelligence - CECOIA 2, pp. 233-238, 1990.
5. Broek, J., Daniels, H.: A Constraint Logic Programming Approach to ALM Modeling in Banks. Comp. Sci. in Economics and Management, Kluwer Acad. Press, 4(2), 1991.
6. Colmerauer, A.: An Introduction to PROLOG-III, CACM, Vol. 33, N. 7, July 1990.
7. Kuipers,B.: Qualitative Simulation. AI, V. 29, pp.289-338, 1986.
8. Jaffar, J., Michaylov, S., Yap, R.: A Methodology for Managing Hard Constraints in CLP Systems. Procs. of the ACM SIGPLAN PLDI Conf., pp. 306-316, June 1991.
9. Lassez, C., McAloon, K., Yap, R.: Constraint Logic Programming and Option Trading. IEEE Expert, 2(3), 1987.
10. Leveson, N.G., Stolzy, J.L.: Safety Analysis Using Petri Nets. IEEE Trans. on Software Engineering, Vol. SE-13, No.3 (1987), pp.386-397.
11. M. Nussbaum and L. Slahor. Production Planning and Scheduling: A Bottom-up Approach. Procs. of the 1st PROTOS Workshop, September 1989.
12. ESPRIT-Project 5246: PRINCE. Report 17: Specification of the FMEA application.
13. ESPRIT-Project 1106: Further Development of Prolog and its Validation by KBS in Technical Areas. Final Report Part 2: Validation. 1990
14. Weld,D.S.,de Kleer,J.: Readings in Qualitative Reasoning about Physical Systems. Morgan Kaufmann, 1990.

A (Gentle) Introduction to Deductive Databases

Shalom Tsur

University of Texas System, Center for High Performance Computing,
Austin, TX, USA

Abstract. This paper is intended as an introduction to deductive databases for the practitioner. After a brief historical overview it discusses the tradeoff between declarative and procedural programming and introduces the topic of deductive databases as an instance of this issue. The paper proceeds to expose some of the features of DD technology by means of the LDL system, shows some typical DD applications and concludes with a discussion of the state of the art and the relation of this technology to that of Object Oriented systems.

1 Introduction

The term *deductive database* is used to denote a technology that extends relational database technology and combines it with certain features of logic programming. The deductive database technology provides a powerful query system that subsumes the current practice of application development by means of a combination of procedural programs and embedded queries. Instead, it provides one uniform language that is powerful enough to specify the problem in its entirety and thus, enables global optimization, both over the application as well as the query portions of the program.

Deductive Databases (DD) evolved during the 80's. The background that led to this development was the maturing of the relational database technology and an appreciation of its strength as well as its weakness. The strength of the relational approach is clearly the *declarative method* of query specification as embodied in such query languages as SQL. The user specifies *what* subset of data needs to be retrieved from the database and leaves the *how* of it to the system itself. A user is thus absolved from the need to plan and optimize the access to the dataset of interest and this function is executed on her behalf by the database system itself. The result is a significant increase in the data independence and maintainability of the data over an extended time. On the downside, users became painfully aware of the limited expressive power of relational query languages and their inability to express certain classes of queries, in particular recursive queries.

Another influence on the evolution of DD technology was the parallel development in the field of logic programming, particularly the use of Horn clauses as a programming device. This lead to the idea of using logical rules as an extension to the relational data model: rule heads express derived relations that can be used to express queries. The abstract queries that can be expressed using these rule extensions are compiled and optimized by the system into extended programs and queries against the underlying data. The rule extensions allow e.g., to express recursive queries. The seminal example is that of the ancestor relation shown in section 3.

This paper is intended as a introduction for the practitioner to the topic area of deductive databases. We provide an overview and illustrate some of the issues by means of examples. No attempt is made here to treat the formal aspects of the theory underlying the technology or the compilation/optimization methods that have been developed for it. After a general discussion of issues pertaining to procedural and declarative programming we move to an exposition of one particular DD system. The other two topics that will be briefly discussed are typical applications of the DD technology and the relation to Object Oriented systems. The LDL++ system, presently under development, combines logical rules with external data types defined as C++ classes.

The present status of technology today is that some of the prototype systems have reached a strength of pre-commercial deployment and at least one vendor has serious plans to develop it into a commercial product.

2 Declarative vs. Procedural Programming

Deductive database technology is one aspect of the larger issue of declarative vs. procedural programming. The latter style is the more common and is embodied in such programming languages as C, FORTRAN, COBOL and others. Essentially, procedural programming is a style of specification that requires *how* a problem is to be solved. In contrast, declarative programming is a style that requires *what* problem is to be solved and leaves the "how" of it to the system itself. Consequently, the procedural programming languages require the programmer to specify the problem of interest in terms of a *sequence* of steps, leading from the data to the result. These languages are featured with *control statements* that enable the programmer to modify the order in which the steps are executed. Using this style, the meaning of the program is determined by the order of specification and different sequences of the same steps have different meanings.

Declarative programming is perhaps best known in the form of query languages to relational database systems such as SQL. Using this language, a programmer specifies which *set* of the data is to be retrieved from the database and leaves the creation of an access plan and its optimization to the DBMS itself. To illustrate the differences in these programming styles consider the following example.

Example: File search

Given, a dataset of the form: {*Author, Title, Year_of_Publication*}.
Problem: "Retrieve all publications of a given author as recent as a given year" e.g., *All publications of A.A. Turing since 1930*.

The procedural solution to this problem can assume at least two forms from which the programmer has to select:

1. Store and search the data sequentially by the *Author* attribute. When *Author* = "A.A. Turing" compare the *Year_of_Publication* with 1930 and add to the result if True.

2. Alternatively, build an index on *Year_of_Publication*. Collect all items for which Value > *1930*, check to see if corresponding Author is "A.A. Turing".

Clearly, of the two strategies listed above there is no one that is always superior. The best strategy would depend on such factors as the selectivity of each data element in its domain, the cost of building and maintaining an index as well as other factors. All of these optimization considerations are the responsibility of the programmer. The declarative specification of this problem would be of the form:

$$\{Title \mid Author = "A.A.Turing" \cap Year\_of\_Publication > 1930\}$$

i.e., the problem is specified as the intersection of two sets. The storage and access details as well as the decision to build/not build an index are left to the system. 2

It would seem obvious that the declarative method has advantages over the procedural method. The specification is more concise, it achieves data independence and hence the application needs not be changed when changes are introduced to the underlying system or data and it would be easier to maintain over the (long) lifetime of the data. On the other hand, the declarative method forces the programmer to give up control over the order of execution even when the order of execution forms part of the specification. This is especially true in certain update situations in which the order of insertion/deletion may affect the end result. In reality, therefore, it is often required to combine both styles of programming so as to achieve the desired result.

The interplay between the procedural and declarative styles of programming in problem solving is the central theme of this paper and we will elaborate on this issue in the following sections. We will develop the theme by

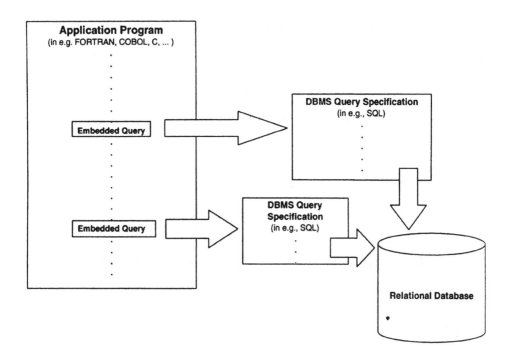

Figure 1: Problem Solving: Existing Order

visiting three worlds in which different orders exist. The present world, an ideal world and finally, a world in-between that comes close to a desired reality.

2.1 Problem Solving: The Existing Order

The existing order in problem solving is schematically described in Figure 1.

A problem to be solved is specified as an application program in a procedural language such as C or FORTRAN. Within this application declarative queries specified in e.g. SQL are embedded. In this order a strict separation between the application program and the DB queries is maintained. The main shortcomings of this order are:

- Limited expressive power of the query language.

- The impedance mismatch problem.

let us elaborate on each of these points.

Limited expressive power of relational query languages

Relational query languages are limited in the class of queries they can express. Certain queries *cannot* be formulated in SQL and such queries are often of interest. Consider the following problems:

- Give a database relation of the form *(Parent, Child)*. For a given child, derive all of its ancestors.

- Given a database relation of the form *(Name, Height)*. Derive the tallest Person(s) in the database.

- Given a network database of the form *(Node1, Node2)* for each pair of nodes connected by an edge. Can we reach node *A* from node *B?*.

Each of these queries require the development of an application program—an activity that we want to avoid if possible—given the expenses involved. Rather, we would like to be able to pose queries of this type directly against the database. To do so however we need a query language which is more powerful than the existing relational languages.

The impedance mismatch problem

Procedural languages utilize *atoms* (integers, reals, strings etc.) as their basic data elements. Query languages, on the other hand, use sets as their basic data elements. This incompatibility of data granularity is the cause of the impedance mismatch problem: data sets retrieved from the DBMS need to be analyzed element-by-element at the application level and need to be (re)constructed as sets at the query level. The impedance mismatch problem causes severe performance problems and prevents the global optimization over application and query programs.

The recognition of the two problems described in this section led to the desire for a new order that we will describe in the next section.

2.2 Problem Solving: New Order

Under the new order, depicted in Figure 2, the need for application programming would be entirely eliminated. The programming system would be endowed with powerful declarative capabilities that would be used to specify both the application and query portions of a problem. Consequently, the two problems that we raised in the previous section viz. limited expressive capabilities and the impedance mismatch problem would disappear. The former problem because the language used would be more powerful than SQL and the latter because no procedural application programming is required. The

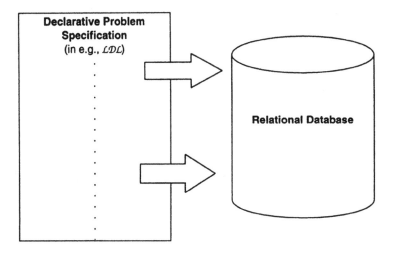

Figure 2: Problem Solving: New Order

declarative specifications produced would be compiled by the system into (procedural) target programs for execution.

Some of the requirements of new-order systems can be summarized as follows:

- At a minimum, the capability to express all of the queries expressible at the underlying relational level.

- A rule-based capability to express super-relational queries.

- A well-defined semantics, consistent with the database paradigm.

- Support for efficient compilation methods.

- An open architecture: the capability to integrate within existing procedural/relational environments.

The last requirement in the list above addresses the fact that we do not operate in a vacuum. We must to recognize the reality in which we operate, which consists of a large legacy of existing programs that cannot simply be eliminated or rewritten. A more realistic approach, therefore, would enable us to combine these elements.

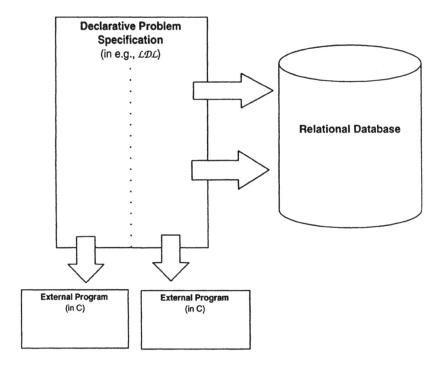

Figure 3: Problem Solving: Actually

2.3 Problem Solving: Actually...

Figure 3 depicts an order in which we combine declarative specifications with procedural ones. We retain the advantages of declarative programming but allow for procedural extensions. Either because we want to utilize existing programs such as libraries for statistical routines, or window display programs or because we want to use procedural constructs such as *if-then-else* or *forever* as a matter of convenience. We also mentioned the need to specify an order in certain update operations.

Ideally therefore,

> We desire a system in which the declarative and procedural capabilities blend seamlessly. We want to be as procedural or as declarative as the problem needs dictate! We desire a well-defined semantics with a system of this type.

While such systems do not exist we will describe one system—LDL (Logical Data Language)—as an approximation to this ideal.

3 LDL

LDL was the focus of a research program at the Microelectronics and Computer Technology Corporation (MCC), Austin, TX., during 1985-91. This effort has resulted in the development of a theory underlying deductive database systems as well as a number of advanced prototypes that implement these principles. In addition, a significant effort has been spent on the identification of suitable application areas for this technology. Because of the author's involvement and familiarity with this project it will be used here for expository purposes. Other deductive database projects include the NAIL! project at Stanford University (Ullman [1989], Phipps [1990], and Phipps, Derr and Ross [1991]), CORAL at the University of Wisconsin (Ramakrishnan [1990]), Aditi at the University of Melbourne (Vaghani et al. [1990]), LOGRES at the University of Milan (Cacace et al. [1990]) and the DedGin project at ECRC (Lefebre and Vielle [1989]). The reader is referred to these sources for further reading.

3.1 Some LDL features

An LDL program consists of four components:

1. A *Schema Description* declaring the relations stored in the database and their argument types. Synonymously we refer to this component as the *Extensional Data Base (EDB)*.

2. The *Dataset* containing the stored data.

3. A *Rule-set* providing by means of logical rules definitions of *derived relations*. We also refer to this component as the *Intensional Data Base (IDB)*.

4. One or more *query forms*. These are generic queries, in which the programmer declares which of the arguments are given and which are to be returned as the result. Each of these query forms is a unit of compilation and is translated into an underlying access program w.r.t the stored data. The target program generated is in C.

As an example consider the "ancestor" program mentioned earlier, as an example of a super-relational query, beyond the capability of SQL.

Example: Ancestor

1. Schema:

```
database({parent(Parent:string,Child:string)}).
```

2. Data:

```
parent('Jill','Joe').
parent('Joe','Jim').
parent('Joe','Sarah').
...
```

3. Rules:

```
(1) ancestor(X,Y) <- parent(X,Y).
(2) ancestor(X,Y) <- ancestor(X,Z),parent(Z,Y).
```

Expressed in English: rule 1 states that X is an ancestor of Y if X is a parent of Y or, as rule 2 states, X is an ancestor of Y if there exists a Z such that X is an ancestor of Z and Z is a parent of Y.

4. Query form: `ancestor(X,$Y)`. For all ancestor queries, all values of the first argument are returned for each value of the second argument given.

5. A query instance: The query `?ancestor(X,'John')` derives all of John's ancestors, direct or indirect. Note that we instantiate the query form by means of substituting a specific value for the second (bound) argument.

2

In this example the ancestor relation is defined by means of rules describing the intention to the extensional parent data. By comparison, solving this problem within the existing order would entail extracting *all* of the parent data by means of an SQL query, the development of an application program that would compute *everybody's* ancestors and a post-selection of those results relevant to John. Clearly this is inefficient! The deductive database system will transmit (via the query form) the knowledge that we require only stored facts pertinent to John's ancestors and hence global optimization can be performed.

Another example of a super-relational query is an extermum.

Example: Tallest person

1. Schema:

```
database({person(Name:string,Height:real)}).
```

2. Data:

```
person('Jill',5.5).
person('Joe',6.2).
person('Jim',6.2).
person('Janie', 6.5)
```

3. Rules:

```
person_taller_than(Height) <-
          person(_,Height1), Height1 > Height.
tallest(X) <-
       person(X,Height), ~person_taller_than(Height).
```

Expressed in English: X is(are) the tallest person(s) if there is no other person taller than X. We have used negation, expressed using the $\sim$ symbol, to derive these extrema data.

4. Query forms: `tallest(X)` derives all of the tallest persons in the database. `tallest($X)` confirms/denies the assertion that the given person is/ is not the tallest person.

2

The power of LDL is in its capability to express by means of rules, very abstract relationships that are derived from the stored data. Powerful compilation methods translate these rule-based specifications into optimized C programs that are executed against the stored data. For a complete exposition of this system the reader is referred to Naqvi and Tsur [1989], and Tsur and Garrison [1991]. The LDL system subsumes thus the normal activity of application program development by the automatic generation of these programs.

3.2 Comparison with Prolog

Although the syntax of LDL resembles that of Prolog, the meaning of the rules is different: in Prolog the rules are interpreted top-down from the goal via the rules to the facts. The interpretation uses SLD resolution and unification as its principal computation and inference mechanisms. Thus, goal resolution in Prolog may produce *an answer*: a proof that there is an instantiation of the variables such that the goal can be derived via the rules and facts. On the other hand, the computation may be non-terminating and the termination may be dependent on the ordering of the rules or the goals within a rule body. In this respect Prolog is procedural: the ordering determines the meaning of the program as well as its termination.

LDL is interpreted bottom-up: from the data via the rules to the query, using matching and fixpoint computation as the computational mechanisms. The result is that LDL computes *all* of the consequences that can be proven from the data and the rules and thus, as in relational systems, provides a complete answer to the query. An LDL program is globally analyzed at compile time: specifications that may lead to non-termination during query execution are detected in advance and unsafe programs will not be executed.

Another distinction is in the orientation of LDL towards data intensive applications. LDL maintains, by means of a schema declaration, a strict separation between rules and data. Consequently, the same rule-set may be applied to multiple data sets or conversely, multiple rule sets may be applied to the same data set. The latter capability is especially important in the emerging application area of scientific databases. Prolog does not maintain this distinction between rules and data and e.g., an update to the data (in form of an **assert**) renders the compiled version of the program invalid.

The different computing paradigms embodied by Prolog and LDL is also reflected in their structures for knowledge representation. The principal data structure in Prolog that can be used for recursion is a list. In LDL by contrast, recursion is performed on the relational data as in the **ancestor** example above. Thus, the data stored in the database can directly be used for the evaluation of recursive queries. In Prolog the data needs to be converted from relational to list format and back. This is another example of the impedance mismatch problem that we mentioned earlier.

4 Selected Applications

A detailed discussion of DD applications can be found in Tsur [1991]. In this section we will briefly comment on some of these. Generally speaking, the deductive database approach lends itself best to those applications requiring ad-hoc querying in which a large number of sophisticated queries need to be posed. Another area of DD applications is as a repository of abstract descriptive data such as description of software modules or objects in an object oriented system. This repository is used then to reason about the underlying software components and to enforce certain rules that pertain to the system as a whole.

Scientific databases

In many scientific areas large volumes of data are collected as part of the empirical routine. Examples include data from Astrophysical observations, DNA data, Nuclear reactions data and others. The data are collected for the purpose of scientific discovery and the formulation of a theory. The discovery process entails the formulation of many ad-hoc queries over the data. Furthermore, the nature of the process is such that queries cannot be planned in advance and are typically posed as a result of previous discoveries. The ability to formulate different scientific theories (in form of rule sets) over the same data is a distinct advantage of the DD technology.

Data dredging

Data dredging is the process of extracting useful knowledge from a (typically massive) volume of low-level data. The knowledge is typically in form of time-varying patterns. Examples include the discovery of market-trends from cash-register data at supermarkets, the discovery of financial market trends from stock-market and other exchange data, computer security violations from computer network logs and many others. Again, the ability to rapidly formulate ad-hoc queries is a major asset in this area.

Data validation

Data validation can assume many forms. At an elementary level it involves the enforcement of integrity constraints in databases. More abstractly it involves the enforcement of inter-relational constraints such as referential integrity constraints. Other types of constraints are often business-specific and can be formulated as rules. A typical example would be the rules that govern the expiration dates of stock options in a stock exchange; e.g. "the last Friday within the month of expiration. If this Friday is a Holiday then the last Business day prior to this Friday". Concepts such as "Holiday", "Business day" and others need to be defined. Presently these are included as part of the application program and hence are encoded over and again (and not necessarily in a consistent way) by different programmers. Instead, these concepts can be defined by formulating the right rules and should be a part of a shared knowledge base. Using this methodology business rules will be used in one consistent form by all applications. Deductive databases offer the right technology for the formulation of such rules and their enforcement.

5 Relation to Object Oriented Systems

Object oriented concepts have been a topic of intense research over the last decade. They have been incorporated in modern programming languages such as C++ and Smalltalk and in parallel have been studied as data models for modern database systems. The systems that have resulted from this effort are known as object-oriented database systems. Let us briefly review some of the salient features of systems of this type:

- Object identity: each object has a unique identity. This identity is used to denote an object or any of its components and serves to distinguish identical objects having the same value(s). Id's can also be used to construct complex objects out of simple objects.

- Classes: OO systems employ a type system which allows for the definition of new types by the user. A related property is that of encapsulation: with each of the class types procedural attachments, known as methods, can be defined. The scope of applicability of a method is defined by the class they are associated with.

- Inheritance: The defined class system forms a hierarchy in which class attributes and methods are inherited to subclasses and can be referred to in subclass instances.

The advent of OO systems allows for the user declaration of sophisticated data models and a richer knowledge representation facility compared with relational systems. This facility is supposed to replace the schema facility of relational systems and, as said, allows for the declaration of user defined types as opposed to the fixed set of types that relational systems employ. This added power that OO systems provide to the user comes however with a price:

- Operations are procedurally defined as methods. As such they suffer from the same shortcomings as the traditional procedural languages such as C or Pascal: their intention is intertwined in their implementation. It would be e.g., impossible for an OO system to recognize that a certain method has the same intention as the predefined join found in relational systems. The system would thus be unable to take advantage of certain properties of joins such as associativity since it would not recognize the operation as such.

- The problem of hidden intentions is magnified when, as the current trend is, programmers define even more abstract objects and methods with the intention of sharing these as building blocks in larger systems. Programmers find it difficult to infer the intention of objects they have not developed themselves and hence, tend to redefine their own rather than reuse others. The result is an uncontrolled proliferation of new object types that largely replicate the functionality of others.

- Another problem is that the semantics of object Id's forces us occasionally to consider sets of result that are much larger than our (value based) intention is. An example is that of the path relation in a graph. Suppose that all we are interested in is to derive the relation between pairs of nodes in a graph that is true for each pair of nodes A,B if a path exists in the graph between these nodes. The declarative formulation of this problem would be similar to the ancestor relation shown earlier and would generate a finite set of tuples assuming that the underlying graph is finite. The OO solution to this problem, on the other hand, would force us to distinguish between different path objects even when they connect the same nodes. If the graph therefore has closed circular paths, the OO solution would generate an infinite set of answers whereas a finite value base set of node pairs is all we need. In other words, it would be the responsibility of the OO programmer to introduce the right termination conditions in her application while in the DD world the problem would be taken care of by the system.

We have only pointed here at some of the issues. For a more detailed discussion the reader is referred to Ullman [1991]. The real issue therefore is: how do we combine the best of the OO and DD technologies? An attempt in this direction is in the design of the LDL++ system, currently in progress. In this system the basic principle is that user defined types, offered by the OO technology, can be incorporated at the schema level in form of external classes and their methods that would complement the schema declaration inherent in the LDL system. Thus, relations are declared and their arguments can assume any of the internal types–simple or complex (relations can be external and can reside in any of a number of server databases). In addition, the external types defined via the C++ class system and its methods can be incorporated in LDL rules and the methods can be used to enforce type restrictions over the logical variables that assume these types. Thus, for example, we can import external types such as "time" or "currency" into the database. We can use the declarative rule based framework of LDL++ to explicate the intention of the underlying object base. We could e.g., declare that a certain method in a certain class type is in fact a relational join and define the join properties by means of the appropriate rules. This would enable us to reason about methods and take advantage of the optimization methods offered with the relational systems. The LDL++ organization comes thus close to the desirable world order that we have postulated in section 2.3.

6 Conclusion

In this paper we have introduced the concept of a deductive database and have argued that this emerging technology is a manifestation of the larger issue of procedural vs. declarative programming. We have argued that the feasible approach towards problem solving lies in an order that combines procedural with declarative programming and have introduced one system, LDL, that attempts to do so. We have also briefly touched upon the relationship with Object Oriented system technology and suggested that, despite its improvement in data modelling capabilities, this technology suffers from problems that can be alleviated by the DD technology. A blend of these seems thus to be the right approach towards problem solving. As we have stated, this paper is intended as an introduction for the practitioner. It is hoped that the suggested reading list that follows will be a source of further information and that the actual problems, encountered by the reader in her workplace will be recognized and analyzed within the framework suggested in this paper. Since this author believes that further progress in the area will come primarily from actual application development, the reader is encouraged to contribute and further the state of the art by means of the description and publication of applications that fall within the declarative framework.

Suggested reading:

1. Cacace, F., S. Ceri, S. Crespi-Reghizzi, L. Tanca, and R. Zicari [1990]. *Integrating object-oriented data modelling with a rule-based paradigm,* TR-90.008 Dipartimento Di Elettronica, Polytechnic University of Milan.

2. Lefebre, A., and L. Vielle [1989]. "On deductive query evaluation in the DedGin system," *Proc. First International Conf. on deductive and Object-Oriented Databases,* Kyoto.

3. Naqvi, S. and S. Tsur [1989]. *A Logical Language for Data and Knowledge Bases.* Computer Science Press, New York, NY.

4. Phipps, G. [1990]. "GLUE: a deductive database programming language," *Proc. NACLP Workshop on Deductive Databases* (J. Chomicki, ed.), Aug. 1990.

5. Phipps, G., M.A. Derr, and K.A. Ross [1991]. "Glue-Nail: a deductive database system," *ACM SIGMOD International Conf. on Management of Data,* pp. 308-317.

6. Ramakrishnan, R. [1990]. "The CORAL deductive database system," *Proc. NACLP Workshop on Deductive Databases* (J. Chomicki, ed.), Aug. 1990.

7. Tsur, S [1991]. "Deductive Databases in Action". *Proc. 10th ACM SIGACT-SIGMOD-SIGART Symposium on Principles of Database Systems,* pp. 142-153.

8. Tsur, S., and N. Garrison [1991]. LDL Users' Guide *MCC Technical Report* STP-LD-295-91, Aug. 1991.

9. Ullman, J.D. [1988] *Principles of Database and Knowledge-base Systems, Vol I: Classical Database Systems*, Computer Science Press, New York, NY.

10. Ullman, J.D. [1989] *Principles of Database and Knowledge-base Systems, Vol II: The New Technologies*, Computer Science Press, New York, NY.

11. Ullman, J.D. [1991]. "A Comparison Between Deductive and Object-Oriented Database Systems" *Proc. Second Conf. on Deductive and Object-Oriented Databases (C. Delobel, M. Kifer and Y. Masunaga Eds.)*, Munich, 1991, pp. 263-277. Springer- Verlag Lecture Notes in Computer Science #566, Springer-Verlag, Berlin.

12. Vaghani J., K. Ramamohanarao, D.B. Kemp, Z. Smogyi, and P.J. Stuckey [1990]. "Design overview of the Aditi deductive database system," TR-90/14, Dept. of CS, University of Melbourne.

Knowledge Based PPS Applications in PROTOS-L

Christoph Beierle

IBM Germany, Scientific Center, Institute for Knowledge Based Systems
P.O. Box 80 08 80, D-7000 Stuttgart 80, Germany
beierle@ds∅lilog.bitnet

Abstract: The basic concepts of the logic programming language PROTOS-L extending ordinary logic programming are briefly described, focussing on types, modules, and database access. Two of the knowledge based PPS applications that have been realized in PROTOS-L are presented. One deals with single step fiber production, and the other one supports distributed planning in a three-level PPS model of global, distributed, and local planning.

1 Introduction

The objective of the EUREKA Project PROTOS[1] [1] is twofold: The development of knowledge based systems in in the area of production planning and scheduling (PPS), as well as the development of various extensions of logic programming suitable for building such planning systems. The logic programming language PROTOS-L [3], [4] that is an outcome of the PROTOS activities, has a powerful type concept with sorts, subsorts and polymorphism, and a module concept that allows the separate compilation of modules as well as the transparent integration of external databases. It has a deductive database component, and a high-level interface to the window management system OSF/Motif. A finite domain constraint solver is currently being integrated into the PROTOS-L system.

These features of PROTOS-L have been used successfully in various application prototypes [2]. In particular, knowledge based production planning systems for the chemical / pharmaceutical sector have been developed. One deals with multi-step production planning in an environment where every apparatus to be used in the schedules can be taken for a number of different purposes. The HoPla system [16] deals with a single-step production planning problem. One of its main objectives is to reduce the resetting costs that arise when changing e.g. from one colour to be produced to another one. For both problem situations purely mathematical approches proved to be unfeasible, whereas the knowledge based approach turned out to be very promising. Especially powerful replanning facilities as they are provided by the PROTOS-L prototype, are vital for a planning support system that aims at modelling directly the reasoning process of a planning expert.

Another dimension of the PPS application area investigated in PROTOS is the extension to distributed and global planning. A three-level planning approach is taken in PROTOS, and a PROTOS-L prototype representing a support system for distributed planning is available [15].

The purpose of this paper is to present a brief overview of the PROTOS-L concepts extending ordinary logic programming, focussing on types, modules and database access (Section 2), and to present two of the knowledge based PPS applications that have been realized in PROTOS-L (Sections 3 and 4). Both systems were developed in close collaboration with the project partner Hoechst.

---

[1]The project partners in PROTOS are BIM, Hoechst AG, IBM Germany, Sandoz, and the Univ. of Bonn and Oldenburg

2 Concepts of PROTOS-L

2.1 Types

Although types play an important role in most modern programming languages, Prolog is essentially an untyped language. However, from a software engineering point of view, types can be vital in the development of reliable and correct software. A type concept with static type checking makes it possible to avoid meaningless expressions and terms, provides a means for better structured programs, and makes explicit the data structures used in a program. For instance, in the type concept of PROTOS-L that has been derived from TEL [14], we could declare the relation

    rel process_order: machine x order.

to take two arguments, the first being of type machine, the second of type order. All uses of process_order are then subject to type checking, so that subgoals with incompatibe argument types would be rejected as being ill-typed, for instance subgoals with incorrectly swapped arguments.

The **subtyping** possibility of PROTOS-L greatly increases the representation facilities since the universe of discourse can now be subdivided and structured in a flexible way. For instance, the type declaration

    vehicle := airplane ++ car ++ train.

introduces the type vehicle as the union of its subtypes airplane, car, and train. Moreover, the deduction process can exploit the subtype relationships when testing for subtype membership or when restricting variables to subtypes. For instance, in a relation with the declaration

    rel travel: vehicle x town x town.

the applicability of clauses could be restricted according to the kind of vehicle given in its first argument. A clause like

    travel(V,TownDep,TownArr) :- V:airplane, ...

is only applicable if its first argument is of type airplane. In the case when the incoming argument V is still a free variable, the subgoal V:airplane restricts the variable V to the subtype airplane and would cause immediate failure and backtracking if V's previous type restriction was already incompatible with airplane, like car or train; in fact, in this case the indexing mechanism on typed free variables used in PROTOS-L would exclude this clause as an alternative to be considered right from the beginning. Thus, V represents the whole *set* of airplanes instead of a particular instance of this set. Thus, the deduction process uses the more abstract level of set-denoting types rather than the level of individuals. This yields not only more compact intensional answers but it may also save a lot of expensive backtracking.

Often, one wants to express data structures in a parameterized way, and therefore, PROTOS-L also offers **parametric polymorphism**. For instance, the type of lists is given by the polymorphic type definition

    list(S) := {[], [_|_]: S x list(S)}.

where the variable S ranges over all types and can be substituted by any type or type term built from the type constants and the polymorphic types. Such a parametric definition makes available all list instances, e.g. list(vehicle), list(airplane), list(list(train)), or list(pair(train,time)) where pair(S1,S2) is another polymorphic type with two type arguments. Whereas list is a built-in type, the PROTOS-L user may also define polymorphic types himself.

As an example for polymorphic type checking and inferencing consider the Prolog predicate definition

    reverse([], []).
    reverse([H|T], R) :- reverse(T, T1), append(T1, H, R).

intended for reversing the order of the elements in a list where **append** is the usual predicate for appending two lists. At first sight the definition might look all right and indeed it can be executed in Prolog without difficulty, but it does not give the intended behavior. In the presence of appropriate type declarations, however, the error can again already be detected at compile time. Adding the relation declaration

> <u>rel</u> reverse: list(S) x list(S).

and using the declaration <u>rel</u> append: list(S) x list(S) x list(S) , the automatic type inferencing mechanism can deduce the type information H:S and H:list(S) for the variable H. Since these are incompatible, the second clause of the **reverse** definition is rejected as being ill-typed. Its correct version (also in the Prolog case) is of course

> reverse([H|T], R) :- reverse(T, T1), append(T1, [H], R).

In order to support the type checking facilities, currently every predicate in a PROTOS-L program must be declared, i.e. the number and the types of its arguments must be given. Among the advantages of this type system are the advantages gained in traditional programming languages (like static consistency checks at compile time, avoidance of non meaningful expressions, explicit data structures and better structured programs), and the importance of these features grows with the size of the program. An additional advantage in logic programming is that computations on types can replace otherwise necessary deductions, which sometimes greatly increases the speed of the computation by the avoidance of backtracking. Thus, where it might seem cumbersome to give the typing information for each predicate in a small prototype, this information might be vital in a large application. Moreover, there are three features of PROTOS-L which greatly reduce the typing effort compared to more traditional approaches: the subtyping facility, the availability of both polymorphic types and predicates, and the type inference capabilities for variables.

2.2 Modules and abstract data types in Prolog

In order to support separate type checking and compilation each PROTOS-L program consists of a set of modules. Each module in turn consists of an *interface* and a *body*. The purpose of the interface of a module is to define the set of imported names and the names that are introduced newly in this module and that are also exported. The user of a module only sees its interface, and not the body, and a compilation unit is either an interface or a body. Thus, PROTOS-L has essentially a Modula-2 like module concept, and here we restrict ourselves to comment on two significant abstraction possibilities that are realized in PROTOS-L by using its module concept.

The first is the availability of abstract data types. In an interface we might have the declarations

> <u>interface</u> planner.
>   time_table := <u>abstract</u>.
>   <u>rel</u> insert: meeting x time_table.
>   <u>rel</u> cancel: meeting x time_table.
>   <u>rel</u> free_time_slot: date x time x duration x time_table.
> <u>endinterface</u>.

where **abstract** is a reserved word in PROTOS-L. Thus, any user of the module **planner** does not know the representation of the abstract type **time_table**, but he may only use the exported relations like insert, cancel etc. to access **time_table**. Of course, in the module body **time_table** can be represented in various ways, and in particular the representation can be changed to another one without having to change any module that is using the module `planner`. This abstraction mechanism corresponds to the *opaque* types in Modula-2 and the abstract types in TEL.

The second abstraction possibility enabled by the module system of PROTOS-L is the transparent access to external databases. For the user of a module it is not visible whether an exported relation is implemented by a sequence of program clauses or by a relation in an external database.

2.3 Deductive database access

We first give an example how transparent access to external relational databases is realized. Consider the interface

```
interface products.
    rel needs: string x string x int.
                % product   product   amount
    rel depends_on: string x string.
                % product   product
endinterface.
```

which exports a relation `needs(A,B,M)` such that for the production of product `A` one needs the amount `M` of product `B`, and a relation `depends_on(A,B)` such that the production of `A` depends on the product `B` via the relation `needs`, possibly involving intermediate products.

One possible implementation of these relations could be in an ordinary *program body*:

```
module products.
    rel needs: string x string x int.
        needs("product1", "product2", 50).
        needs("product1", "product3", 100).
        needs("product2", "product9", 80).
        ...
    rel depends_on: string x string.
        depends_on(P1,P2) :- needs(P1,P2,A).
        depends_on(P1,P2) :- needs(P1,IM,A), depends_on(IM,P2).
endmodule.
```

However, another possible implementation could be to state that the fact relation `needs` is defined by the tuples of a relation `Product_needs` in an external relational database `Product_DB`. This is achieved by the concept of *database bodies*:

```
database_body products using Product_DB.
    rel needs: string x string x int.
        dbrel needs is Product_needs(Product,UsedProduct, Amount).
        ...
endmodule.
```

where the definition of `depends_on` is exactly the same as above. This simple example already shows several aspects of database access:

- **Base relations**: In order to access base relations in an external database one just has to state the correspondence between the PROTOS-L predicate and its arguments, and the database relation and its attributes; in the example, the predicate `needs` corresponds to the database relation `Product_needs` (in the `Product_DB` database), and the attributes `Product`, `Used_Product`, and `Amount` correspond to the first, second and third argument of `needs`, respectively. Note that since not all attributes must correspond to an argument of the predicate, this concept already allows views in the form of projections.

- **Views**: Using ordinary Prolog syntax, one can define database views by clauses in the same way a Prolog predicate is defined. (In fact, since there are no nested terms in relational databases, there is the obvious restriction to function-free logic programs, usually called Datalog). These are automatically dealt with by the system, and translated into appropriate SQL queries to be executed at run time.

- **Recursive queries**: As in the given example where the definition of depends_on is recursive, one can formulate arbitrary recursive queries. This goes beyond the power of SQL systems (which do not allow recursion) and plays a central role in deductive databases.

- **Language uniformity**: The user does not have to use a second language, like SQL or another database query language, but uses just one language.

In the PROTOS-L system there is a database interpreter (DBI) [13] which realizes all three levels of the database access: base relations, non-recursive views, and recursive views. The DBI evaluates rules in a deductive database manner, using a mix of top-down and bottom-up evaluation techniques and reusing intermediate results. It is well-known that on large relations (as they are typical for database relations) this can result in a significant efficiency gain over the top-down evaluation approach of Prolog using backtracking. The latter, however, is also applicable in the presence of complex terms, and it may cause less overhead in situations with smaller relations. Thus, two different evaluation mechanisms, each having its own merits, are combined, and the PROTOS-L module concept abstracts away from the evaluation strategy used inside the body of a module. By choosing where to put e.g. the recursive depends_on definition (in a program body or in a database body) the user has the choice which evaluation method will be used.

Similar to the transparent access to external relational databases, access to the deductive database LILOG-DB [12] has also been integrated into PROTOS-L. In the context of database updates PROTOS-L offers a transaction concept as the underlying database management system. For instance, update operations within a transaction are made permanent only if the transaction can be completed successfully; thus backtracking inside a transaction undoes every insert and delete operation.

2.4 Constraint handling and further extensions

Currently, a a finite domain constraint solver (c.f. [8]) is being integrated into the PROTOS-L system and will be available soon. Among the further extensions of logic programming that have already been realized in PROTOS-L is the availability of functions which allow for functional nesting. There are also various new built-in relations and functions related to types, e.g. for testing, instantiating, and generating typed variables. All built-ins are type-safe, including file input and output [4]. Type-safe is also the interface to OSF/Motif that was developed for the PROTOS system [11]; through a collection of a few built-in predicates and types it provides an object-oriented access to the powerful window handling facilities which have already been used extensively in the PROTOS-L applications. The implementation of the PROTOS-L system prototype is based on an extension of the Warren Abstract Machine to polymorphic order-sorted resolution [5] and it is currently available on RS/6000 under AIX 3.1, on RT/PC 6150 under AIX 2.2.1, and on PS/2 under AIX 1.2.

3 Single step production planning

The following PPS application deals with fiber production as it occurs at the PRO-TOS partner Hoechst AG. The problem is to assign a set of orders (typically between 100 and 500 per month) to a number of production lines (say, for instance, 15 lines). Once the production of an order has been started on a production line it will also be finished before anything else is produced on the same line. Thus, this problem situation is an instance of a single step production planning problem. Its main characteristics are:

- Each **order** specifies a certain type of fiber, the amount to be produced, a bobbin type, whether the fiber should have a profile or not, a diameter, and a due date. Furthermore, it may specify a certain colour or it may specify colourless.

- The **production lines** may differ in their characteristics. There are slow and fast production lines, each line can produce only a subset of all types of fibers, etc. From the **production line characteristics** it is possible to determine which order can be produced on which production line. This may depend on all parameters of an order like its fiber type, its bobbin type, its diameter, etc.

- Producing an order S after an order P on the same line will cause a certain amount of **resetting costs**. The amount depends on the line and P's and S's fiber type, colour, diameter, and bobbin type.

- Apart from the production restrictions already mentioned, there is one major restriction which is called the **colour constraint**. There are special fiber types which form a subset of all possible fiber types. The constraint is that no two orders for these special types can be produced at the same time on any of the production lines if the orders specify two different colours. Therefore, orders for Coloured Special fiber types (called CS-orders, for short) must be handled especially careful in the planning process. For instance, a green CS-order running on line 4 from time T1 to time T2 will prohibit the production of *any* blue, yellow, red etc. CS-order on *any* other line between T1 and T2.

All data mentioned so far is stored in an external database. The goal of the planning process is to find a schedule such that all orders are produced in time and all restrictions and constraints are satisfied. Since the resetting costs can be rather high (e.g. when changing from one colour to another) a important objective is to reduce the overall resetting costs of a plan.

According to the experiences of Hoechst, this PPS problem could not be solved successfully with purely mathematical programming techniques. Tow main reasons for this were that the resulting plan could not be modified easily (which had to be done frequently due to modified orders), and the plan often did not reflect additional constraints the human experts knew and obeyed but that were not represented in the program (which would also require a plan modification).

Therefore, a knowledge based approach was taken at Hoechst where the planning knowlegde of the experts could be expressed directly. When entering the PROTOS project the Hoechst group had already built the EXAMPL planning system described in [10]. It was decided to model this approach in PROTOS-L (HoPla system, [16]) and to extend the work in the direction of further re-planning and re-scheduling facilities. The main objective of the system is to reduce the set-up costs that arise when changing e.g. from one colour (or a specific diameter, bobbin type, etc.) to be produced to another one, while providing a *scheduling support system* in which high-level user interactions and flexible replanning options are available.

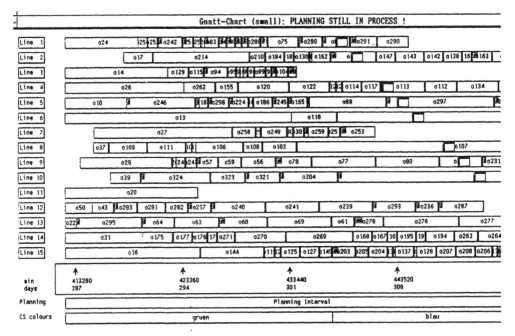

Figure 1: Gantt Chart poduced by the HoPla system

Certain heuristics to select a "best next" order cut down the search space significantly. For instance, there are rules like

- A best next order should have the same parameters as its predecessor.
- If one has to switch colours *or* diameters, take the same colour and switch the diameter.
- If one has to switch colours, it is better to switch from a lighter colour to a darker one.
- If one has to switch diameters, it is better to switch from a smaller to a greater one.

Each of these rules expresses a planner's knowledge how to keep the resetting costs low, and can ideally be modelled directly within the logic programming paradigm.

In addition to the data given above, the planner gives a time interval for the planning period, and a list of colour intervals. Each of these colour intervals specifies a sub-interval of the planning interval in which the system may plan CS-orders of that colour. The system generates a Gantt chart displaying the produced plan that is coloured according to the orders' colours (Fig. 1).

After an order has been scheduled in the plan it is immediately visualized in the Gantt chart. Already during the still ongoing planning process, every order that has been planned so far can be clicked with the mouse. A new window appears and shows the details of the clicked order, the resetting costs it has caused, and the reason why it has been placed into this position in the plan. Thus, an explanation facility is available for the planner.

After completing a schedule proposal, the HoPla system generates a window showing all non-scheduled orders, and by clicking on a non-scheduled order, detailed information about this order is given, which includes the list of all production lines where this order could be processed. The user now has a range of replanning facilities. In particular, he may require modifications of the plan by

- changing an order's priority,
- moving a (scheduled or non-scheduled) order to a particular production line, or
- chaining a (scheduled or non-scheduled) order S to another order P so that S will be the direct successor of P.

All these modifications are collected as requirements for the next plan proposal that will be generated by the system when the user starts replanning. However, sometimes the planner also wants to keep certain parts of the old plan, e.g. the first week, or everything up to a particular order on a specific production line. To this end, the system provides the possibilities to

- fix the whole plan up to a time to be given by the user,
- fix the plan on a production line up to a given allocated order.

Given the full set of replanning options, the user has complete control over the system. For instance, when he wants to ensure that a non-scheduled order N is included into the next plan proposal, he could select an apropriate order P scheduled on a production line L, fix the plan on line L up to P, and define N to be a direct successor of P. Moreover, the user is free to fix also the rest of the plan, or to allow further plan modifications. The latter could be caused by the scheduling knowledge rules since P might now be a "best successor" on another line L', causing orders previuosly scheduled on line L' to be allocated to other lines, etc.

When starting replanning, the system first makes several consistency checks, e.g. that no order is both in the fixed part of the plan and also is requested to be allocated to another production line. Moreover, the planner may save his scheduling decisions and also backtrack over them by

- undoing the last replanning step; this may be done repeatedly until the first planning in the current session,
- saving the current plan under some name to be given,
- restoring a saved plan from a previous planning session.

As with other PROTOS-L applications, also in the HoPLa system there was a smooth transition from a small test version using only main memory to the full version with the external database relations containing the orders information, the production requirements, etc. The typing concept helped to discover many programming errors already in an early phase of program development which might have otherwise caused much more difficulties in locating and eliminating them.

The modification and addition of the knowledge base is quite easy. In addition, the high-level window based user interface made it much easier to work with the system, e.g. to see what effects a change of the colour intervals or changes or additions of planning rules have.

4 Global, distributed, and local planning

4.1 A hierachy of planning levels

Whereas the system described in the previous section operates on the level of a single plant, other aspects of the PPS problem area deal with the bill of materials, lot size

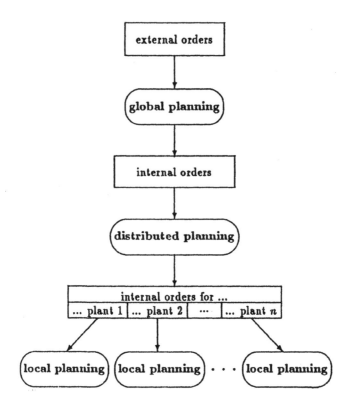

Figure 2: Hierarchy of planning levels

determination, stock levels in warehouses, assignment of orders to different plants, etc. In order to structure the whole problem into better manageable parts, three levels of planning are distiguished in the PROTOS project: global, distributed, and local planning [9] (c.f. Figure 2). A similar 3-level approach is taken in [7].

Input to the whole system is the set of *external orders* for finished products with given amounts, due dates, etc. Purpose of the **global planning** phase is to generate from these external orders a set of internal orders (for finished products or intermediates) that can be produced in one production step. This is done by respecting the bill of materials for all involved products, taking into account current stock levels, etc. Thus, this work could be part of a standard MRP II system. However, MRP II systems usually do not take into account production capacities which is necessary in order to reduce conflicts in later planning refinement steps. Output of the global planning phase and thus input for distributed planning is a set of (internal) orders, and a set of precedence constraints on these orders. Each order has a time window with earliest possible starting date and latest possible finishing date which is determined by the global planning phase.

Purpose of the **distributed planning** phase is to assign to each order a plant where this order should be produced, and to adjust - if necessary - the time windows for each order.

Purpose of the **local planning** phase is the generation of a concrete production schedule where each order is produced within its given time window and where all precedence constraints are met.

At each of the three planning levels, different levels of abstraction and various approximations are used. So on each level conflicts may arise when not all goals can be met. Thus, in addition to the tasks sketched above, each level might produce also conflict reports (e.g. "Order O2 can not be produced within ist given time window") which are then dealt with at the next higher level.

4.2 A distributed planning prototype

The three-level planning model sketched above can be applied to many production situations, for instance to the Hoechst fiber business unit [9]. For distributed planning in this fiber unit a PROTOS-L prototype has been developed (DPla system [15]). Similar as with the HoPla system, its main objective is to provide a flexible planning *support* system where the planning expert can easily choose automatic planning options, but can also interfere manually at every stage. Therefore, the DPla system realizes a distribution cycle which consists of the following four steps:

1. The planer first determines the **working set** of orders to be distributed. This can be done by selective criteria like orders that have no predecessors, or orders that belong to a certain product class. Usually, the most critical orders (e.g. w.r.t. time or production location constraints) will be selected first.

2. In the **distribution step** plants are assigned to the orders in the working set, where again various selection criteria can be used. The available capacity information for each plant is updated continuously according to this distribution.

3. If there is a precedence constraint "O1 must be produced before O2" between orders O1 and O2 that are distributed to *different* plants, then the distribution step has produced a so-called **global dependency**. In the DPla system global dependencies are resolved by restricting the orders' time windows set by the global planning phase accordingly. I.e. if the latest possible finishing date of O1 is greater than the earliest possible starting date of O2, an appropriate time t is selected and both values are set to t. The reason for this is that local planning in each plant should be independent from the actual scheduling decision in another plant; this would not be the case if the plant producing O2 depended on the scheduling of the plant producing O1. The system offers different strategies for restricting the time windows, and they can be applied to several of the involved plants or only w.r.t. a particular one.

4. In the fourth step, the local planning components of one or more plants are activated. Their results - actual production schedules and conflict reports - are obtained and evaluated. The time slots for all scheduled orders are propagated, and the non-scheduled orders are dealt with, e.g. by reallocating them to another plant.

The planner can now go on to the next distribution cycle. He has the option to start again with Step 1 and determine a new working set, or to continue with Step 3, focussing e.g. on another plant for global dependency resolving and subsequent local planning activation in Step 4.

In all four distributed planning steps of the DPla system it is very easy to integrate heuristics ("It is usually a good idea to distribute orders for red fibers first"), special cases ("make sure order O3 for customer XY is produced before the end of the month"), algorithmic approaches ("resolve global dependencies always by constraining the involved time intervals proportionally to the expected production times"), and any combinations thereof. The current version of the system which includes an OSF/Motif based user interface was realized in less than three person months which was certainly enabled by the use of a high level logic programming language like PROTOS-L. Among the extensions of the DPla system which are currently being carried out is the use of built-in constraint satiasfaction in the new version of PROTOS-L which can be used directly e.g. for representing the precedence constraints between orders.

Acknowledgments: For their contributions, support, and discussions towards the development of the PPS applications described above I would like to thank H. Wittmann, G. Urban, M. Lenz, O. Wauschkuhn, as well as G. Meyer, and the Hoechst PROTOS team J. Jachemich, R. Augustin, and A. Rose.

References

[1] H.-J. Appelrath, A. B. Cremers, and O. Herzog. *The Eureka Project PROTOS*. Stuttgart, 1990.

[2] C. Beierle. An overview on planning applications in PROTOS-L. In R. Vichnevetsky and J. H. Miller, editors, *Proceedings 13th IMACS World Congress on Computation and Applied Mathematics*, Dublin, Ireland, July 1991.

[3] C. Beierle. Types, modules and databases in the logic programming language PROTOS-L. In K. H. Bläsius, U. Hedtstück, and C.-R. Rollinger, editors, *Sorts and Types for Artificial Intelligence*. LNAI 418, Springer-Verlag, Berlin, 1990.

[4] C. Beierle, S. Böttcher, and G. Meyer. *Draft Report of the Logic Programming Language PROTOS-L*. IWBS Report 175, IBM Germany, Scientific Center, Inst. for Knowledge Based Systems, Stuttgart, 1991.

[5] C. Beierle, G. Meyer, and H. Semle. Extending the Warren Abstract Machine to polymorphic order-sorted resolution. In V. Saraswat and K. Ueda, editors, *Logic Programming: Proceedings of the 1991 International Symposium*, pages 272–286, MIT Press, Cambridge, MA, 1991.

[6] S. Böttcher. A tool kit for knowledge based production planning systems. In *Proc. Int. Conference on Database and Expert System Applications*, Vienna, 1990.

[7] P. Burke and P. Prosser. A distributed asynchronous system for predicative and reactive scheduling. *Artificial Intelligence in Engineering*, 6(3):106–124, 1991.

[8] P. Van Hentenryck. *Constraint Satisfaction in Logic Programming*. MIT Press, Cambridge, MA, 1989.

[9] J. Jachemich. *Global production planning and scheduling at the HOECHST fibres business unit*. PROTOS Working Paper, Hoechst AG, Frankfurt, 1990.

[10] J. Jachemich. Rule based scheduling in a fibre plant. In R. Vichnevetsky and J. H. Miller, editors, *Proceedings 13th IMACS World Congress on Computation and Applied Mathematics*, Dublin, Ireland, July 1991.

[11] H. Jasper. A logic-based programming environment for interactive applications. In *Proc. Human Computer Interaction International*, Stuttgart, 1991.

[12] T. Ludwig and B. Walter. EFTA: A database algebra for deductive retrieval of feature terms. *Data and Knowledge Engineering*, 6, 1990.

[13] G. Meyer. *A Poor Man's Deductive Database*. IWBS Report, IBM Deutschland GmbH, Stuttgart, 1992.

[14] G. Smolka. *TEL (Version 0.9), Report and User Manual*. SEKI-Report SR 87-17, FB Informatik, Universität Kaiserslautern, 1988.

[15] O. Wauschkuhn. *Untersuchung zur verteilten Produktionsplanung mit Methoden der logischen Programmierung*. Studienarbeit Nr. 1088, Universität Stuttgart und IBM Deutschland GmbH, Stuttgart, March 1992.

[16] H. Wittmann. *An Example for Knowledge Based Production Planning with PROTOS-L*. Diplomarbeit, Universität Stuttgart und IBM Deutschland GmbH, Stuttgart, 1991. (in German).

The SECReTS Banking Expert System from Phase 1 to Phase 2

Carlo Chiopris

ICON srl
34, L.ge Rubele
37121 Verona, ITALY
tel. ++39 45 8004767
fax. ++39 45 8004767
AppleLink ITA0282

Abstract. "SECReTS" is an expert system used by several Italian banks to support the analysis of client specific data. SECReTS performs a monitoring task, observing a well-defined data flow that takes place on a monthly basis between each bank and the Bank of Italy. As is typical in monitoring expert systems, SECReTS' main task is that of identifying the occurrence of meaningful situations, and in such cases to produce a detailed report to a user. SECReTS is implemented in Prolog, and is built around a meta-interpreter. To interact with the user, SECReTS employs a HyperText-like graphic user interface. A brand new SECReTS architecture is currently under implementation, that incorporates the original architecture in a Computer Supported Cooperative Workgroup framework.

1 Introduction

"SECReTS" is an expert system used by several Italian banks to support the analysis of client specific data. Unlike most banking expert systems, which are Mycin-like consulting and diagnostic systems that gather input data mainly from the user, SECReTS performs a monitoring task, observing a well-defined data flow that takes place between each bank and the Bank of Italy. As is typical of monitoring expert systems, SECReTS' main task is that of identifying the occurrence of specific situations, and in such case to produce a detailed report to a user. SECReTS is implemented in Prolog, and is built around a meta-interpreter. To interact with the user, SECReTS employs a HyperText-like graphic user interface.

We found it rather easy to design and implement a system capable of performing smartly and efficiently. It was much more difficult to move it into the main stream of a user's everyday operations. We also found that monitoring tools should be given a wider independence of initiative and should be deeply integrated in the working of the user. For example, an expert system might decide when to employ e-mail to send a report to a user signalling a specific event about a client, and then remain pending for an answer before forwarding another message to a different user.

For this reason we are now extending the SECReTS architecture in order to incorporate it into a GroupWare environment. Instead of dealing with a single user that uses a specific (though well-integrated) system for a single task, we now deal with a network of "Actors" that cooperate on a much larger set of inter-related tasks. Actors are both the human users of the system and the so called Artificial Actors: Knowledge Based Systems with varying capabilities, from pure DB querying to typical expert system tasks.

Actors can be created to detect the occurrence of specific events, for sending alarms to other responsible actors, for decision support or for the coordination of complex tasks.

2   SECReTS Phase 1

SECReTS is an expert system originally developed by ICON srl with the participation of the "Banca Popolare di Sondrio" (BPS), an Italian credit institute. It later became a product sold on the Italian market.

2.1   The Specific Problem

Every month, each Italian banking institute is asked to send information about client loans to a section of the Bank of Italy named "Centrale dei Rischi" (Risk Centre, in the following "CR"). The bank must specify the loan granted and the effective usage of the client on different credit types. A month later, the Bank of Italy sends this data back, integrating it with loan data on every single client of the whole banking system (that is: all other Italian banks).

While the main goal of the CR is to monitor risk, it is possible to use CR data in a marketing-oriented manner, as one can guess clients' behaviour with competing banks and then decide appropriate marketing actions. The availability of this information becomes an important opportunity for the banks.

The typical user of CR data is the branch manager, who is expected to look at CR data every month. Central Offices are other possible users, as they may need to inspect the data on a much higher level of abstraction in order to identify general market trends so that they can take strategic decisions.

The main problem here is the amount of data involved. Each client is described monthly by about 70 numbers, and a reliable analysis should take into consideration a history of at least 6 months. All this amounts to hundreds of numbers every month for each client. Because of the problem size, CR data is often underused.

The typical solution provided by the DP Centre is to implement filter programs that for each client compute a given set of indicators. The resulting output is printed in a tabular form and lists those clients whose indicators suggest an anomalous situation.

While this solution is helpful in reducing the number of clients to inspect, it is not of much help in performing the real analysis that must take into consideration the effective data, and not just the computed indicators. It takes at least ten minutes to access and read the CR data of a client, and much longer to produce a written comment on it.

We decided to solve the problem using Logic Programming and Artificial Intelligence techniques, since we felt that this approach would be feasible and would yield a high return. We also felt that it was a kind of application general enough to be extended to different fields and to different data bases.

2.2 The General Problem

The CR data analysis problem is a typical case of a mixed Monitoring / Diagnostic task.
In a monitoring task a human or artificial actor must identify the occurrence of specific conditions and report them. Typical instances of monitoring systems are found in industrial process control (e.g. alarms). They are maybe less widespread in financial domains, but they are nevertheless very important. Though a monitoring task can be in principle very rich (see [Hickman] and [Breuker]), monitoring actors most often needn't have a great deal of intelligence or deep knowledge, since the process of identifying interesting events is usually strictly time constrained and so bound to be simple. For this reason, monitoring applications are usually implemented using simple ad hoc programs. If a data base is used as the source of the analysis, query languages can be used.

What can Logic Programming and Artificial Intelligence offer beyond the traditional tools used in this framework?

We chose to use Logic Programming because the naive approach has two main weaknesses:

- identifying a meaningful event may not be that simple (that is: requires simple reasoning);
- the boundary between monitoring and diagnosis is not very clear.

The CR case is a good example of these weak points:

- some interesting situations require non-trivial reasoning that cannot be expressed easily in a language like SQL but require more complex logic and functional expressions (see [Parker]);
- raising an alarm is not enough, the user wants to understand the general framework, not the simple indicator, and this means classification and diagnosis.

2.3 Logic Programming as a Tool

The last two points suggest that it is useful to employ powerful tools, capable of performing both simple and complex tasks.

Another interesting point is that a monitoring task usually employs a behavioural model of some system. As it is the evolution in time of such a system to be under consideration, there is likely to be some kind of history recorded in a data base. This gives a strong data base flavour to many monitoring systems: querying a DB is usually part of a monitoring system's activities.

Logic Programming is the ideal tool in such cases. The relationship between Logic Programming and Relational Data Bases is well understood and documented. Though it is true to say that there is an "impedance mismatch" between SQL and Prolog, it is a fact that this mismatch is much smaller than with procedural languages. A loose coupling of Prolog and a RDBMS is a technically feasible task, and products of this kind are already available on the market (e.g. ProDBI by KeyLink).

Prolog coupled with a RDBMS is an ideal tool for data analysis applications in many aspects:
- integration, since it allows a system to interact with the real DB, used by all other existing applications;
- speed, since many DB intensive operations can be downloaded on the RDBMS;
- flexibility, since different coupling strategies between Prolog and the RDBMS can be easily implemented.

SECReTS was developed on a Macintosh using LPA MacProlog, which is an extraordinarily rich environment providing all the graphic features needed.

It was later ported to other Prolog implementations: MVS Prolog by IBM, LPA Prolog for MS-DOS, Quintus Prolog 3.1 on HP9000/400 machines running HP-UX and Olivetti LSX machines running System V.4. In all cases the porting was very simple, apart from well known problems in the user interface area.

We found Prolog to be a very effective development environment and a reasonably efficient language. Being a real programming language, and not just an ES tool, Prolog gives complete control on all parts of a system.

2.4 SECReTS goals

SECReTS' main goal is that of supporting the person responsible for inspecting CR data.

This goal can be divided into three subgoals:

- perform all needed data abstractions, formulate consistent hypothesis and data interpretations;
- present the results to the user in a friendly way, integrating data and interpretations with explanations and graphics;
- operate in a DB-like manner to allow selection of clients that show given conditions.

2.5 SECReTS functions

SECReTS goals are satisfied by the three main functions of the system.

Data Analysis. Data Analysis is performed according to the contents of a Knowledge Base.

As is typical of this type of system, knowledge is represented in terms of frames. There are about one hundred frames in the KB. Each frame represents the occurrence of a specific "situation". Each frame can be connected to many others by different kinds of links (e.g.

specialisation links, similarity links). These links are used by a generic algorithm to build a connected graph of correlated frame instances, which is the real output of the system.

Frames can represent very simple data abstraction tasks, like computing averages or trends, or they can be rather complex when performing real diagnostic functions.

The Inference Engine is implemented as a Prolog Meta-Interpreter. It operates on frames instead of rules, has explanation capabilities and implements the interface to the DB and also to the User Interface.

User Interface. Presenting the results of Data Analysis can be a very complex task. The problem lies in the fact that in a typical run of the Inference Engine, hundreds of frame instances can be created, along with their links. Not only are there too many to be presented to the user, but they are mostly uninteresting, corresponding to "nothing to say" situations. The user wants to be shown only 'interesting' information.

We provided two solutions to this problem.

The first is a very interactive one: the connected graph which is the output of the Inference Engine is used to build a HyperText stack. Each frame represents a kind of HyperCard card, with its explanation, while other connected frame instances can be reached by clicking buttons which in turn represent frame links. This is a very general and efficient solution, that allows the user to freely navigate in the possibly large explanation space. Furthermore, it is possible to use a graphic representation for all the frames performing given kinds of data abstraction (like computing trends).

The second solution is based on producing written reports. The reports are produced by traversing the explanation graph, and by linearizing it into a textual format. Though it is much less flexible, it is preferred by the user. An interesting feature of this module is that it can export to different file formats, and so the report can be read by different tools such as a Word processor with outlining capabilities, or a spreadsheet to produce high quality graphics.

Selection. Inspecting the CR data of a single client is only one aspect of the problem. Another aspect is identifying the interesting subset of clients to inspect. No fixed criteria can be given in this field and so it becomes very important that the user be able to select from a set of clients only those satisfying given conditions.

This is implemented by a simple query tool, where the user specifies his/her requirements. All clients matching the conditions are then extracted and an appropriate report is generated.

2.6 Lessons learnt

SECReTS has been a rather successful ES. Unlike many other experiments and prototypes it went straight into production and it is regularly used by tens of users in different Italian banks.

Four years later we feel that SECReTS as a system still suffers from a lack of integration with the rest of the information world of a bank user. It uses its own DB, it runs on an unusual machine (Macintoshes' are rare in Italian banks), it cannot take any initiative, it focuses on a single and limited, though important, data stream. For a system of this kind the combination of all these factors can be very dangerous, since the user may simply forget to use it, being out of his/her everyday scope of operations.

For this reason we decided to integrate the SECReTS paradigm into a larger framework, with the SECReTS Phase 2 project.

3.   SECReTS Phase 2

3.1.   Background and goals of Phase 2

Banca Popolare di Sondrio decided to launch a medium term project with far reaching architectural and organisational implications to investigate the integration of recent technologies like: distributed RDBMS, Computer Supported Cooperative WorkGroup, Expert Systems.

The reasons behind this project are many:

-    the number of branches keeps on growing, and they become more and more geographically spread;
-    the mainframe is becoming increasingly loaded;
-    coordination between different parts of the bank is becoming increasingly difficult;
-    the bank needs to formalise and control organisational procedures;
-    know-how is difficult to spread;
-    simple decisions can be delegated to a Knowledge Based System.

The architecture proposed to face all these problems is the following:

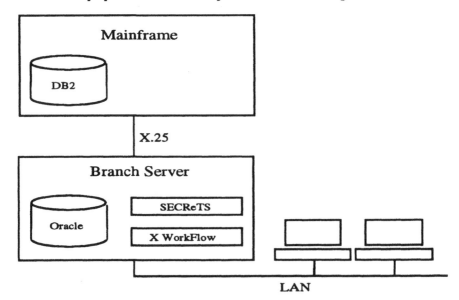

Each branch has a Unix server and is connected through X.25 lines to the EDP centre. On each Unix machine (an Olivetti LSX) there is a local Oracle RDBMS containing all the information of interest to the branch; branch databases are partial copies of the larger DB on the mainframe; alignment between the central DB and those in the branches is guaranteed by update procedures that download new data on local DB's.

The main goal is to allow all decision support functions to be run on the local machine, instead of on the mainframe. This reduces mainframe loading and provides greater freedom to sophisticated users.

Apart from these considerations that are typical of downsizing, a major element is introduced: all decision oriented activities are integrated in a GroupWare framework and supported by a GroupWare tool. In our architecture, coordination functions are performed by X_Workflow, a tool developed by Olivetti Systems & Networks which is part of the larger "IBISys" product, an office automation and integration tool.

The SECReTS architecture is implemented using Quintus Prolog 3.1 for Olivetti LSX.

Different partners take part in the project:
- BPS;
- Olivetti Systems & Networks;
- RSO Futura;
- ICON.

3.2 The Framework

Instead of considering a single task, like CR data analysis, we aim at modelling more and more of the decision oriented activities that are performed in a bank. This is a very ambitious goal, that can never be fully accomplished. All these activities, or tasks are correlated, and are performed by actors that communicate through messages. A basic task is performed by just one actor, while a complex task may involve many smaller tasks, to be performed by many different actors.

The structure of complex tasks can be modelled as Petri Nets, and can be represented using the X_Workflow programming language "COPLAN-S". Actors can be either human or artificial. Artificial actors are Knowledge Based Tools implemented using the SECReTS architecture.

An interesting point is that the concept of cooperating actors found in CSCW literature is very close to the Analysis of Cooperation in the KADS methodology (see [Wielinga]).

We devised three actor typologies:

- monitoring actors;
- decision support actors, that can support decision oriented activities by writing reports, running simulations etc.;
- communication actors, that monitor the communication flow between actors.

3.3 A Scenario

An overdraft is identified on a client's current account. It is signalled to an artificial actor "A1" capable of simple reasoning about overdrafts. A1 decides that it is a non trivial overdraft, sends a message to the branch manager responsible for the client, and waits for an answer. The branch manager reads the message, and decides to perform a deeper case analysis with the support of another artificial actor "A2" that will gather all meaningful information about the user and then prepare a detailed report. The answer given back to A1 is that the situation cannot be solved in the branch, and that a higher manager from the central office must intervene. A1 then sends the appropriate message to the higher manager.

In the meanwhile one can query the system:

- what events are related to this client/branch/manager?
- what procedures have been pending for more than ... days?
- what is the state of this practice?

The goal of this scenario is to stress the following points:

- any anomalous situation is signalled by an appropriate actor, human actors can trust that they will be alerted when needed;
- events may produce chains of other events, through message passing;
- many actors are artificial, but they can anyway be given limited responsibilities;
- it is possible to supervise the global state of communication between actors.

3.4 Integrating SECReTS

In projects of this kind problems do not usually arise from the sophisticated features, but from the basic technologies. In our case, much work was spent on the data base side, on procedures maintaining alignment of the data bases, on the data model, on data distribution and telecommunications, on efficiency problems of mainframe procedures. By comparison the integration of SECReTS with the rest of the architecture was rather simple.

Integration with the RDBMS. Integration with the Oracle server is obtained using ProDBI by KeyLink, a software library to interface Quintus Prolog to different RDMS. From Prolog it is possible both to execute arbitrary SQL calls and to access tables as if they were clauses.

Integration with X_Workflow. This integration is very simple to implement, as X_Workflow is programmed using text files and also communicates using text files.

User Tools. Users interact with the system from PC's. Quite often they want to use their own productivity tools. For this reason the actors implemented produce files in text formats that can be opened by Word Processors or Spreadsheets. A tool to browse on explanations was implemented by BPS on the PC .

4 Conclusion

We think that data analysis (monitoring and diagnostics) applications can be very useful in financial domains. Given the appropriate tool, that we identified in the coupling of Prolog and a RDBMS, they are easy to implement and can offer a very high return. Furthermore, applications of this kind are not difficult to integrate in an Office Automation or GroupWare framework. This integration can give a definite impulse in the effectiveness of their use, changing them from passive to active and autonomous members of a complex coordination structure.

References

J. Breuker, B. Wielinga et al.. *Model-driven Knowledge Acquisition: Interpretation Models*. Esprit Project P1098, Deliverable D1, University of Amsterdam and STL Ltd., 1987.

F.R. Hickman et al.. *Analysis for Knowledge Based System - A Practical Guide to the KADS Methodology*. Ellis Horwood, 1989.

Olivetti Systems & Networks. *X_Workflow Author's Guide*. 1992.

D. S. Parker. *"Stream Data Analysis in Prolog"*. In L. Sterling, *The Practice of Prolog*, MIT Press, 1990.

P. Torasso, L. Console. *Diagnostic Problem Solving*; North Oxford Academic, 1989.

B. Wielinga, G. Schreiber, P. De Greef. *KADS Synthesis Report*. Esprit Project P1098, Deliverable Y3, University of Amsterdam, 1989.

Logic engineering and clinical dilemmas[*]

John Fox
Advanced Computation Laboratory
Imperial Cancer Research Fund, London

Abstract

There are many challenges to designers wishing to build computer systems to help doctors in their everyday decision making. Systems must be flexible, easy to maintain, robust, sound and safe. It is not easy to meet all these requirements simultaneously. General medical practice exemplifies these problems in quite an extreme form; we discuss how the use of logic programming techniques has facilitated the design and implementation of decision support systems for this medical setting. Although classical logic provides a strong formal foundation for the work, practical decision making requires significant extensions. Principled extensions for constructing arguments under uncertainty, and for framing and taking decisions, are outlined.

Computer support for clinical decision making

The increasing healthcare needs of an ageing population are creating a demand for general medical practitioners (GPs) to play a more central role in the provision of high quality health services. This is necessary in order to limit calls on high-cost hospital facilities. Information and decision support technologies will be crucial in increasing the range of services that GPs can offer. Decision support systems (DSSs), such as expert systems, are under development in many areas of specialist medicine, but systems which are appropriate for the general practitioner are less well advanced. General practice raises practical, technical and theoretical problems that are even more severe than those encountered in designing technologies for hospital settings. Among these are the following.

a. The system must cover a large proportion of the clinical cases routinely faced by the GP. Systems that can only assist in narrow specialities of medicine have little value, and experience shows that computer systems that are infrequently required quickly become derelict. A useful GP system requires a large and comprehensive medical knowledge base; this raises major questions about systematisation and compilation of medical knowledge.

b. Support for a wide range of decisions must be provided. Diagnosis has been the main focus of research in medical expert systems. This type of decision is now quite well understood, but diagnosis represents only a moderate part of general practice. GPs make decisions about investigations and tests, treatment selection and planning, drug prescribing, risk assessment and referral, contingency planning, patient monitoring and follow-up, and so on.

c. Practical systems need to be able to play multiple roles. The needs and expectations of individual GPs vary widely. One doctor may have considerable specialist knowledge in some medical areas, while in others his knowledge is more limited. A DSS should be able to offer simple database facilities (eg what are the contraindications of a little-used drug? what countries require innoculation for yellow fever); aide-memoires (what diseases or information sources are relevant to this case?), and sometimes a "second opinion" about the preferred treatment or action may be required.

d. Software must be sound and safe. Ensuring that software performs well and reliably, and continues to do so throughout its development and evolution, is a central problem of software engineering. It is crucial for safety-critical fields like medicine. Building and maintaining a reliable system on the scale required for general practice is probably

---

[*] This work has been supported by the European Commission under its Advanced Informatics in Medicine and Esprit Basic Research programmes, and by the UK Science and Engineering Research Council.

intractable using current knowledge engineering techniques.

Logic engineering for knowledge engineering

We describe our approach to the design and implementation of DSSs as a logic engineering approach. The term is loose, intended only to indicate the importance we attach to logic and deductive techniques in developing flexible yet sound knowledge based systems, particularly where the size of the application is likely to be large. The following techniques have proved valuable:

a. A declarative style of knowledge and task representation to yield clarity, ease of maintenance, provision of views and reuse of knowledge etc;

b. Deductive database techniques for efficient inference on large bodies of knowledge;

c. Reactive (event-driven) database techniques for invoking high priority procedures;

d. Formalisation of decision making and other procedures in first-order logic;

e. Implementation of specialised theorem provers (eg the argumentation theorem prover described later);

f. Use of formal specification languages to assist verification and validation of knowledge models and safety critical procedures.

The Oxford System of Medicine

Logic engineering techniques which are intended to meet the requirements of general practice have been under development for some years. *The Oxford System of Medicine* (OSM), is DSS designed for use in medical general practice. The OSM acknowledges the complexity and unpredictability of a typical GP-patient encounter by providing a range of services within a uniform environment. The user can freely review or update a patient record, retrieve information from the OSM's knowledge base, invoke various kinds of assistance during decision making, and obtain case summaries, data analyses or explanations of advice offered by the system. Development of the OSM has clearly benefited from these techniques, and has provided an interesting platform for investigating logic engineering as a general development discipline.

The techniques which have been developed for this system draw heavily on ideas from logic programming and deductive databases. The general architecture of the OSM is a layered deductive database. Patient data and medical knowledge, covering the symptoms and causes of diseases, investigations, treatments etc are encoded as ground Horn clauses. Strategic knowledge of how to carry out particular tasks (such as diagnosis decisions, treatment decisions) is encoded as meta-knowledge which guides reasoning and decision making. This is stored as a further database layer. Finally, a generalised decision procedure which reasons over these data is encoded as a theory in first-order logic.[*]

This decomposition helps to meet many requirements. The separation of task knowledge from domain knowledge simplifies the maintenance and extension of the large medical knowledge base the OSM will require; the simple factual encoding facilitates the construction of multiple views of data and knowledge bases; the generalised decision procedure can be applied to many types of decision, with additional decision functions provided by adding new definitions to the task database. The ease with which meta-procedures can be implemented in logic languages like Prolog facilitates development at all levels, and the logical basis of the decision procedure assists formalisation of the design

---

[*] Not all of these facilities are to be found in all the systems we have developed; for brevity we shall not distinguish these systems in the description.

and analysis of safety issues.

Representation of patient data

A record of patient illnesses, tests and results carried out, and drugs and other treatments given, has to be maintained over time. These things are encoded as a set of facts stamped with the date and time the events occurred or data were recorded. This simple representation has proved to be adequate for complex medical tasks. Alternative views of the patient data (eg disease oriented views; treatment oriented views) and predicates for temporal reasoning over the data (event or interval-oriented) are easily provided.

Representation of domain knowledge

The *medical* knowledge base also consists of ground facts representing medical objects (diseases, observations, investigations, treatments, drugs etc). Object class information (eg cancer is a kind of disease, chemotherapy is a kind of treatment) and the attributes associated with classes (eg symptoms and signs of diseases; indications, contraindications and side-effects of treatments etc), and the values for these attributes for specific instances, are straightforwardly represented as relation tuples. Procedures for inheriting information between classes and instances are implemented by first-order deductive rules.

A demonstration knowledge base of 20,000 "core" facts covering such medical topics as joint pain, nausea and vomiting and breathlessness (a fragment of the eventual knowledge requirement, which is estimated to be about 10M facts) has been implemented. These are stored explicitly; additional "virtual" facts can be derived when needed or generated in advance and stored explicitly. (The choice depends on the acceptable tradeoff between space required for fact storage and the time to generate facts during inference. A medical data model specifies the relations which can hold between objects, and constraints on those relations; it is used to guide knowledge base construction and the generation of virtual facts (Glowinski, Coiera and O'Neil, 1991). In order to support the clause database a specialised clause compiler/indexer and retrieval procedures have been implemented in C. As efficient, robust deductive database products become available we may make use of these in preference to specialised utilities.

Decision Support.

Knowledge about particular types of decision task are represented in an explicit specification in the knowledge base. The specifications are declaratively defined as ground facts. Their function is to guide the decision procedure in carrying out particular tasks, such as diagnosis, by indicating what information is relevant to making the decision, what kinds of reasoning are appropriate, and so forth. The logical decision procedure provides functions for:

(a) identifying candidates for a specific decision (possible diagnoses, treatments etc) and relevant information sources (eg symptoms, test results);

(b) generating "arguments" for and against the candidates as findings are entered, by reasoning from causal, statistical and other medical knowledge;

(c) identifying important relationships between candidates (eg subsumption or inconsistency relationships);

(d) carrying out symbolic or quantitative assessments of the relative merits of the alternative candidates.

The decision making process creates a task specific interpretation of the current patient data. The logical forward closure of new data added to the patient record is computed on arrival, added to a "volatile" database, and truth-maintained as changes are made to the record. This database is complete except for certain numerical and other inferences which are computed on request. A variety of views of the volatile database, providing subject checklists, decision summaries and comparisons, and a number of subsidiary reports are constructed as required.

The stratification of the database provides simplicity of representation, aiding performance optimisation and reusability of knowledge. Traditional expert system representations based on mixed rules, frames, attached procedures etc are relatively difficult to extend and maintain, particularly when the knowledge base becomes large.

Practical reasoning and decision making

Although a logical approach offers many benefits, there are important issues in the design of DSSs that it does not address directly. It provides a powerful and sound approach to categorical reasoning, but it does not provide a general theory of decision making, particularly for situations which are normal in medicine where there are high levels of uncertainty, and the consequences of actions taken as a result of decisions involve subjective values.

The development of an appropriate theoretical framework for knowledge based decision making is a surprisingly neglected topic. There is currently only one widely accepted theoretical framework for decision making, the theory of expected utility maximisation (Lindley, 1985) which was developed in statistics and economics rather than AI or knowledge based systems research. Expected utility theory dominates much thinking about medical decision making, but the framework is restrictive. The theory requires that all decision situations be wholly predefined and, further-more, that a comprehensive set of probabilistic and cost/benefit (utility) parameters is established for these situations.

In general practice it is effectively impossible to predefine all the decisions and problems the GP may face or to estab-lish all the necessary quantitative parameters. The expected utility approach also does not provide a formalism or cal-culus for exploiting qualitative knowledge (such as knowledge of causality, anatomical structure or physiological function). Furthermore, numerical formalisms provide no basis for automating important decision making functions, such as deciding whether a decision is needed and what its objections are or what candidates should be considered. The OSM project has therefore required development of a theoretical framework which addresses these problems, providing ways of composing decisions dynamically and coping with circumstances where facts are uncertain and knowledge is incomplete and imprecise.

Coping with uncertainty: argumentation

In the classical propositional or predicate calculus an argument is a sequence of inferences leading to a conclusion. The conclusion is true or it is false. The interest of the logician is in procedures by which arguments may be judged valid or invalid. An argument

$$A_1 ... A_{n-1} \vdash_L A_n$$

is syntactically valid in the logic L iff the formula A_n can be derived by $A_1...A_n$ and the axioms of L, if any, by the rules of inference of L (Haack, 1978). In the reasoning system we have developed an argument may not only prove or refute a proposition but also, more weakly, support or oppose it. Standard inference treats such weak qualifiers as undistinguished strings in a sentence ie mere linguistic decorations (of a unitary structure which as a whole can only be true or false. If we are to permit reasoning about the qualifiers as well as the base proposition we must distinguish them in some way.

Argumentation can be characterised as a process of applying sources of domain knowledge, or "theories", within a specific context to deduce a qualified conclusion:

[1]   Context $\cup$ Theory $\vdash$ (Proposition, Qualifier, Grounds)

Informally, a decision maker may argue "the patient possibly has gastric cancer because he is elderly and has recently lost weight and I know this is a classical presentation of an advanced malignancy". Here the *Context* equates to all the things the agent knows about the situation and the *Theory* corresponds to the decision maker's knowledge about typi-cal presentations, relationships between pathologies, development of disease etc. The theory is so-called because a theory embodies a generalisation about the world which, *ipso facto*, justifies the argument as a special case. *Proposi-tion* is the base sentence "the patient has cancer" but this is *Qualified* as "possible", which can be construed as "has arguments in favour but not definitely true". (There is nothing particularly radical here; the inclusion of the qualifier is analogous to the annotation of facts with numerical coefficients of belief (such as "the patient has cancer, with probability 0.3".) The *Grounds* include the specific facts that were the basis of the conclusion (the patient is elderly and has lost weight).

Constructing the arguments

Different kinds of argument can appeal to individual experience, authoritative opinion, the use and interpretaton of observations made using specialised devices, and domain theories with generalised schemas for reasoning about causality, structure, function, and so on. A general approach to implementing different forms of arguments is described in Fox and Clarke (1990). A specialised logic, LA, based on intuitionistic logic, sanctions the construction of lines of reasoning for hypotheses, and a meta-logic permits us to derive reasons to doubt arguments (Fox and Krause, 1992). Current work is extending these ideas to provide a scheme for reasoning over symbolic uncertainty terms (cf numerical utilities) and constructing arguments about actions. A rigorous theorem prover for LA is described by Krause, Ambler and Fox, 1992.

Labelling the arguments

Intuitively, arguments have "strength" or "force", which can be captured by the qualifier in [1]. A labelling scheme for qualifying an argument with its uncertainty is summarised by the mapping:

[2] L-Function_R: Proposition x Grounds $\longrightarrow$ Proposition x Grounds x Label_R

L-Function is a class of functions which maps arguments to arguments with labels representing the uncertainty associated with them. Uncertainty labels can be quantitative or qualitative, and in various representations, R.

In traditional expert systems rules are supplied by a knowledge base designer along with an explicit probability, certainty factor etc; here the L-Function is a human designer, who objectively or subjectively estimates these values. However, as observed earlier, it is often impractical in medicine to estimate precise uncertainty coefficients for facts, and hence for arguments that use those facts. In argumentation we can finesse this by using qualitative labels; the L-Function simply labels each argument with a sign indicating whether it increases or decreases belief in P. The semantics of such labels can be viewed as the sign of the change in belief in P that is warranted by the argument. Such labelling schemes can be defined for any uncertainty representation, R; probability, possibility, belief function etc.

Since the grounds for an argument are represented explicitly, we can reason about arguments as well as their conclusions. One use of this capability is to construct "meta-arguments" about the credibility of other arguments, which provides a basis for labelling the relative strength of arguments on purely logical grounds (Fox and Krause, 1992).

Arguments and beliefs

Given an exclusive set of possible decison options the decision maker needs to be able to weigh up their relative merits. The recommendation for how to do this from classical decision theory is clear: "Numbers have ... been associated with the events and with the consequences. The final stage in the argument [*sic*] is to associate numbers with the decisions, in such a way that the best decision is that with the highest number" (Lindley, 1985, p 57).

Although argumentation is based on qualitative reasoning it does not exclude a quantitative result. In classical FOL a single proof makes further proofs redundant, but in argumentation each proof provides further information by yielding new grounds for increases or decreases in belief. More formally

[3] A-Function_R: $P(\text{Proposition x Grounds x Label}_R) \longrightarrow (\text{Conclusion x Merit}_R)$

A simple function for computing the merit of a possible decision option simply determines the proportion of supporting arguments in the total set of arguments. In other words all arguments are treated as having equal weight. On the surface this seems unsatisfactory but there is a considerable literature indicating that decision procedures which ignore weights of evidence can be very effective (eg O'Neil and Glowinski, 1990).

Progressive decision making and clinical dilemmas

We now turn to the second problem mentioned, that of developing a decision procedure that does not require prior definition of all the options, relevant information sources etc. for all possible decisions that the GP (or DSS) may be required to assist with. We require a procedure that can do this progressively as information about the patient is acquired. The capabilities required can be illustrated with a simple medical scenario.

A young patient, Fred Smith, is complaining of stomach pain after meals and loss of weight. A DSS recognises that this is an abnormality and raises goals to explain and remedy the problem. After consulting the knowledge base, the DSS argues that the symptoms could be caused by gastric cancer or a gastric ulcer; it has a "dilemma", and raises a goal to make a decision which will resolve the dilemma. Since the dilemma concerns the aetiology of abnormal observations it is classified as a diagnosis dilemma. Knowledge of diagnostic dilemmas indicates that relevant sources of information include personal and life-style information, symptoms that can be caused by any diseases that are under consideration, and tests that can confirm the presence of a disease. Using this information as a guide the DSS establishes the patient's age from the patient record, and notifies the user that Fred's appetite, and an endoscopy test, are relevant to distinguishing the two diseases. The fact that Fred is young argues against both gastric cancer and gastric ulcer, but also justifies an argument in favour of duodenal ulcer which is more common among young patients. In due course information is added to the patient record that Fred's appetite is normal and the endoscopy test is negative; the DSS concludes that there is no organic disease and suggests antacids to relieve the symptoms.

Dilemmas can arise in any problem that a medical DSS is required to deal with - dilemmas about the treatment to use, whether to refer a patient to a specialist or not, whether to carry out tests and so forth. A decision procedure which is capable of dealing with such dilemmas must have a sound procedure for marshalling and evaluating relevant information. This must include:

a. identification of all relevant decision options
b. identification of information sources that bear on the options
c. construction of arguments for and against the options
d. assessment of the relative merits of the options

A symbolic decision procedure

Symbolic methods allow us to explicitly represent decisions, knowledge sources, reasoning strategies, representations etc. Decisions can be treated as objects in a conventional class hierarchy. Distinct classes of decision can be defined by a set of attributes, which are inherited by subclasses, such as diagnostic decisions and treatment decisions in medicine, but the values of the attributes are specific to each class. (The attributes and specific values correspond to the OSM "task specifications" described above.) In order to diagnose Fred Smith's stomach pain, for example, we consult general knowledge of how to make a decision and more specific knowledge of how to make a diagnosis decision.

More formally the framework can be summarised by two schemas, which define the data-structures to be maintained when any specific decision is being taken and the functions to do this. The first schema defines the data space associated with a decision:

$$(Dilemma, Topology, Labelling, Merit, Decision)$$

where:

Dilemma is a description of the decision, ie a goal for which there is more than one possible solution that could satisfy it.

Topology is a directed acyclic graph consisting of: (a) a set of input or information nodes (eg test results, symptoms) (b) a set of output or option nodes (eg diagnoses, treatment options) (c) a set of arguments linking input and output nodes.

Labelling is a set of coefficients, one for each argument in the *Topology*, representing the strength of arguments as discussed above.

Merit is a set of pairs *{(Node, Value)}*, one for each output node of *Topology*. *Value* is the result of aggregating the coefficients of arguments that lead into the node (for example the overall belief in a diagnostic hypothesis).

Decision is either the value "open", meaning no decision can yet be made, or an identifier indicating that the arguments/merit of an output node satisfy criteria associated with the class of decision being taken.

The second schema identifies a set of functions that operate over the data-structures:

$$(T\text{-}Function, L\text{-}Function, A\text{-}Function, D\text{-}Function)$$

where:

T-Function is a function whose domain is the current dilemma and the application knowledge base. It adds arguments to *Topology* on the basis of facts retrieved from the knowledge base, deductions, or other forms of computation.

L-Function is a function with domain *Topology* $\cup$ *KB* and range *Labelling* (discusssed above).

A-Function is a function with domain *Topology* $\cup$ *Labelling* and range *Merit* which aggregates the arguments for each option to yield a measure of the relative merit of the options. There are various aggregation functions, including classical probabilistic and utility functions, as well as simpler arithmetics and semiqualitative methods (Parsons and Fox, 1991)

D-Function is a function with domain *Topology* $\cup$ *Merit* and range *Decision*. Its role is the central one of determining when a specific option satisfactorily resolves the dilemma, or taking the decision is otherwise justified.

Taking the decision

Construction and aggregation of arguments for the set of decision options yields a total ordering on the set (eg the degree of belief in each possible diagnosis). However there is more to decision taking than simply ordering the alternative options and picking the one with the highest value. Decisions, particularly important decisions, should not be taken simply on the basis that some figure of merit has been exceeded. When we are accumulating information over time we may, by chance, obtain a run of findings that favour a certain option and our threshold of merit is exceeded, yet further data acquisition may reverse the position. We appear to need stronger criteria than simple ordering on the basis of some numerical measure of merit.

A more satisfactory class of decision functions is easy to appreciate but the precise properties such functions should have are open to discussion. Among the functions we are investigating are:

a. Committing to a decision option when we can prove from medical knowledge and patient data that there are no sources of information that could yield arguments which would change the most preferred candidate.

b. Demonstrating that the expected costs of not committing to an option exceed the expected costs of seeking further information.

There may also be criteria for making decisions which are restricted to specific classes of decision, as when we only accept a diagnosis if it can explain ("cover") all the observed symptoms, and criteria which are associated with safety critical decisions (eg proving that an action will be effective and safe).

Towards a design theory for decision support systems

Our emphasis on methods based on logic arises partly because of benefits in designing and implementing knowledge systems, but we believe logical methods also allow us to address a problem of long term importance for medical information technology. This concerns how we expect to evaluate and certify new clinical technologies.

Wyatt and Spiegelhalter (1990) have reviewed evaluation methods used in knowledge based system development, drawing an analogy with empirical trials of new drugs. They point out that there is much to be learned by designers of medical DSSs from the vast methodological experience accumulated in conventional drug testing. We agree that empirical evaluation of DSSs needs improvement, but we think this cannot be our only methodological goal. We draw a second analogy with mature engineering disciplines, many of which have moved beyond purely empirical testing (such as test-flying aircraft prototypes) because they have established strong design theories (eg for modelling stress in airframe structures and laminar flow over flying surfaces). The consequence is that designers can confidently predict failure modes, performance boundary conditions and so forth before the systems are implemented.

We do not currently know how to do this with decision support systems but we believe that it is a vital aim for medical engineering. Well understood formal theories, such as first-order logics and technologies based on them, offer an important direction for improving the soundness and safety of designs.

Work in progress

Although the logic engineering approach has taken us a considerable way towards our goal of building flexible, easy to maintain, robust, sound and safe decision support systems for medicine there is clearly much still to be done. A number of projects are in progress that are intended to address some of the more important issues.

DILEMMA (DIstributed Logic Engineering in Medicine: Methods and Architectures)

As mentioned earlier we are carrying out development work to field and evaluate DSSs based on the ideas described in this paper. The work is funded by the European Commission and involves a number of clinical and commercial organisations. The main clinical objectives are to design advanced DSSs for use in general practice and in specialist cancer and cardiology management clinics, and to integrate these into existing practice management and hospital information systems. Technical problems being addressed include the design of an efficient and general toolset for configuring specialist DSSs and methods for constructing complex arguments based on "deep" medical theories. A particular challenge is to develop a model of distributed decision making, to permit communication between remote medical centres and to support coordination of patient care by professionals in multiple disciplines.

RED (Rigorously Engineered Decisions)

Much of the theoretical work on decision making has been aimed at the development of a sound logical and mathematical basis for the design of advanced DSSs. The work has been stimulated by problems in medicine but, in common with classical decision theory, the theory of argumentation and symbolic decision procedures are believed to be generally applicable. A problem which is not well addressed by either classical decision theory or our work is that of decision making in safety critical situations. "Soundness is not safety"; software which is formally sound can still have consequences involving death, injury or damage to property. In the project RED (funded by the UK Government Department of Industry) we hope to clarify technical and legal issues of safety and develop techniques for eliminating or minimising decisions whose consequences may be unsafe.

ODD1 (Oncology Deductive Database)

The use of data models and integrity constraints in defining the OSM knowledg base helped to ameliorate problems of building and maintaining the large knowledge base, but current techniques for knowledge engineering on a large scale are not really satisfactory. We are participating in an ESPRIT funded project (IDEA: Intelligent Database Environment for advanced Applications) which is aimed at development of a database technology combining deductive and object-oriented techniques. Our role is to develop a substantial application for the new database system. The goals of the application will be to inform design and assist with testing by building a knowledge base covering a substantial fragment of knowledge about cancer. A central element of the ODD1 application will be an integrated data model for oncology together with tools for simulation, decision making and scientific theory formation. The technology could open up the possibility of providing cancer scientists and clinicians with a powerful resource for managing, sharing and interpreting research data, as well as providing insights into developing high integrity knowledge bases for medical and other safety critical applications. Achieving a consistent knowledge model with a comprehensive semantics for domain concepts is likely to be a necessary condition of success.

Acknowledgements

Thanks to Andrzej Glowinski, Colin Gordon and Mike O'Neil whose work has been central to the OSM project, and to Paul Krause and Simon Ambler who have contributed a great deal to theoretical developments.

References

Fox J, Clarke M "Towards a formalisation of arguments in decision making" *Proc. Stanford Spring Symposium on Argumentation and Belief*, Stanford: AAAI, 1990.

Fox J, Glowinski A J, Gordon C and O'Neil M "Logic engineering for knowledge engineering: the Oxford System of Medicine" *Artificial Intelligence in Medicine*, 2, 323-339, 1990.

Fox J, Krause P and Ambler S"Arguments, contradictions and practical reasoning" *Proc. European Conference on Artificial Intelligence*, 1992.

Glowinski A J, Coiera E and O'Neil M "The role of domain models in maintaining consistency of large medical knowledge bases, *Lecture notes in Medical Informatics 44*, Berlin Springer, 1991

Haack S *Philosophy of Logics*, Cambridge: Cambridge University Press, 1978

Krause, P Ambler S and Fox "The development of a logic of argumentation" *Proceedings of IPMU '92*, Majorca, 1992.

Lindley D V *Making decisions (2nd edition)* Wiley, 1985.

O'Neil M and Glowinski A J "Evaluating and validating very large knowledge based systems" *Medical Informatics*, 1990.

Parsons S and Fox J "Qualitative and interval algebras for robust decision making under uncertainty" in M G Singh and L Trave-Massuyes (ed) *Decision support systems and qualitative reasoning*, North Holland, 1991.

Wyatt J and Spiegelhaoter D J "Evaluating medical decisions aids: what to test and how" in J Talmon and J Fox (eds) *Knowledge engineering in medicine: Methods, applications and evaluation*, Heidelberg: Springer, 1992.

A Knowledge-based Approach to Strategic Planning

Edward H. Freeman

U S WEST Advanced Technologies
6200 South Quebec
Englewood, Colorado 80111
efreeman@uswest.com

Abstract. The Strategic Planning System (SPS) is a knowledge-based causal modeling and analysis technology. It has been designed to provide an organizing framework within which planners can (1) identify the key goals, actions and environmental variables which potentially contribute to the success or failure of a particular plan of action, (2) express the underlying causal relationships between these critical business factors, (3) obtain automatic identification of all feedback loops in a given model, including the determination of the positive or negative polarity of each loop, (4) obtain structural information on how these positive and negative loops interact to produce what are often counterintuitive effects from strategic system inputs, (5) determine the effects and side-effects of a given action plan or strategy, and (6) through ad-hoc queries, conduct "what-if" analyses and obtain various sorts of advice on manipulating action and goal variables. In group planning situations, SPS provides a model building methodology and an automated environment within which specific scenarios can be defined, and strategic plans can be iteratively refined based on group feedback and consensus.

1 Introduction

Corporate strategic planning maybe defined as an attempt to set a strategic direction for a corporation based upon an implicit or explicit model of how "critical" business factors might potentially interact to produce target outcomes. Strategic planning, by its very nature, is an exercise in navigating, hypothetically, a future that is undependable and unknowable [18]. While the strategic plans of a corporation typically describe its long term direction, a major portion of the planning process may be characterized as an exercise in developing a shared understanding between decision makers [16]. One of the difficulties in planning is that each planner's mental model of the world, and the articulation of that model as it relates to a particular business problem may be varied and inconsistent. Lack of complete understanding and poor communication can be fatal to the planning process and can lead to the development and execution of plans based on erroneous assumptions, unforeseen consequences, and unsuspected side-effects.

In this paper, we describe a knowledge-based tool that supports an interactive approach to planning and system modeling. We first provide a brief summary of causal modeling and its use as a framework for knowledge representation, analysis and inference. We then describe the salient features of the Strategic Planning System (SPS) including its knowledge acquisition, model analysis and knowledge presentation capabilities. SPS system effectiveness is then touched on briefly, and we close with a discussion of system status and future development plans.

2 Causal Modeling

Causal analysis and causal modeling methods have a long history of development and use in a variety of different scientific disciplines. Some examples include: Genetics [20], Econometrics [15], Electrical Engineering [12,10], Sociology [5], Political Science and Decision Analysis [1], Artificial Intelligence [13,8,2,6], Systems Dynamics and Causal Loop Analysis [14], and Social Psychology [7,11]. More recently, strategic planners have described causal modeling and causal loop diagramming as a mechanism for providing a natural, formal framework by which plans can be developed, articulated, justified and evaluated objectively [9,16].

Unlike any of the other systems listed here, the SPS causal modeling framework is based on a first order predicate calculus (fopc) formulation of causality, which has been used to define an "executable" causal theory, implemented as a knowledgebase of axioms and theorems. As an example, the following fopc sentence:

$$(1)\ \forall x\ \forall y\ causes(x,y) \Leftarrow \quad direct\_cause(x,y) \lor$$
$$indirect\_cause(x,y).$$

asserts that for all variables x, and y from the relevant domain of discourse,[1] the relation x causes y holds if either x directly causes y or if x indirectly causes y. As a second example, we can further specify what "indirect cause" means by the fopc sentence:

$$(2)\ \forall x\ \forall y\ \forall z\ indirect\_cause(x,y) \Leftarrow \quad (direct\_cause(x,z) \land$$
$$direct\_cause(z,y)\ ) \lor$$
$$(direct\_cause(x,z) \land$$
$$indirect\_cause(z,y)\ ).$$

This is a recursive definition which states that for all variables x, y and z from the relevant domain of discourse, x can be said to indirectly cause y if either x directly causes z and z directly causes y (i.e., one intervening variable), or x directly causes z and z indirectly causes y (i.e., two or more intervening variables). A large number of such assertions, describing path tracing rules, effect propagation rules, loop finding algorithms, etc., have been defined and the entire knowledgebase is available for "what-if" analysis, path finding, causal explanations, and ad-hoc queries. Within this framework, interactive model analysis is supported in a very rich and flexible manner.

It should be noted that the most primitive causal relationship in the SPS representation framework is that of direct cause. This is a simple pairwise relationship between variables of the form "X directly_causes Y," which means that a change in one variable (say X) will directly produce a change in another variable (say Y) without having the causal effect flow through any other variable in the model [5,3]. In their most generic form, complex causal models can be viewed as networks built from a collection of such simple atomic relations.

---

[1]In this case, the "domain" would be the domain of relevant causal variables.

The analysis of causal models can help strategic planners predict effects and unsuspected side-effects from possible system inputs, determine and explain the inherent feedback loops underlying the structure of the business, provide explanation for promoting certain actions based on their potential effects on goals, and perform "what-if" analyses of plausible scenarios.

3 Overview of System Implementation

The Strategic Planning System is based upon the OMNI/MOD causal modeling, expert system shell [4]. Because of its extensive knowledgebase of the conceptual structures and behavioral characteristics of causal systems, SPS can be viewed by the user as "understanding" the essential characteristics of causal relationships, causal topology, and causal variables. Besides acyclic structures, SPS has also been designed to handle cyclic graphs, (i.e., causal networks containing feedback loops), a feature that is used extensively to model dynamic systems [6,17]. SPS has been implemented in MacPROLOG™[2] and currently runs on Macintosh™[3] computers making extensive use of the Macintosh graphical interface. Source code for the entire system consists of 15,000 lines of Prolog, with approximately 2,000 lines of supporting C code, which is used to improve the runtime efficiency of the system's numerical algorithms.

4 SPS Model Building

SPS users typically begin their strategic planning effort by creating a causal model representing the context within which one or more strategic issues are to be analyzed. Creating a model consists of naming and arranging relevant causal variables in a network editing window, and then drawing connections between those factors to indicate a causal relationship. Variables and causal links may be created or removed in any sequence that the user finds convenient. SPS also provides facilities for documentation of both variables and relationships, an important part of model building for explanation purposes. In fact, the system has been designed to use this documentation as an integral part of its explanation facility, during the model analysis phase of the strategic planning process.

The SPS modeling framework is based on a rather radical notion that there are only four generic types of variables that a strategic planner must consider when developing corporate strategy: *Action* variables, *Environmental* variables, *Goal* variables and miscellaneous *Intervening* variables. Surprisingly rich and accurate models can be developed by judiciously choosing causal variables from these classes and by specifying their causal interrelationships. A short description of each class of causal variable is presented in table 1 below.

SPS allows the user to represent causal relationships in four different ways. The simplest form of causal relationship supported by SPS is the "+/-" model, which describes each causal influence in terms of its positive or negative impact. Figure 1 illustrates a simple "+/-" model of customer service relationships. SPS also supports *fuzzy weights* (e.g.,

---

MacPROLOG™ is a trademark of Logic Programming Associates, Ltd.

Macintosh™ is a trademark of Apple Computer, Inc.

"strong positive", "weak negative", etc.), *interval weights* (e.g., [.3, 1.2], etc.), and *crisp numbers* (e.g., .325, 12.29, etc.), as the user acquires more accurate data

![person with A box]	Action Variables are business factors that are under the direct control of decision makers. Unilateral decisions can be made to increase or decrease Action variables (e.g., increasing the size of the sales force or reducing R&D budgets).	![FACTOR hexagon]	Intervening Factors are intermediate factors over which business decision makers and/or environmental factors have only indirect control (e.g., employee morale, market share).
![E cloud]	Environmental Variables are external factors that affect the business in some way, but over which the business has no direct nor indirect control (e.g., the prime interest rate, the weather, etc.).	![GOAL triangles + and -]	Goal Variables are "evaluative" factors representing positive goals that a business would like to increase or maximize (e.g. cash flow, customer satisfaction) and negative goals, which a business would like to decrease/minimize (e.g., customer complaints).

SPS Causal Variable Categories
Table 1

In the current version of SPS causal effects propagate linearly, and the only restrictions on model building are that no other causal variables in the model can affect an environmental

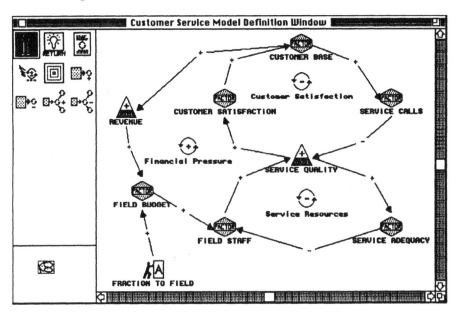

A model of customer service
Figure 1

variable. Environmental variables are considered exogenous in nature, and are by definition uncontrollable, as far as other variables in the model are concerned.

5 Knowledge Acquisition

Typically group strategic planning exercises are oriented towards making key business decisions, and are not often designed to acquire knowledge and insights from participants, as a first priority. Far too often, these meetings degenerate into antagonistic discussions, in which participants feel that they must attack the perceived weak points in competing strategies, to win acceptance for their own strategy proposals. SPS promotes a more refined level of dialog, whereby the focus is not on generating a discussion about which plan is best, but instead the focus is on developing group consensus about "how the world works" (i.e., develop an understanding of the "physics" of the business). Once this causal model has been developed, SPS can compute the logical consequences of various strategic alternatives (i.e., compute the logical implications of the dynamics associated with a set of proposed actions). Strategic decisions, then, are viewed as logical outcomes of knowledge acquisition, group consensus formation, and model analysis.

In addition to group model building exercises, SPS knowledge engineers also conduct individual interviews to facilitate iterative model refinement, and to make sure that all relevant aspects of individual "mental models" are captured in the group model. During the interview process, achieving the proper level of granularity or model simplicity has been critical in developing meaningful strategic models. At too coarse a level, analytic results from the model become obvious and contentless. At a level that is too deep, the important results are lost in a myriad of detail. What is desired is a level of granularity that leads to a deep understanding of strategic alternatives, with an adequate amount of supporting detail when necessary. The graphical nature of the SPS knowledge representation scheme has proven to be very useful in helping the knowledge engineer obtain the best level of model granularity for the problem under investigation [9].

6 Model Analysis

Once an initial model has been developed, SPS provides a variety of analytic tools to support structural exploration, model interrogation, and the analysis of strategic alternatives. Figure 2 illustrates a simple model designed to explore some of the dynamics associated with order processing and the potential effects on market share and corporate revenue.

6.1 Structural Exploration

The window in the upper left-hand corner of the screen presents the basic order processing model and a tool pane providing easy access to tools for structural exploration. These tools include mechanisms for highlighting direct and indirect causes of a given variable, direct and indirect effects of a given variable, summary *roll-ups* indicating the total effects of a unit change in any variable on all goal variables, summary *roll-ups* indicating the total effects of a unit change in action variables on a given target variable, etc. The upper right-hand window illustrates a roll-up of the total effects of advertising on the model's two goal variables, revenue and market share. Each arrow summarizes the total impact of all causal paths from advertising to each goal variable. The second roll-up window summarizes the total impact of all actions on market share. For very complex models,

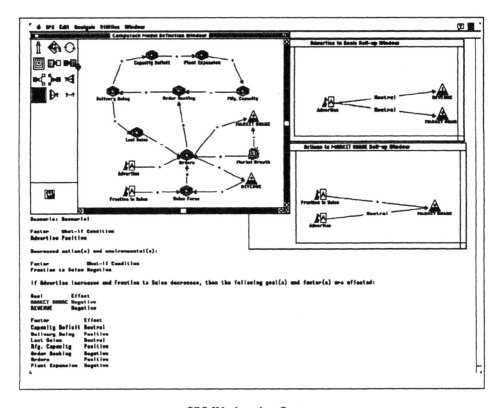

SPS Workstation Output
Figure 2

6.2 Model Interrogation

In addition to exploring various structural relationships, SPS also supports various model interrogation mechanisms. These tools can be accessed from the "Analysis" menu and include the ability to ask questions about

- How to increase or decrease any intervening variable or goal in the model. The corresponding answer can be given within the context of actions that can be performed by corporate decision makers, or within the context of both actions that can be performed by decision makers and potential impacts associated with external events generated from outside the corporation.
- Pro arguments for doing more of a particular action.
- Con arguments against doing more of a particular action.
- Pro arguments for doing less of a particular action.
- Con arguments against doing less of a particular action.

Model interrogation has proven to be very useful in helping strategic planners understand some of the subtleties associated with a given strategic domain or in preparing for presentations or debates.

6.3 Analysis of Strategic Alternatives

SPS defines a *strategy* as a particular set of increases and decreases in action variable inputs. It defines a *strategic scenario* as the combination of a strategy and a set of assumptions about how environmental variables will behave. Perhaps one of the most powerful features of SPS is its ability to generate the logical consequences of a given strategic scenario. This feature is called *what-if analysis*, and can be accessed from the "Analysis" menu. The textual output in the lower left-hand portion of figure 2 illustrates part of the output of a what-if analysis based on a particular strategic scenario, which assumes an increase in advertising, a decrease in fraction to sales (i.e., the percentage of revenues reinvested in the sales force), and no change in the environmental variable market growth. SPS provides the results of this what-if analysis, in the form of a summary of the net effects of these inputs on all goal variables, on all intervening factors and on all action variables(if there is feedback in the model associated with any of the action variables). Traditionally what-if analysis has been limited to presenting the results of, best case, worst case and most likely scenarios. With the speed and flexibility of SPS, many more scenarios can be explored and analyzed, resulting in a much richer set of strategic alternatives from which to choose.

7 Knowledge Presentation

For the most part, planners present their assumptions and models of the world in textual, spreadsheet and graphics forms, to explain and justify their strategic plans. SPS supports this type of output, with English-like explanation facilities that explain each factor, connection and path in the causal network. Corporate planners routinely generate automatic reports, which includes complete model specification, all user defined system documentation, and any roll-ups, and what-if analyses requested by the analyst. This output can be transferred to a word processor for further editing and formatting.

Users, however, are most attracted to SPS's interactive graphical presentation capabilities. This visual network-oriented presentation of knowledge helps planners gain critical insights into the most important relationships between business factors, within the context of their direct and indirect effects flowing through acyclic paths and feedback loops. During model building activities, this interface expedites knowledge acquisition. With the ability to view a complex model from different perspectives, previously elicited knowledge is remembered, and new pieces of knowledge can be easily related to the existing model. When a factor is added to a model, users typically employ structure exploration to find all points of causal impact for that factor. Color-coded highlighting is used to help planners see the connectivity of models by showing cones of influence and positive and negative feedback structures. As pointed out earlier, this graphic-oriented approach also helps planners recognize what level of granularity is relevant to a planning situation and appropriate for a given audience.

The use of alternate communications "channels," textual/linguistic and visual/diagrammatic helps users to remember acquired knowledge, to structure upcoming knowledge elicitation, to see areas of knowledge that need further development and filling-out, and to relate emerging ideas to the framework of existing knowledge.

8 System Effectiveness

Both U S WEST planners and MetaLogic Inc. consultants have found the SPS tool and methodology to be extremely useful, across a gamut of activities from quick single person brainstorming to full-blown strategic planning exercises. They have also found the tool very useful as an educational technology. Whenever new members come into a group that uses the SPS technology, a fully documented SPS model can serve as an interactive guide and tutorial of the group's shared vision of how their portion of the world "works." The new person can read model documentation, investigate causal relationships, perform what-if analyses and generally interrogate the model until he/she has become familiar with the group's perspective. Debates about model structure and strategic alternatives can then occur wherever necessary and the new person's point of view can be incorporated into the strategic model, as required.

Overall, users have found SPS to be an unusual, but very powerful approach to corporate strategic planning. They have been impressed with the system's ease of use and analytic capabilites and have used SPS to advantage in over fifty planning exercises.

9 Status and Future Enhancements

SPS was first released on May 1, 1989. Most recently, version 2.12 was released on September 1, 1990. We are currently experimenting with a number of laboratory versions of the system, designed to test various extensions to system functionality. Possible future enhancements include integrating multiple models, providing a model building "assistant" that will interactively help new users construct their SPS models, providing more mechanisms for representing time explicitly, simulating the effects of feedback loops and comparing the results to key model variables, and handling model uncertainty and input probability distributions (to support Monte Carlo simulation).

Acknowledgments

I would like to thank Russell McGregor and Gary Mischke for their unfailing support of the project. We are indebted to Mitchell Smith and Michael Epstein for providing architecture and algorithm development support to the project.

References

1. R. Axelrod, Structure of Decisions: The Cognitive Maps of Political Elites, Princeton University Press, Princeton, NJ., (1976).
2. J. de Kleer and J. Brown, "A qualitative physics based on confluences," Artificial Intelligence 24, 7-83 (1984).
3. E. Freeman, "The Implementation of Effect Decomposition Methods for Two General Structural Covariance Modeling Systems," Ph.D. Dissertation, Dept. of Psychology, UCLA, Los Angeles, 1982.
4. E. Freeman, "A Logic-based Prototyping Environment for Process Oriented Second Generation Expert Systems," SPIE Applications of Artificial Intelligence VI, 937 (1988), pp. 126 - 134.
5. D. Heise, Causal Analysis, New York: Wiley, 1975.
6. Y. Iwasaki and H. Simon, "Causality in Device Behavior," Artificial Intelligence, 29, 3-32(1986).

7. H. Kelley, "Perceived causal structures," in Attribution Theory and Research: Conceptual and Developmental Dimensions, J. Jaspars, et. al. eds., pp. 343-369, Academic Press, London, (1983).

8. B. Kuipers and J. Kassirer, "Causal reasoning in medicine: Analysis of a protocol," Cognitive Sciences 8, 363-385 (1984).

9. J. F. Ledbetter, "The Strategic Planning System User Interface Evaluation," U S WEST Document, 1989.

10. C. Lorens, Flowgraphs for the Modeling and Analysis of Linear Systems, McGraw-Hill, NY. (1964).

11. P. Lunt, "A causal network of loneliness," in Attribution Theory and Research: Conceptual and Developmental Dimensions, J. Jaspars, et. al. eds., pp. 343-369, Academic Press, London, (1983).

12. S. Mason, "Feedback theory -- some properties of signal flow graphs," Proceedings of the Institute of Radio Engineers, 41,1144-1156 (1953).

13. C. Reiger and M. Grinberg, "The Declarative Representation and Procedural Simulation of Causality in Physical Mechanisms," Proceedings of the Fifth International Joint Conference on Artificial Intelligence, William Kaufmann, Inc., 1977, pp. 1080 - 1085.

14. G. Richardson, A Pugh, Introduction to System Dynamics Modeling with DYNAMO, Cambridge, MA, MIT Press, 1981

15. H. Simon, "Spurious Correlation: a Causal Interpretation," Journal of American Statistical Association, 49, 467 - 479(1954).

16. R. Stata, "Organizational Learning—The Key to Management Innovation," Sloan Management Review, 30 (1989), pp. 63 - 74.

17. J. Sterman, "Modeling Managerial Behavior: Misperceptions of Feedback in a Dynamic Decision Making Experiment," Management Science, 35, No. 3, 321-339 (1989).

18. B. Tregoe, J. Zimmerman, Top Management Strategy, New York, Simon and Schuster, 1980.

19. S. Weiss, C. Kulikowski, and A. Safir, "A Model-based Consultation System for the Long-term Management of Glaucoma," Proceedings of the Fifth International Joint Conference on Artificial Intelligence, William Kaufmann, Inc., 1977, pp. 826 - 832.

20. S. Wright, "Correlation and Causation", Journal of Agricultural Research, 20, 557-585 (1921).

Expert Systems in Mining

Lutz Plümer

Rheinische Friedrich-Wilhelms-Universität Bonn

Institut für Informatik III, D-5300 Bonn 1, Römerstr. 164

lutz@uran.informatik.uni-bonn.de

In consequence of difficult geological circumstances and heavy competition on the world market the German mining industry is facing strong economic pressure for several decades. In Germany hardcoal mining is mainly done by the Ruhrkohle AG (RAG), a company with about 86 000 employees. The chance of survival of German mining activities rests on strong emphasis on technological innovations and rationalization measures. Some years ago the company started a project developing computer based work benches for the several engineering disciplines involved in the mine planning process. In this context the need for an overall information management was realized. A data model for the technical aspects of the mine was designed (1), (2), an information system based on this model is now being implemented. In parallel, the applicability of knowledge based systems for several aspects of the planning process was explored.

In the last few years we have developed several expert system prototypes in cooperation with RAG and DMT*, the research center of the mining industry:

- an expert system for checking electrical devices to be used underground (firedamp-proof) (3)

- an expert system for machine diagnosis (4)

- an expert system for the planning of underground illumination in coal mines (5), (6)

- an expert system for the correlation of stratigraphic sequences (7), (8), (9)

These systems have been implemented in LPA Prolog and are running on PCs. Several features are common to all of them: A knowledge-based approach was required in order to solve the given problem. Much conventional, partly numerical programming was necessary as well. The needs were not always very precise from the beginning, 'rapid prototyping', however, helped to identify them. Access to a relational database - which does not exist yet

---

* Deutsche Montan Technologie - Gesellschaft für Forschung und Prüfung GmbH

- must be possible. With regard to acceptability, convenient, interactive, graphics oriented user interfaces turned out to be of great importance.

The lecture will concentrate on two systems, which are near to an application in practice: SCHIKORRE and BUT.

BUT - an expert system for the planning of underground illumination in coal mines

Appropriate lighting minimizes the risk of accidents in hardcoal mining. In order to identify objects, sufficient light intensities are required, the value of which depends on the special task to be performed. Gradual changes in light intensity are desired in order to ease the accommodation of the eyes. Recently the European Coal and Steel Community has adopted guidelines which are an abstract of different research projects and define the requirements of underground illumination in coal mines. These guidelines relate the different tasks to be performed with minimal light intensities. These requirements will be obligatory in near future. They also suggest patterns which allow to find out which tasks may be performed on a certain place.

At present each mine employs a group of experts which plan the lighting of a scenario manually based on their experience with previously planned scenarios. The judgement of ergonomic properties of a planned installation is too complex to be performed by the planning experts in advance without any computer support. Until now the only way to verify whether a proposed solution satisfies the ergonomic requirements has been to measure the average lighting intensity and regularity after the installation has been finished. When the guidelines for underground illumination become obligatory, it will be necessary to check in advance whether or not a proposed installation satisfies these requirements. This was the main motivation to develop BUT*, a knowledge based system for the planning of underground illumination in coal mines.

The planning process realized by BUT consists of the following steps. First, the user has to specify the actual mine layout. In order to minimize the amount of information which has to be entered, we have built a knowledge base containing a taxonomic hierarchy of mine layout types. The various types, for example, contain information about possible equipment and possibly installed objects. The inheritance mechanism in the taxonomic hierarchy is top-down, although it is possible to specify exceptions to this rule. Below we give a small extract from the specification of 'stone drift' and 'drift':

---

* BUT is an acronym for 'Beleuchtung unter Tage', in English: underground lighting.

```
frame('stone drift', is_a, drift).
frame('stone drift', 'conveyor system', 'band conveyor').
frame(drift, equipment, ramp).
frame(drift, 'conveyor system', 'rail transport').
```

The next step in the planning process is the identification of tasks to be performed by the miners. To identify such tasks, spatial reasoning about the equipment of the mine layout is combined with knowledge extracted out of the guidelines and heuristics acquired from the planning people. In order to implement the guidelines, we formulated them as first order sentences before we transformed these sentences to clausal form. Inspired by (11) this technique was already successfully applied to build up the knowledge base of the diagnostic expert system described in (3).

The guidelines and heuristics often contain vague phrases such as: 'Material is transshipped at the boundary of two transport systems.' The concept 'boundary' is vague and not defined precisely. A heuristic to identify a place of transshipment is that there are at least two transport systems close to each other and, additionally, at least one of both systems begins or ends at that particular place. Whether an object is a transport system is derived from its type specification. The relative distance between two objects is derived from their resp. geometric attributes.

Generally it is possible to deduce several different tasks for a certain point in the mine layout. In this case BUT sorts the tasks according to the priorities implied by the guidelines. This is the basis for the calculation of the minimal light intensities.

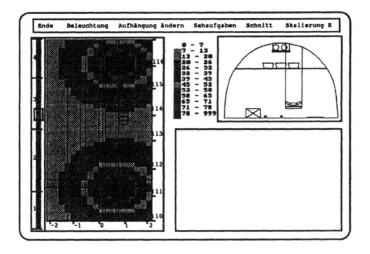

Fig. 1. Planning a stone drift with BUT

In a final step graphical simulation of the lighting configuration resulting from a generated plan is displayed. This feature has been identified as a key component for user friendliness and interactivity. The user can directly assess the pros and cons of a given proposal. He can focus on arbitrary parts of the scenario to analyze critical regions in more detail, and on those regions where only emergency lighting is provided. In the actual implementation, BUT allows a top view including the lighting intensity and a sectional view. An interpreter draws all objects, which are represented by vector graphics, in a particular mine layout according to the actual view.

Figure 1 shows a screen dump displaying the result of a session. The scenario includes a stone drift containing a ramp, a rail, a band conveyor, explosion barriers and conduits. The illumination resulting from the planning process is shown in the top view.

Schikorre: A Knowledge Based System for the Correlation of Stratigraphic Sequences

Schikorre* is an expert system for the correlation of stratigraphic sequences. It facilitates the identification of geological seams, especially coal seams, based on boreholes. Mining aims at the extraction of minerals, in our case hardcoal. The whole structure of the mine depends on the location of the coal seams. Knowing the exact location of these seams is a basic precondition of mine planning. Wrong predictions imply heavy costs. Boreholes, however, give only evidence on the local situation, so the best what one can get is a good model of the global situation. Prediction of coal seams is the task of geologists. Their work starts with an exact description of the different segments which have been extracted by the drillings. In order to get a prediction for a certain area, several drillings are grouped as polygons. Such a polygon is illustrated in figure 2, taken from (12). It shows the correlations between segments of different boreholes generated by a geologist aiming at the prediction of the location of a coal seam named 'Flöz Wilhelm 1/2'. Appropriate forming of polygons facilitates correlation and involves already some global knowledge. Major problems are caused by geological (tectonic) irregularities. They may be small but are very common in our area. Correlations are complicated if irregularities occur between two related boreholes. Thus it is normally a good idea to start with polygons which are located on either sides of (known) irregularities.

Geological reasoning comprises different aspects. There are general physical laws, but important in our context are also observations which may be true only for a special area like the Ruhr area. There are rules which say how to proceed best (procedural knowledge), and

---

* *SCHIKORRE* is an acronym for "Schichtenkorrelation", in English: correlation of seams

there are statements saying for instance which segments may correlate (declarative knowledge).

A geologist for instance states that one should start with a segment which is highly significant. A segment is significant if it has a high number of attributes which occur rarely. Significant are occurrences of several minerals (kaolinites). The first statement is procedural. The second is declarative and an example for something which might be true only for a certain geological area. From a formal point of view, stratigraphic correlation is the task of finding proper mappings between pairs of lists.

This is a combinatorial problem, since the number of possible mappings between two lists is exponential in the size of these lists. A proper mapping must respect geological and geophysical constraints. Segments which are to be associated have to be similar with regard to fundamental geological properties. With regard to coal seams, important properties are thickness and relative distances, ash and sulphur content, occurrences of significant kinds of kaolines etc. Correlation is complicated by the fact that seams may carve or disappear. Apart from very rare cases - which can be identified by other considerations - it can be assumed, however, that seams never cross each other. From a problem solving point of view this is the most important constraint and the starting point for Schikorre's approach.

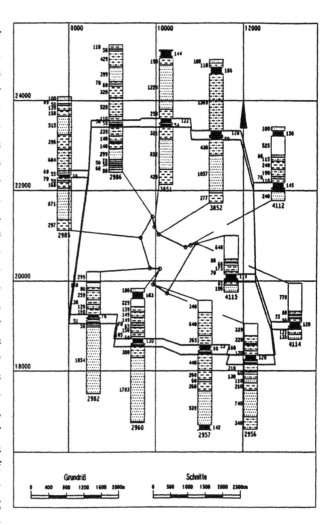

Fig. 2: Polygon of boreholes

This approach can roughly be described by the following clauses:

```
correlate([],B₂,[]).
correlate(B₁,[],[]).
correlate(B₁,B₂,Edges) ←
                find_significant_seam(B₁,S₁),
                find_partner_seam(S₁,B₂,S₂),
                user_accept(S₁,S₂,X),
                split(B₁,S₁,B₁_upper,B₁_lower),
                split(B₂,S₂,B₂_upper,B₂_lower),
                correlate(B₁_upper,B₂_upper,K_upper),
                correlate(B₁_lower,B₂_lower,K_lower),
                append(K_lower,[(S₁,S₂)|K_upper],Edges).

user_accept(S₁,S₂,X) ←
                output(S₁,S₂),
                read(X),
                X = yes.
```

The procedure *correlate* gets two boreholes B_1 and B_2 as input and generates a set of edges. Boreholes are - at least on this specification level - described by lists of (identifiers of) segments, and edges are pairs of (identifiers of) segments related to the two boreholes. Geological and spatial attributes of the segments are defined in the database of the system which is implicitly accessed by the predicates *find_significant_ seam* and *find_partner_seam*. The specification of these two predicates comprises (the main part of) the knowledge base of the system. Important aspects for both of them are, among others, occurrences of significant kinds of kaolinkohlentonstein, typical fossils in the seam's roof, ash content of the seam, sulphur content of the seam, seam thicknesses and seam distances.

For given B_1 *find_significant_seam(B_1,S_1)* may have several solutions, and again *find_partner_seam(S_1,B_2,S_2)* may have several solutions for given S_1 and B_2. However, a correlation which is appropriate from a local point of view may be inconsistent from a global point of view. Above, as the call user_accept(S_1,S_2,X) indicates, the user may reject this correlation after carefully inspecting it. Thus the first two calls are backtrackable. For the moment the correlation (S_1,S_2) defines an upper and a lower part for the resp. boreholes B_1 and B_2 and allows to analyze the upper and lower parts separately. Thus the basic problem solving approach of Schikorre is *divide and conquer with backtracking*. The knowledge base of Schikorre aims at a high reliability of local correlations, such that backtracking is needed rarely, if ever. The specification of the predicates find_significant_seam and find_partner_seam consists of 700 facts and 30 rules.

Interleaving the correlation of subsequent pairs of boreholes is another mean to focus search on highly reliable local correlations. At a time Schikorre processes up to 4 boreholes B_1, B_2, B_3 and B_4 forming a polygon. Having identified an appropriate partner-seam B_2 from S_2 for a significant seam B_1 in S_1, correlation of the pair (B_1, B_2) is delayed and (B_2, B_3) is considered. S_2 is now taken as significant seam of B_2, thus the next step is to find an appropriate partner in S_3. This step is iterated once again, giving S_4 until finally the pair (B_4, B_1) is considered. If B_1 is now derived as appropriate partner seam for S_4, then there is another evidence for the correlations derived so far, otherwise failure occurs and backtracking is enforced. We found that in our context *divide and conquer with backtracking* is more powerful if applied to polygons rather than to pairs of boreholes. Moreover this approach is quite natural for the experts.

Schikorre is highly interactive and has a graphic user interface. Correlations generated by the system are explained by the main geological and geometrical attributes of the involved segments. The rules applied are given on demand in a special explanation window via language templates. Correlations proposed by the system can always be rejected by the user. With the mouse he can require all relevant information of the geological attributes of the segments of the boreholes, thus getting a very convenient database access. He can also enforce *(assert)* correlations on his own.

In the context of the European coal mining industry, the topic of knowledge based stratigraphic correlation has been studied for nearly a decade. Schikorre, based fundamentally on logic programming and constraint handling techniques, was developed within 15 months. It is the first running prototype, producing plausible correlations and offering an interactive, user-friendly environment.

Conclusions

We have described two expert systems which have been developed in cooperation with the mining industry. They are implemented in Prolog and are running on PCs. In these projects we have been cooperating with engineers who are interested in solutions rather than programming techniques. Logic programming convincingly demonstrated its ability to solve complex problems which occur in a real industrial environment. A major point has been flexibility: supporting several phases of software production, starting with (rapid) prototyping and ending up with the final implementation of applicable systems. Another important point is that an expert system based on the logic programming paradigm fits well to - and can can even be regarded as a natural extension of - an information system which is based on a relational database management system.

Acknowledgements:

A. B. Cremers (University of Bonn) and R. Grevé (Ruhrkohle AG) initiated the project developing BUT, W. Burgard, S. Lüttringhaus-Kappel (University of Bonn), A. Kappel, J. Grebe (University of Dortmund), F. Mücher and F. Ende (Ruhrkohle AG) made substantial contributions. A. B. Cremers and J. Leonhardt initiated the project developing Schikorre, U. Baumbach, U. Freudenberg and F. Lehmann made substantial contributions.

References:

(1)   A. Baumewerd-Ahlmann, A. B. Cremers, G. Krüger, J. Leonhardt, L. Plümer, R. Waschkowski: An Information System for the Mining Industry, in: Karagianis, D. (ed.), Database and Expert Systems Applications, Proceedings of the International Conference in Berlin, Springer-Verlag 1991

(2)   W. Keune, L. Plümer: Ein Informationssystem für den Bergbau, Glückauf-Forschungshefte (erscheint demnächst)

(3)   Arnold, G., Burgard, W., Cremers, A.B., Lauer, M., and Lüttringhaus, S. Ein Expertensystem zur Prüfung elektrischer Betriebsmittel für explosionsgefährdete Bereiche. In Proceedings 2. Anwenderforum Expertensysteme, Cremers, A.B. and Geisselhardt, W., 1988, pp. 57-64, in German

(4)   Danielczyk, P., Einsatz tiefer Modelle zur Fehlerdiagnose in technischen Anlagen, Diploma thesis (in German), University of Dortmund, 1990

(5)   W. Burgard, S. Lüttringhaus-Kappel, L. Plümer: A Prolog-based Expert System for Underground Illumination in Coal Mines, Proceedings of the International Conference on the Practical Application of Prolog, London 1992

(6)   Grebe, J. Wissensakquistion für ein Expertensystem zur Planung von Beleuchtung unter Tage, Diploma thesis, (in German) University of Dortmund, 1991

(7)   U. Baumbach, L. Plümer: Schikorre: A Knowledge Based System for the Correlation of Stratigraphic Sequences in Prolog, Proceedings of the International Conference on the Practical Application of Prolog, London 1992

(8)   U. Freudenberg, F. Lehmann, L. Plümer: Schikorre: Wissensbasierte Korrelation geologischer Schichten, Zeitschrift für das Markscheidewesen (erscheint demnächst)

(9)   Baumbach, U., Ein Expertensystem zur Korrelation geologischer Schichten, Diploma thesis (in Gertman), University of Dortmund, 1990

(10)   Guidelines on the Ergonomics of Underground Illumination in Coal Mines. Tech. Rept. 15, Community Ergonomics Action, European Coal and Steel Community, Luxembourg, 1990

(11)   Sergot, M.J., Sadri, F., Kowalski, R.A., Kriwaczek, F., Hammond, P., Cory, H.T., The British Nationality Act as a Logic Program, Communications of the ACM 5 (1986)

(12)   Lehmann, F., Über die Ermittlung zu erwartender Lagerungsverhältnisse aus benachbarten Aufschlüssen. In: Das Markscheidewesen 98, (1991), H. 1, S. 9 - 16

Natural and Formal Language Processing

Michael Hess

ISSCO, University of Geneva, Switzerland

Abstract. Natural Language Processing, one of the most important branches of
Artificial Intelligence, has close links with Logic Programming. It is one of the
two roots from which Logic Programming developed, and still shares important
core concepts with it (e.g. unification). The concepts and techniques used for two
of its main tasks, viz. syntax analysis and semantic analysis, are illustrated with
concrete examples. For syntax analysis, it is the notations of Definite Clause
Grammars and Extraposition Grammars, for semantic analysis, the concepts of
Montague Grammar and Generalized Quantifier Theory. Some of these techniques
can also be applied to the analysis of formal languages.

1. Goals of Natural Language Processing

Natural Language Processing (NLP) is one of the most important branches of
Artificial Intelligence . It is also a remarkably varied field, mainly because natural
language is used by humans for quite different purposes. Language is, of course,
used to *communicate* information between humans, but it is also used to *store and
retrieve* and, to some extent, to *process* information in our minds. The systems
designed in the field of NLP try to model one or several of these components but
hardly ever all of them. Some of the main goals of NLP are

1. translation by computer

2. interaction with computers in natural languages rather than in special query
   languages

3. accessing information represented in natural language

"Machine Translation", i.e. fully automatic high quality translation of human
languages, was one of the first goals researchers set themselves when they began,
shortly after the end of the Second World War, to use computers for tasks normally
requiring human intelligence. Ambitions have been reduced considerably over the last
decades, and most researchers now prefer to think in terms of "Machine *Aided* Translation", as a fair amount of pre-editing and/or post-editing of the raw translations is
necessary even for the best systems available today. Nevertheless, some of these systems are now commercially viable for large companies with a high volume of technical documents to be translated.

Of equal importance is the second task, i.e. interaction with computers in natural language. While modern query languages used for contemporary relational Data Base Management Systems (DBMSs), such as SQL, are very powerful they are also fairly complex and unforgiving, and therefore intimidating to the casual user. Moreover, the more demanding the information requirements of users become, the more complex must be the query languages required for the job. This becomes particularly clear when interactions with computer systems other than DBMSs are considered. Take interaction with expert systems. It is well known that users trust the advice given to them by an expert system only if they are given clear explanations of the reasoning that resulted in a particular piece of advice, and if they are allowed to ask the system what it knows about the subject domain in general. This kind of interaction requires a much richer type of language than what is needed for querying a DBMS. While richer formal languages could, of course, be developed to this end it is doubtful whether users would accept them, given that even the relatively simple query languages of today's DBMSs turn out to be a major obstacle for the average user. It would be far more intuitive to use natural language for the interaction with information systems.

Finally, the need to automatically access information stored in the form of texts becomes increasingly pressing. In almost any given organisation, the amount of information stored in the form of texts still surpasses that stored in other forms, *and* by a large margin. It will be sufficient, in the present context, to simply refer to the almost exponentially growing number of scientific papers and monographs published in all fields of research. Storing these vast amounts of texts in electronic form no longer poses any fundamental problems but locating relevant information in this ocean of data becomes increasingly more difficult. It is generally recognized that the standard methods of information retrieval have reached the limit of their usefulness when applied to natural language texts. The main reason is that existing automatic indexing and retrieval methods treat texts on a basically statistical basis. In particular, all non-content-bearing words, the so-called "noise words" (function words, and extremely frequent non-function words) are removed from the documents to be indexed. The retrieval of documents requires in most cases that the users cast their queries in terms of Boolean combinations of search terms. This method retrieves, in many cases, so much irrelevant material that it is prohibitively time-consuming to locate the relevant information among the irrelevant documents (i.e. lack of precision). Any attempt to increase precision results in an often unacceptably high loss of relevant information (i.e. lack of recall). One of the reasons for this disappointing performance lies in the fact that this method ignores all the linguistically interesting information in documents. By ignoring function words, morphology, and word order, all the syntactic structure in the sentences of a text is lost, and consequently all the semantic information that is encoded in natural language through syntax becomes unavailable. It is, therefore, not really surprising that this method does not perform very well. Several attempts have been made in the mid-seventies to use the syntactic structure in the language of documents to increase the performance of full-text retrieval systems. Less often one tried to use the syntactic structure of the users' original questions (i.e. before their reduction to a combination of search terms) to the same end. Results have, on the whole, been rather disappointing. The main reason seems to be that the linguistic methods applied in this period were, by today's standards of NLP, fairly simplistic. In recent years, renewed efforts have been made to use the more sophisticated methods of NLP available today to access the content of texts in document retrieval systems, and even to develop fully-fledged question answering systems operating on a knowledge basis derived from natural language texts.

2. The Relationship between NLP and Logic Programming

The choice of a programming language for the design of some piece of software is normally made on the basis of purely pragmatic considerations (efficiency, availability, ease of integration with pre-existing software products etc.). The fact that a growing number of projects in the field of NLP are written in one of the Logic Programming languages (so far mostly Prolog) is, however, due to an intrinsic relationship between NLP and Logic Programming. Historically speaking, NLP is one of the two roots from which Logic Programming grew, while automatic theorem proving is the other. Colmerauer and Kowalski discussed, in 1971 in Marseille, different ways in which theorem proving could be applied to natural language processing, and the result of these discussions, together with contributions by Roussel, also at Marseille, resulted in the first implementation of Prolog, complete with an implementation of a French-language question answering system (Kowalski in: Shapiro 1987:546). It is therefore not surprising that Prolog in particular is superbly suited for NLP tasks. In fact, even the syntax of present-day Prolog is, in part, fashioned after that of natural language: "Sentences" (clauses) end with a period, and the "phrases" of a "sentence" (terms) are separated by commas. But much more significant is that the concept of unification is as basic to Logic Programming as to an entire class of modern linguistic theories that have sprung up in the seventies, and that rival today the descendents of transformational grammar in importance and influence. In a stunning case of historical convergence not only the *concept* of unification was developed independently in the disciplines of automatic theorem proving and linguistics at around the same time but even the *term* "unification" was invented independently in the two fields, for the same basic concept. If you use one of the unification based linguistic theories as the basis of your NLP system it is, therefore, far more than a mere convenience to use a unification based framework for the actual implementation.

On the other hand, the techniques that were developed on this background for the analysis of natural languages can, of course, also be applied to formal languages, for instance in the design of compilers. The priorities are, however, a bit different from those set in the analysis of natural languages. Formal languages tend to be much simpler than natural ones since they are *designed* to be processed efficiently. Analyzing them in a Logic Programming framework does not therefore require many of the more subtle techniques used for the analysis of natural languages. There is, for instance, no need for a treatment of context sensitive constructions in formal languages. On the other hand, some other properties of formal languages make it possible to use sophisticated program transformation techniques such as partial evaluation and unfolding, which are less useful for natural languages. The reason is that programs are, by definition, always completely explicit. They can be subjected to a full static analysis of their text in order to determine suitable transformations. For the correlate of programs in the field of natural language, viz. for their grammars, it is much less clear how exactly such concepts can be used. The grammars of natural languages must be inferred laboriously from the language behaviour of native speakers, and we can never be sure that we have a correct, let alone complete, grammar available. One of the contemporary unification based linguistic theories, Generalized Phrase Structure Grammar (GPSG), is based on the notion of grammatical meta-rules that are unfolded into object-level grammar rules but, for the reasons given, it is still unclear how adequate this kind of combined system is for a description of natural languages.

3. The Core Problems of NLP and their Treatment in Logic Programming

Although each specific NLP application is, of course, a tailor-made program, there are certain core tasks that must be performed in most NLP systems. We will not describe them all with the same degree of precision, not least because some of them are of little relevance for the processing of formal languages.

3.1 Morphological Analysis of Natural Language

One of the things that cannot be avoided in any NLP system, in one form or another, is the analysis of word forms, i.e. a morphological analysis of the linguistic material processed. In any but the most primitive applications we want to derive, from a given word *form*, its *root*, and the *inflectional* information (in English: number, gender, case) which not only is often semantically relevant (e.g. number) but has also important syntactic repercussions (agreement phenomena). But morphological analysis is also important for other reasons. It is only when the root of a word form has been determined that additional information about it can be retrieved from the lexicon: Syntactic information, e.g. how many and what kind of complements can be combined with a word, or categorial information (e.g. "animate", "concrete"), either about the word itself, or about its complements, etc., or even factual knowledge about the object denoted by it.

Although the problem of morphological analysis is truly central we can, in certain circumstances, get away with a conceptually very primitive solution. We simply list, exhaustively, all the word forms occurring in a language, together with their analysis and any additional information required. Needless to say, this leads to the creation of huge dictionaries very fast. Even for languages like English, which has precious little in the line of morphology, this is feasible only for smallish applications. For highly inflectional languages such as French, German and Russian or, as fairly extreme cases, Finnish and Turkish, this soon becomes unmanageable. We will, however, not go into any details here as the problem is fairly irrelevant for non-natural languages, and Logic Programming has nothing particular to offer for its solution. This does not hold for the next core problem of NLP, syntax analysis.

3.2 Syntax Analysis of Natural Language

Finding the syntactic structure natural language utterances (normally: of entire sentences), i.e. *parsing* utterances, is of paramount importance in NLP, and there are no quick fixes here to circumvent the problem. How central the task of parsing is, becomes clear if we consider the three applications mentioned above:

- In Machine Translation, it turned out to be extremely hard to map sentences of a source language directly into sentences of a target language. The earliest attempts at Machine Translation were in fact based on this concept. Soon it turned out that the resulting programs became monstrously big, monolithic pieces of software that were nearly incomprehensible and prohibitively expensive to maintain, not least because of the need to employ staff equally competent in at least two languages and software engineering. Moreover, one needed a separate piece of software for each pair of source language and target language, without the possibility to re-use much of the software used for another pair. In the following years it became clear that the computation of the syntax structures of the source language sentences is an indispensable intermediate step. These structures can then be transformed into the corresponding syntax structures of the target language from where, in a third step, the surface forms of the target sentences are generated. That way the analysis and generation modules of each language can be used for different language pairs, and only the middle module, achieving the transfer between syntax structures, must be written specifically for each language pair. Moreover, the staff maintaining such a transfer based system, need not be fluent in both languages of a given language pair. It is sufficient for them to know one language well.

- Natural language interfaces to DBMSs are, in a way, specialised Machine Translation systems. The difference is that the target language is not another natural language but an artificial query language, such as SQL. Again, it is possible in theory but not advisable in practice to do the translation in one step. It is far easier to first compute the syntax structure of the natural language query and then to transform it into a corresponding expression in the query language. This last step is, however, much simpler than in the case of a translation between two natural languages. Since query languages are much simpler than natural languages, the syntax structure of the natural language query can often be translated *directly* into the surface form of the query in the formal language, without the need to first create a "deep" syntax structure of the formal query and then to generate from it the surface form.

- In order to make retrievable the information contained in natural language texts, even for simple document retrieval at least some parts of the text (e.g. the noun phrases) must be transformed into complex index terms expressed in a suitable knowledge representation language (e.g. a variety of logic). For fully-fledged question answering, the requirements are much more demanding, as whole sentences must be translated in their entirety. This problem resembles the task of translating natural language into query languages in that the target language is, again, a formal language, not a natural one. However, the range of linguistic phenomena to be covered is much wider (documents consist, after all, of text in *unrestricted* natural language). Accordingly more demanding is the translation job, and experience in NLP has shown that it is a fairly hopeless task to translate natural language into any kind of logic without the intermediate step of a more or less complete syntax analysis.

In a sense, then, these three problems all reduce to a problem of translation. In all cases, it was said, the intermediate step of syntax analysis is indispensable, and more will be said about parsing shortly. But while Machine Translation can go a fairly long way without performing any further, i.e. semantic, analysis, the same thing does not hold for natural language interfaces to DBMSs, and even less for intelligent text retrieval. In these two fields, semantics is of central importance.

Before going into some of the details of how we can parse utterances in a Logic Programming framework it will be useful to give a concrete example how it is in fact often indispensable to compute parse structures as intermediate structures for further linguistic processing. This example will be from the field of automatic translation of natural languages.

An Example: Syntax Analysis for Machine Translation

Let us consider the following simple nominal constructions in English:

the configuration
the preliminary configuration
the installation configuration
the preliminary installation configuration

Let us now assume that we want a computer program to translate them into the corresponding French phrases, viz.

la configuration
la configuration préliminaire
la configuration d'installation
la configuration d'installation préliminaire

We notice two well-known things: In English, adjectival modifiers of nouns (and nominal complexes) are put *in front* of the modified expressions, and in French, *after* them (in most cases, anyway). Second, nominal complexes in English are formed by simple juxtaposition of nouns but in French, by means of a "de" and a different ordering of the simplex nouns.

If one wanted to translate these phrases from English into French *directly*, one would need four translation rules, one for each type of phrase (the meaning of the functors and predicates should be self-explanatory):

```
translate([e_art(E_Art),e_n(E_N)],
         [f_art(F_Art),f_n(F_N)]) :-        translate(E_Art,F_Art),
                                            translate(E_N,F_N).


translate([e_art(E_Art),e_adj(E_Adj),e_n(E_N)],
         [f_art(F_Art),f_n(F_N),f_adj(F_Adj)])
                            :-              translate(E_Art,F_Art),
                                            translate(E_Adj,F_Adj),
                                            translate(E_N,F_N).


translate([e_art(E_Art),e_n(E_N1),e_n(E_N2)],
         [f_art(F_Art),f_n(F_N2),prep(de),f_n(F_N1)])
                            :-              translate(E_Art,F_Art),
                                            translate(E_N1,F_N1),
                                            translate(E_N2,F_N2).


translate([e_art(E_Art),e_adj(E_Adj),e_n(E_N1),e_n(E_N2)],
         [f_art(F_Art),f_n(F_N2),prep(de),f_n(F_N1),f_adj(F_Adj)])
                            :-              translate(E_Art,F_Art),
                                            translate(E_Adj,F_Adj),
                                            translate(E_N1,F_N1),
                                            translate(E_N2,F_N2).
```

We would, of course, also need translation rules for the individual words, and rules to determine the syntactic category of each word plus a morphology component.

This proliferation of translation rules for nearly trivial phrases is, in itself, extremely worrying. It is easy to imagine what the set of translation rules would have to look like even for a small subset of English and French. Worse than that is the fact that both core constructions of these phrases, i.e. adjectival modification of nominal complexes as well as the formation of nominal complexes itself, can be recursively extended, without a theoretical upper limit:

the preliminary configuration
the complete preliminary configuration
the first complete preliminary configuration
...
the installation configuration
the installation configuration design
the installation configuration design manual
...

or to take a famous (and it seems, real) example

1) Airport long term car park courtesy vehicle pickup point

The approach using direct translation rules does not allow a truly general solution to this problem. If, however, we take the path via syntax structures, we can solve this particular translation problem with remarkable ease. In the following we will ignore the possibility of multiple adjectives. Under this simplifying assumption, the English noun phrase

the preliminary installation configuration design

will get the following syntax analysis (ignoring number)

```
e_np(e_art(the)
    e_adj(preliminary)
    e_cnp(e_cnp(e_cnp(e_n(installation))
               e_n(configuration))
          e_n(design)))
```

The corresponding French phrase (ignoring, in addition, gender)

le dessein de configuration d'installation préliminaire[1]

will get the analysis

```
f_np(f_art(le)
    f_cnp(f_n(dessein)
          f_prep(de)
          f_cnp(f_n(configuration)
                f_prep(de)
                f_cnp(f_n(installation))))
    f_adj(préliminaire))
```

Although the two structures may look, at first, fairly different, at closer inspection it can be seen that the differences are at least of a systematic character. The embedding of the nominal complex, for instance, runs from top left to bottom right in French, and exactly the other way in English. Both complexes form one coherent block, and the adjective is arranged on the same level of embedding, either before or after this block. Due to these systematic differences, merely three very simple transfer rules are sufficient to transform the first structure into the second. One of the rules takes care of the different ways adjectival modification is done in the two languages, and the other two take care of the different ways of (recursively) forming nominal complexes:

---

1. the contraction of "de installation" to "d'installation" is not taken into account below

```
transfer(e_np(E_Art,E_Adj,E_CNp),f_np(F_Art,F_CNp,F_Adj))        :-
                                            transfer(E_Art,F_Art),
                                            transfer(E_Adj,F_Adj),
                                            transfer(E_CNp,F_CNp).

transfer(e_cnp(E_CNp),f_cnp(F_CNp))      :-       transfer(E_CNp,F_CNp).
transfer(e_cnp(E_CNp,E_N),f_cnp(F_N,f_prep(de),F_CNp))
                            :-              transfer(E_CNp,F_CNp),
                                            transfer(E_N,F_N).
```

We only need to add rules for basic word categories, i.e. the entries where lexical entries are translated:

```
transfer(e_n(installation),f_n(installation)).
<etc.>
transfer(e_adj(preliminary),f_adj(préliminaire)).
<etc.>
transfer(e_art(the),f_art(la)).
transfer(e_art(the),f_art(le)).
```

Invoking 'transfer(E,F)' with the variable 'E' bound to the English syntax structure will now result in the creation of the corresponding French syntax structure, from which the French surface phrase can be generated. These rules will, of course, work for any depth of recursion of nominal complexes.

While this example had to be, by necessity, extremely simple it should have shown why syntax structures are, in the case of Machine Translation, an indispensable step. However, it can be shown, in a similar way, that performing a syntax analysis of the natural language input is crucial for almost any of the other types of task mentioned above. We can now turn to the question of *how* parsing can be done in the framework of Logic Programming.

Some Basic Techniques of Syntax Analysis in Logic Programming

Computing the syntax structures of natural language utterances is, in principle, surprisingly simple in a logic based programming environment. At first, the task looks like an intrinsically procedural problem: We must consume one word form of a sentence after the other, from left to right, analyze each word form morphologically, determine the syntax rules according to which the word is combined with the preceding words, and incrementally build up the syntax structure in the process. However, this task can be re-formulated in a purely declarative way. Then it is not only easy to implement the solution in a logic based programming language but, in addition, the resulting solution is much more general than what was needed for the original task.

The first key to this re-formulation lies in the concept of *parsing as deduction*. The second key is the use of *difference lists* as the representation scheme for phrases to be analyzed syntactically. Consider the following grammar of a tiny fragment of English, represented in BNF:

```
<Sent>        ::=    <Np>
                     <Vp>

<Np>          ::=    <Det>
                     <N>

<Vp>          ::=    <V_tr>
                     <Np>
<Vp>          ::=    <V_intr>

<Det>         ::=    the
<Det>         ::=    a
<Det>         ::=    an

<N>           ::=    man
<N>           ::=    dog
<N>           ::=    rock

<V_tr>        ::=    loves
<V_tr>        ::=    beats
<V_intr>      ::=    sleeps
<V_intr>      ::=    runs
```

It defines sentences such as the following as grammatically correct:

The dog loves the man.
The man beats the dog.
The man sleeps.

If all we want from a system is to perform a *check* whether a given string is a legal sentence with respect to this grammar, i.e. if we want to design a *grammar checker* (or an "acceptor"), all we have to do is to *prove* that this first part of the string consists of an 'Np', and the rest of the string of a 'Vp'. In order to do that we will have to prove that the first part of the string consists, in its turn, of a 'Det' followed by a 'N', and similarly for the 'Vp' part. This procedure goes, recursively, down to the level of terminal elements. If the proof succeeds, then the string is a legal sentence, otherwise it is not. What we need to do is, in other words, to recursively descend a logical program of this general form:

```
sent            :-      np,
                        vp.

np              :-      det,
                        n.

vp              :-      v_tr,
                        np.
vp              :-      v_intr.

det             :-      the.
det             :-      a.
det             :-      an.

n               :-      man.
n               :-      dog.
n               :-      rock.

v_tr            :-      loves.
v_tr            :-      beats.
v_intr          :-      sleeps.
v_intr          :-      runs.
```

But how do we represent the string of words to be analyzed? We must somehow extend the program so that we can, for instance, prove that 'the dog' is a noun phrase, maybe along the following lines:

```
np('the dog')           :-      det('the'),
                                n('dog').
```

But how do we get the original string from the outside into the program and, moreover, how can we split the string exactly so that each of its parts ends up precisely in that component of the program that is "responsible" for it? This is, after all, one of the main problems to be solved, so we cannot assume it will be solved beforehand. This is where the notion of difference lists enters the picture. Instead of referring to a word string by representing it as such (as done just above), we now refer to a string by representing the difference between two lists. Thus we refer to the string "the dog" by means of the *difference* between, for instance[2], the list '[the,dog,loves,the,man]' and the list '[loves,the,man]'. If we represent difference lists as two adjoining simple lists, i.e. as "A,B", this gives us, for the same fragment of the program above and for the example just used

---

2. there are, of course, infinitely many ways of representing the same string by means of difference lists

```
np([the,dog,loves,the,man], [loves,the,man])
         :-      det([the,dog,loves,the,man], [dog,loves,the,man]),
                 n([dog,loves,the,man], [loves,the,man]).
```

If we generalize this notation, the first part of the program above will look like that:

```
sent(S0,S)      :-      np(S0,S1),
                        vp(S1,S).

np(S0,S)        :-      det(S0,S1),
                        n(S1,S).

vp(S0,S)        :-      v_tr(S0,S1),
                        np(S1,S).
vp(S0,S)        :-      v_intr(S0,S).
```

This representation has also several procedural interpretations, which are very help-ful. Let us first consider the procedural interpretation for the recognition of a string of words. We can imagine that a word list is fed by the user into the variable 'S0' of the top predicate 'sent', from where it flows automatically into the same variable 'S0' in the right hand side predicate 'np'. After a successful proof of this predicate, the *rest* of the word list (i.e. everything *not* being part of the noun phrase) will pop up in the variable 'S1' in 'np'. From there it will flow into the variable 'S1' in 'vp'. Again, a successful proof will result in the unused rest of the string popping up in the second part of the difference list, i.e. in the variable 'S' in 'vp', from where it will be passed back to the variable 'S' in the predicate 'sent'. If the entire word list has been "used up", this rest will be the empty list.

What we still need is a way of cutting off one individual word form from the incom-ing list, and determining whether it is the word which we need. This is done by means of a predicate, here called 'c' (for "connects"), whose definition is

```
c([Word|Rest],Word,Rest).
```

This can be paraphrased in procedural terms as follows: If supplied with a word list in its first argument, this predicate cuts off the first element, tries to unify it with what it finds in the second argument, and returns the rest in the third argument. We now sup-plement the program fragment from above with the following clauses, using this predicate:

```
det(S0,S)       :-      c(S0,the,S).
det(S0,S)       :-      c(S0,a,S).
det(S0,S)       :-      c(S0,an,S).
```

```
n(S0,S)         :-       c(S0,man,S).
n(S0,S)         :-       c(S0,dog,S).
n(S0,S)         :-       c(S0,rock,S).

v_tr(S0,S)      :-       c(S0,loves,S).
v_tr(S0,S)      :-       c(S0,beats,S).
v_intr(S0,S)    :-       c(S0,sleeps,S).
v_intr(S0,S)    :-       c(S0,runs,S).
```

We can now phrase our queries as theorems of the following form:

```
?- sent([the,dog,loves,the,man],[]).
```

In words: Can it been shown that the difference between '[the,dog,loves,the,man]' and '[]', i.e. the string "the dog loves the man", is a 'sent' with respect to the grammar used? Note that the standard proof procedure of Prolog can now be used to perform the proof, without the need to write a special parsing program. Naturally, the restrictions on the form of Prolog programs now also apply to the form of such grammatical analyzers (among others the restriction which forbids left-recursive rules).

Since the correspondence between the original grammar of the language, given in BNF, and the acceptor program derived from it, is so straight-forward there is, in most Prologs, a special notation for grammar rules. A set of grammar rules is expanded automatically (i.e. whenever such a program is read in) into the corresponding acceptor program in standard Prolog syntax. This special notation, called "Definite Clause Grammar" notation, uses the special two place infix operator '-->', e.g. the first rule of the grammar will look like that:

```
sent    -->      np,
                 vp.
```

The operator '-->' is the signal for the system to pre-process (i.e. to compile) the expression in which it occurs into a standard Prolog clause with difference lists added in all the appropriate places, i.e.

```
sent(S0,S)      :-       np(S0,S1),
                         vp(S1,S).
```

Terminal elements are represented in square brackets. They are expanded into 'c'-predicates. This gives us, as DCG representation for the above grammar,

```
sent            -->     np,
                        vp.

np              -->     det,
                        n.

vp              -->     v_tr,
                        np.
vp              -->     v_intr.

det             -->     [the].
det             -->     [a].
det             -->     [an].

n               -->     [man].
n               -->     [dog].
n               -->     [rock].

v_tr            -->     [loves].
v_tr            -->     [beats].
v_intr          -->     [sleeps].
v_intr          -->     [runs].
```

Remarkably enough, this looks almost like the BNF representation of the *grammar* but it is, due to this automatic expansion, a fully-fledged acceptor *program*.

Such a simple-minded grammar will, of course, accept a very great number of sentences that are not correct English. Not least, number agreement is not enforced, i.e. sentences like

The dog love the man
The dogs loves the man

would be accepted. We could overcome this problem by extending the grammar with new categories like 'np_sing', 'vp_sing' etc.:

```
sent            -->     np_sing, vp_sing.
np_sing         -->     det_sing, n_sing.
vp_sing         -->     v_tr_sing, np_none.
vp_sing         -->     v_intr_sing.
<etc.>
```

```
sent              -->     np_plur, vp_plur.
np_plur           -->     det_plur, n_plur.
vp_plur           -->     v_tr_plur, np_none.
vp_plur           -->     v_intr_plur.
<etc.>
```

However, this would be an extremely awkward solution. As soon as other types of agreement are taken into account, the situation gets soon out of hand. For pronouns, for instance, where gender and case are formally distinguished, we would need a whole battery of different, completely artificial, categories such as 'pronoun_masc_sing_subj', 'pronoun_masc_sing_obj', 'pronoun_fem_plur_subj' etc., altogether 3*2*2 combinations (and in languages other than English many more). This is not only awkward to write, it also makes it impossible to use some of the most basic generalizations of language. We could, for instance, no longer speak of "noun phrases" as such. We would always have to speak of "plural noun phrases" and "singular noun phrases", even in those numerous cases where a certain phenomenon uniformly applies to *all* kinds of noun phrases.

Luckily, there is a much simpler solution. By using *one* additional variable in some of the terms of a DCG, we can enforce number agreement:

```
sent              -->     np(Number),
                          vp(Number).

np(Number)        -->     det(Number),
                          n(Number).
<etc.>

det(_)            -->     [the].
det(sing)         -->     [a].

n(sing)           -->     [man].
n(plur)           -->     [men].

v_tr(sing)        -->     [loves].
v_tr(plur)        -->     [love].
<etc.>
```

The standard machinery of resolution will now ensure that subject and predicate of a sentence agree in number, as it will for the determiner and the noun of a noun phrase. This solution is more powerful than it may look like at first. More will be said about it below.

However, we still do not know how to *parse* a sentence. All the programs shown so far were mere acceptors. How do we get the syntax structure of a successfully analyzed sentence out of the system? Again, one additional argument, in *every* term of a DCG this time, is sufficient to do the trick. The additional argument is filled, in the left hand side terms, with a complex structure, giving the "outline" of the structure recognized by this particular grammar rule, while the arguments in the right hand side expressions are filled with variables:

```
sent(sent(Np,Vp))              -->      np(Np,Number),
                                        vp(Vp,Number).


np(np(Det,N),Number)           -->      det(Det,Number),
                                        n(N,Number).


vp(vp(V,Np),Number)            -->      v(V,tr,Number),
                                        np(Np,_).
vp(vp(V),Number)               -->      v(V,intr,Number).
```

Pre-terminal elements are structured similarly, only with fully instantiated entries in the left hand side:

```
det(det(the),_)                -->      [the].
det(det(a),sing)               -->      [a].


n(n(man),sing)                 -->      [man].
n(n(man),plur)                 -->      [men].


v(v(love),tr,sing)             -->      [loves].
v(v(love),tr,plur)             -->      [love].
<etc.>
```

This simple expedient will make the machinery of resolution create a structure that faithfully represents the "audit trail" of any proof successfully performed. If we now ask the system prove the theorem

```
?- sent(S, [the,dog,loves,the,man], []).
```

with an additional variable 'S', it will instantiate this variable to the value

```
sent(np(det(the),n(dog)),vp(v(loves),np(det(the),n(man))))
```

This is the desired syntax structure, as can be seen in the easier-to-read representation of a pseudo-tree:

```
sent(
    np(
      det(the)
      n(dog))
    vp(
      v(loves)
      np(
        det(the)
        n(man))))
```

One of the most appealing aspects of the Logic Programming approach to NLP is the fact that the non-directionality of logic programs automatically applies to grammars written in any of the grammar formalism phrased in it (such as DCG). Thus the *same* DCG can be used

- as parser
- as generator
- as syntax checker
- and as grammar verifier.

This is what was meant above when it was stated that the declarative re-formulation of a supposedly procedural problem would result in a more general solution than originally required.

First, let us consider generation. If we ask the system to prove

```
?- sent(sent(np(det(the),n(dog)),vp(v(loves),np(det(the),n(man)))),S,[])
```

it will, without the need for any further changes, *generate* the surface sentence corresponding to this syntax structure. The result will be made available as the binding of variable the 'S' in the theorem, viz. as

```
S = [the,dog,loves,the,man]
```

This is, of course, exactly what we need in a Machine Translation system that uses syntax structures as intermediate representations for the transfer from source language to target language, in the manner outlined above. In fact, the example translation above was generated by a query of the following form:

```
?- e_np(P1,E,[]), transfer(P1,P2), f_np(P2,F,[]).
```

where 'E' was instantiated with the English input sentence, and 'F' got then bound to the French translation. 'e_np' and 'f_np' accessed an English and a French DCG, respectively, and 'transfer' accessed the rules given earlier.

If we make the system prove the theorem

```
?- sent(sent(np(det(the),n(dog)),vp(v(loves),np(det(the),n(man)))),
        [the,dog,loves,the,man], [])
```

it will *check* whether the syntax structure in the first argument is a correct analysis of the sentence given in the other arguments, and reply 'yes' or 'no' accordingly.

Finally, if we ask the system to prove

```
?- sent(P,S0,[]).
```

it will blindly generate the first legal sentence allowed by the grammar plus its syntax analysis. On backtracking, the system will generate the next sentence plus analysis, and this can be repeated until all legal sentences have been generated. This is very useful when one wants to test a grammar for correctness. The main problem when debugging grammars of natural languages is finding sentences which erroneously get accepted (false positives). Humans can easily think of grammatically *correct* sentences but have a hard time coming up with the wildly *incorrect* sentences that a grammar is likely to let go through in the early stages of debugging. When a DCG is used as a blind generator it will, however, be easy to spot all the grammatically aberrant sentences it generates, and correct the grammar.

While the techniques just outlined will be equally useful for parsing natural as well as formal languages, there are some special formalisms that are particularly important in NLP. They will be mentioned next.

Special Techniques for "Quasi Context-Sensitive" Constructions

Although not many natural languages are known that definitely use context-sensitive constructions (Swiss German is one certain case), most natural languages have properties that can be processed only in the most awkward manner if context-free methods are used. One such case has already been mentioned, viz. number agreement in English. Although it is possible to describe and analyze this agreement phenomenon (and all other agreement phenomena) with purely context-free methods, viz. by splitting each basic category into several artificial categories, it is, in practice, impossible to proceed that way. We solved the problem in a very elegant way, by means of an additional argument position in the relevant non-terminals. This additional argument gave the grammar context-sensitive power[3]. As in many other cases in linguistics, an intuitively (and practically) acceptable analysis of a certain phenomenon requires methods that are, strictly speaking, unnecessarily powerful.

There are many more such cases of "quasi context-sensitive" constructions in natural language. While in the case of agreement phenomena, it was possible to deal with them by means of a single additional argument in the appropriate places, this is not so easy in other cases. Wh-questions are a case in point. If we compare the question 2 and the declarative sentence 3 we notice two things: First, the question uses an auxiliary verb ("did") and inversion to express the interrogative character of the sentence and, second, it puts the interrogative pronoun ("what") in sentence initial position:

2)   **What did the man see?**

3)   **The man saw a dog**

How could we cover such interrogatives in a grammar? The first thought might be to add grammar rules such

---

3. In fact, the power of an unrestricted language. The implicit difference lists of DCGs, if used merely as a convenient means to refer to the string to be analyzed, will itself not yet give the grammar this power.

```
sent            -->     qu_prn, aux, np, v_tr(inf).

qu_prn          -->     [what].
aux             -->     [did].
v_tr(inf)       -->     [see].
```

to the pre-existing rules for declarative sentences, i.e.

```
sent            -->     np, vp.
vp              -->     v_tr,
                        np.
```

However, this solution is very unattractive. First, for each rule describing a declarative sentence type we would have to write an additional grammar rule for the corresponding interrogative sentence type. This in itself would result in the same kind of rule duplication as in the case of agreement phenomena. But other types of information would also have to be duplicated. The subcategorization frames of verbs are an example. These frames indicate how many complements a verb expects and of what kind they must be. The verb "to see" requires, syntactically, one direct object which, furthermore, must be of the right semantic type (roughly: a physical object) while "to talk" syntactically requires an indirect object introduced through a prepositional phrase with the preposition "to", and whose semantic type is "animate". It is the syntactic constraint which rules out sentence 5, and the constraint on the semantic type (the so-called "selectional restrictions") which make sentence 6 semantically odd:

4)   The man talked to a boy

5)   * The man talked a boy

6)   ? The man talked to a table

This kind of information is stored in "subcategorization frames" which look, in simplified form, as follows

```
talk:   V, +[ _          - Pp(to)
              +animate    +animate ]
```

which says that "talk" is a verb ('V'), and that both the subject ('_') as well as the (immediately following) indirect object must be animate, and that the indirect object is given in a prepositional phrase ('Pp') with "to".

For each of the declarative sentences above there is an interrogative one, with the same degree of acceptability:

7)   Who did the man talk to?

8)   * Who did the man talk?

9)   ? What did the man talk to?

Intuitively it is quite clear that the same kind of constraints is at work in the interrogative as in the declarative sentences, yet we could *not* use the same subcateogorization frames as for the declarative sentences. The verb form is, after all, *not* followed by a full prepositional phrase in the interrogative sentence. We would, in other words, have to create a complete set of subcateogorization frames for interrogative sentences, and would lose the possibility to make the intuitively obvious generalization over both types of sentences.

This situation was one of the reasons why Transformational Grammar introduced transformations into the linguistic theory. One could use a single "deep structure" for, in this case, both declarative and interrogative sentences. It was on this level that the constraints defined by the selectional restrictions were tested. In a second step one could then apply suitable transformations to the deep structure and derive the surface form of the sentence. For a interrogative sentence, this would involve, among other things, moving the interrogative pronoun to the front of the sentence. For example 7 above this would mean that the "deep structure" would correspond to 7a

7a) The man did talk to who?

Operating on this structure, two different transformations would move the auxiliary verb and the interrogative pronoun into their final positions. Inverting word order is a local kind of transformation but pulling a word to the beginning of a sentence is a kind of movement that can, in many cases, result in a move over an indeterminate amount of intervening material, at least in principle. This is why it is is called "long distance dependency".

Transformations have a series of drawbacks and have fallen, in their original form, into disuse but the problems they were meant to solve must be taken care of somehow or other. Unification based linguistic theories normally use a technique called "gap threading" to this end. Gap threading has the effect of a "movement" but is perfectly declarative and, therefore, static. Moreover, it does not operate on a syntax structure previously generated, like the deep structures of Transformational Grammar, but works on the same level as all other rules of grammar. Theories using this technique are therefore often called "mono-stratal". The basic idea is very simple, and it is reminiscent of the technique of difference lists used for the representation of strings of words. Whenever we encounter a type of word in a syntactic position where we would not "normally" expect it (say, a certain type of noun phrase right at the beginning of what we expect to be a sentence), we make a "note" of it in a special "filler list". When, later in the sentence, we would expect a certain type of word at a certain position (say, a noun phrase) but find there a gap instead, we go and look in the store whether there is a filler of the correct type, meaning that this kind of word had been encountered earlier. If so, the filler is removed, and the gap "filled". Since the distance between the position of the original word and the gap can be arbitrarily long, since the gap can (in principle) appear anywhere later in the sentence, and since there are no global variables in Logic Programming, the store must take the characteristics of a "communication channel" between the non-terminals in the grammar. Since there may easily be *several* "displaced constituents" in the same sentence, the contents of the store must be passed to the next non-terminals once a certain item has been removed. It is therefore sensible to implement the communication channel as a difference list. In procedural terms this means that the fillers enter a non-terminal through the list 'GapIn', the non-terminal is proved, and what remains of the fillers leaves the non-terminal via the list 'GapOut':

```
s(GapIn,GapOut)              -->      np(GapIn,Gap),
                                      vp(Gap,GapOut).
```

We define additional rules for declarative and interrogative sentences:

```
wh_interr(GapIn,GapOut) -->      qu_prn(GapIn,Gap),
                                 s(Gap,GapOut).
assertion(GapIn,GapOut) -->      s(GapIn,GapOut).
```

and make sure that a preposed interrogative pronoun adds a "filler" of type "np" to the list:

```
qu_prn(GapIn,[np|GapIn])-->      [who].
qu_prn(GapIn,[np|GapIn])-->      [what].
```

Finally we must allow a noun phrase, found missing later in the sentence, to be substituted by a suitable filler taken from the filler list. This can be done with a rule

```
np([np|GapOut],GapOut)     -->      [].
```

saying that noun phrases may be preposed. As in DCG, the empty square brackets in the last grammar rule say that *no* terminal element is required for the rule to succeed. Here the additional requirement for success is that there is an entry "np" in the filler list. If so, even a gap will be recognized as a (virtual) element (here: noun phrase). The filler 'np' will then be removed from the list, and the rest of the list returned in 'GapOut'. In a similar way we would have to deal with the displaced auxiliary verb in interrogative sentences.

It will be obvious that we are now in a situation that is very similar to that which led to the introduction of the DCG formalism. Then as now, the human grammar writer is reduced to inserting, by hand, additional arguments to all non-terminal elements of a grammar. This is the sort of repetitive and mindless work that could done much faster by a computer. In an extension of DCGs, called "Extraposition Grammars" (XGs)[4], the operator '--->' is defined so that non-terminals are supplemented by *two* difference lists, one for the word string, as in DCGs, and a second one for the 'gap fillers'. The use in a grammar rule of the second operator which is new in XG, written '...', triggers two things: First, the addition of a filler (of the type named in its right argument, below: 'np') to the filler list and, second, the creation of an additional rule allowing the recognition of a virtual constituent of the same type. The grammar from above, in XG notation, becomes now much simpler:

---

4. strictly speaking, "Extraposition Grammar" is a bit of a misnomer as most linguists understand something slightly different by "extraposition"

```
s                       --->    np,
                                vp.

wh_interr               --->    qu_prn,
                                s.
assertion               --->    s.

qu_prn  ... np          --->    [who].
qu_prn  ... np          --->    [what].
```

The rules that recognize virtual constituents, i.e.

```
np([np|GapOut],GapOut)  -->     [].
```

from above, need not be mentioned explicitly in an XG as they will be created during the expansion of rules containing the operator '...'.

It goes without saying that a whole host of detail problems must be solved before this approach works. The full version of XG[5] offers more possibilities but the basic principles should be clear even on the basis of this simplified version of the theory given here.

Extending the rules given so far, we can write a very small XG that can, nevertheless, cope with a certain range of declarative and interrogative sentences. Its central part is reproduced here:

---

5. described in most detail in Pereira 1983

```
sent(Sent)                      --->    decl(Sent).
sent(Sent)                      --->    wh_interr(Sent).
sent(Sent)                      --->    yn_interr(Sent).

decl(decl(Sent))                --->    s(Sent).
wh_interr(Sent)                 --->    qu_prn(Np,Anim,Num),
                                        yn_interr(Sent).
wh_interr(interr(Sent))         --->    qu_prn(Np,Anim,Num),
                                        s(Sent).

yn_interr(interr(Sent))         --->    fronted_aux,
                                        s(Sent).

fronted_aux   ... verb_form(V,aux,Num)
                                --->    verb_form(V,aux,Num).

s(s(Np,Vp))                     --->    np(Np,_,Num),
                                        vp(Vp,Num).

qu_prn(wh(who),anim,_)   ... np(wh(who),anim,Num)
                                --->    [who].
qu_prn(wh(what),inanim,sing) ... np(wh(what),inanim,Num)
                                --->    [what].

np(np(det(the),Pn),anim,sing)
                                --->    proper_name(Pn).
np(np(Det,N),Anim,Num)          --->    det(Det,Num),
                                        n(N,Anim,Num).

vp(vp(verb(be),Np),Num)         --->    verb_form(verb(be),aux,Num),
                                        np(Np,_,_).
vp(vp(V,Np),Num)                --->    verb(V,main+trans,Num),
                                        np(Np,ObjAnim,_),
                                        {requires(V,ObjAnim)}.

vp(vp(V,Np),Num)                --->    verb(V,intrans,Num).
```

Apart from the lexicon, not shown here, there is (rudimentary) information about the
selectional restrictions (only about the direct object of verbs):

```
requires(verb(employ),anim).
requires(verb(teach),inanim).
requires(verb(pay),_).
```

This grammar wil analyze declarative and interrogative sentences such as
Some department employs a tutor.

Every department employs john.
The department employs the tutor.
John employs a tutor.
Every department employs a tutor.
Does every department employ a tutor?
Who employs a tutor?
Who does every department employ?
Does some department employ a tutor?

One of the advantages of using an XG is that an interrogative sentence gets almost the same syntax structure as the corresponding declarative sentence, despite its, superficially, very different word order. In the case of a yes-no-question, the difference between the parse structures is localised in one place (the outermost layer of the structure), indicating the type of sentence :

```
Some department employs a tutor


decl(s(np(det(some)
          n(department))
       vp(verb(employ)
          np(det(a)
             n(tutor)))))



Does some department employ a tutor?


interr(s(np(det(some)
            n(department))
         vp(verb(employ)
            np(det(a)
               n(tutor)))))
```

In the case of wh-questions there is, in addition, an interrogative pronoun instead of a full noun phrase, and it is in the *same* position as the full noun phrase would be:

```
Every department employs John?


decl(s(np(det(every)
          n(department))
       vp(verb(employ)
          np(det(the)
             proper_name(john)))))
```

```
Who does every department employ?

interr(s(np(det(every)
             n(department))
         vp(verb(employ)
         wh(who)))))
```

When differences that were spread out over the entire sentence in its original form
have been reduced, in the underlying syntax structure, to smaller, more to localised
differences, it is much easier to process the output in subsequent program components. This will be shown in the next section.

3.3 Semantic Analysis of Natural Language

It may seem to be intuitively fairly clear what we mean when we talk about "meaning" in natural language. However, things become murkier if we look at them from
close quarters. At first view, the meaning of an utterance simply is what the utterance
is *about*. To be "about" something primarily means, in its most basic sense, to *refer* to
something. Reference is what we (normally) achieve by the use of proper names or
definite noun phrases, and sentences containing only this type of nouns are called
"singular" statements, such as

10) Miller is a student in class CS 101

where two individually known objects are said to stand in a certain relationship.
Checking whether this sentence is true is quite straight-forward.

When either one of the two proper names is replaced by a general term, such as
"some(thing/body)" or "every(thing/body)", we get a "general" statement such as

11) Someone is a student in class CS 101

12) Miller is a student in every class

Again, we assert that there is a certain relationship between certain objects although
we do not know the identity of all the objects involved. It is still clear how we would
have to test the truth value of these sentences.

This is the kind of general statement that the European medieval philosophers knew
how to deal with. They got, however, into deep problems with the so-called sentences
with "multiple generality", i.e. sentences where more than one general term occurs, as
in

13) Everyone is a student in some class

The medieval logicians felt that this type of sentence was somehow vague, and that it
was very unclear under what conditions it would be true. They devised special rules
to come to grips with this problem, but they never found a general solution.

It was a general solution to this problem that became the starting point for the entire modern logic. As is well known, Frege realised that these sentences are not vague but structurally ambiguous. It can mean either of

13a) Everyone is a student in some class (probably a different one for each student)

13b) There is some class (one and the same) everyone is a student in

There is no denying that one of the readings (viz. 13a) is strongly preferred but it is equally certain that the other reading *is* available (and can be made the preferred reading by stress and intonation). Frege designed First Order Logic (FOL) as a special language where these ambiguities could be made explicit. In FOL, we get two unambiguous statements instead of the one ambiguous statement in natural language:

```
13c) ∀ P: person(P) → ∃ C: class(C) ∧ student(P,C)
13d) ∃ C: class(C) ∧ ∀ P: person(P) → student(P,C)
```

Now we can, again, say under what conditions exactly each of the two readings is true. It is fair to say the whole discipline of formal semantics of natural language centers on the problem how sentences of natural language can be translated into unambiguous statements in FOL which can then be interpreted in a model (in the model theoretic sense of the word). For most of this century, formal semanticists spent their time figuring out how to map the various types of natural language utterances into FOL statements of the same meaning.

One of the people who did so was Russell. The starting point for him was the fact that a great number of sentences in natural language do not contain either any of the obvious quantifying words "every", 'all", or "some", or proper names. Nevertheless, such sentences seem to be perfectly unambiguous, with clear truth conditions. The reason is that they contain *definite descriptions*, with the same function as proper names in the examples given above: They *refer*. In 14

14) My brother is a student in the class

the definite noun phrases "my brother" and "the class" play basically the same role as the proper names "Miller" and "class CS 101" in 10.

Frege had already realised that in many cases the object that a definite noun phrase purports to refer to simply does not exist, i.e. that a definite description may fail to refer. For him, the fact that referential failure is a common thing to occur in natural language was not particularly surprising. He thought natural language was a fairly chaotic affair anyway, and that the main task of semanticists was to correct its shortcomings. Moreover (and more importantly) he thought that reference to real world objects was not the only function that language had to fulfil. Russell thought otherwise. To him, reference to real world objects *was* the only function of language. He therefore considered the occurrence of a referential failure totally unacceptable in a theory of natural language semantics. To him, any sentence resulting in referential failure would be *meaningless*. For this reason he wanted to *force* every natural language statement to have a well-defined truth value. To his end he suggested, in his famous "Theory of Description", to rewrite referential terms as quantified terms. Thus a sentence like

15) The king of France is bald

is re-cast as the conjunction of

1. There is a king of France

2. he is the only king of France

3. he is bald

The result of this procedure is that any sentence containing a definite noun phrase that fails to refer becomes *false* rather than meaningless.

However, this move has considerable consequences. First, the structure of a sentence in natural language and the structure of its logical rendering now become, in most cases, very different. The definite article is a pre-terminal category, i.e. a category located at the *innermost* layer of the syntax structure, e.g. the syntax structure of "The king is bald" is

```
sent(np(det(the)
         n(king))
      vp(aux_v(be)
         adj(bald)))
```

yet in the logical representation of this sentence, viz.

```
∃ X:
∧ king(X)
  ∧ ¬   ∃ Y:
        ∧ king(Y)
          ¬(X=Y))))
      bald(X)
```

the definite article becomes the *outermost* existential quantifier. The verb, on the other hand, which is similarly deep in the syntax structure, creates a term which does not rise to the surface in a similar way but ends up deep in the logical structure. With transitive verbs this becomes even more striking. The sentence

16) The Queen received the Prime Minister

```
sent(np(det(the)
         n(queen))
      vp(tv(receive)
         np(det(the)
            n(prime_minister)))))
```

becomes

```
∃ X:
∧ queen(X)
  ∧ ¬ ∃ Y:
      ∧ queen(Y)
        ¬(X=Y))))
    ∃ U:
    ∧ prime_minister(U)
      ∧ ¬ ∃ V:
          ∧ prime_minister(V)
            ¬(U=V))))
        received(X,U)
```

and here the translation of the verb even got moved inside the existential quantifier translating the second definite article.

And then there is a second problem. Proper names and definite noun phrases fulfil, syntactically, the same function. The syntax structure of

17) Elizabeth II received John Major

```
sent(np(pn(elizabeth_2))
    vp(tv(receive)
      np(pn(john_major))))
```

is basically the same as that of 16 above. In logic, however, these sentences would normally get entirely different representations:

```
16a) ∃ X:(queen(X) ∧ ¬(∃ Y: (queen(Y) ∧ ¬(X=Y)))
      ∧ ∃ U: (prime_minister(U) ∧ ¬(∃ V:(prime_minister(V) ∧ ¬(U=V)))
      ∧ received(X,U) )
```

```
17a) received(elizabeth_2,john_major)
```

The task of mapping syntactic structures into very different logical structures in a principled manner is a central problem of NLP, and one of the basic techniques devoted to this task will be outlined in one of the next sections.

An Example: Semantic Analysis for Natural Language Interfaces to Data Bases

If we assume, for the time being, that this mapping can be performed in a satisfactory manner, what would we do with the resulting logical formulae? A simple yet realistic example will be used to show this: A natural language interface to a data base. The domain of the system are the tutors, students, and departments of a university. As data base we use, for the sake of simplicity, the Prolog data base itself. In order to model a regular relational data base we allow only ground facts to be used. As additional restriction we require the system to merely process yes/no-questions. They must be translated into legal Prolog queries, to be proved over the Prolog data base. The example system is, obviously, not meant to be immediately useful. It should merely serve to illustrate the most basic problems to be solved when designing such a system. These problems would not be much different if we used a commercial relational DBMS and wanted queries to be translated into a query language such as SQL but the design of an interface to a commercial system is obviously a major software engineering task.

Interfaces to relational data bases are a particularly good starting point as we can stick quite closely to the model theoretic view of standard semantics. Since the user is primarily interested in getting information out of the system, the entries in the data base can be treated as the world *itself* (i.e. not as a more or less perfect image of the world). We hope, naturally, that the data in the system are correct but this is not the responsibility of the interface designer. We will, however, not have to deal with questions such as what the basic objects and relationships of the world are, how we can identify them, how we can determine that two proper names actually refer to the same object, how proper names really denote etc. All these questions must have been explicitly answered by those who set up the data base. The interface designer can interpret natural language sentences, or their translation into a logic of his choice, without getting distracted by any of these worrisome problems.

Under this substantial simplification it is relatively straight-forward to design a natural language interface. Let us start with a query like

```
Does some department employ a tutor?
```

which would give the syntax structure (repeated here)

```
interr(s(np(det(some)
            n(department))
         vp(verb(employ)
            np(det(a)
               n(tutor))))))
```

The core part of this structure is the same as that of the corresponding declarative sentence. If we translate this core part into FOL[6], we get

```
exists(D): (department(D) ∧ exists(T): (tutor(T) ∧ employ(D,T)))
```

If we want to turn this into a legal *Prolog query* to be proved we must *negate* the whole expression, due to the basic principle of the refutation proof procedure used by the standard Prolog interpreter, which wants to find a contradiction to the negated theorem. Then we must turn the FOL formula into *clausal form*. This can be done in seven steps with the help of a well-known algorithm, which can be turned into a very

---

6. here, and in many of the formulae below, quantifiers are written as follows

```
all(V): P(V) → Q(V))
exists(V): P(V) ∧ Q(V))
```

fast program (Clocksin 1984, Appendix B).

The same procedure can be used for a considerable number of questions. Sentences with proper names, for instance, are even simpler:

```
Does John employ a tutor?

¬ exists(T): (tutor(T) ∧ employ(john,T))

 :- tutor(T), employ(john,T) .
```

Questions with the predicative and the identity use of the verb "to be" would result in the following queries:

```
Is John a tutor?

exists(T): (tutor(T) ∧ john=T)

 :- tutor(T), john=T .
```

```
Is John Peter?

john=peter

 :- john=peter .
```

The first fundamental problem with this approach surfaces if we try to ask questions like the following:

```
Does every department employ a tutor?
```

which is, in FOL

```
¬ all(D): (department(D) → exists(T): (tutor(T) ∧ employ(D,T)))
```

The conversion of the FOL expression into clausal form will give us

```
department(sk12).
 :- tutor(T), employ(sk12,T) .
```

of which the first clause is not a theorem and can therefore not be proved by Prolog. The reason is that universally quantified statements in the antecedents of implications result in non-Horn clauses.

If we convert the original FOL statement in a different way we can, however, find a way around the problem. The following statements are all equivalent

```
¬ all(D): (department(D) → exists(T): (tutor(T) ∧ employ(D,T)))
¬¬ all(D): (department(D) → exists(T): (tutor(T) ∧ employ(D,T)))
¬¬ exists(D): ¬ (department(D) → exists(T): (tutor(T) ∧ employ(D,T)))
¬¬ exists(D): ¬ (¬department(D) ∨ exists(T): (tutor(T) ∧ employ(D,T)))
¬¬ exists(D): (department(D) ∧ ¬exists(T): (tutor(T) ∧ employ(D,T)))
```

If we now turn, in the last version of this statement, the two negations immediately in front of the existential quantifiers into normal predicates, say 'not', treat them as invisible to the conversion into clausal form, and convert the expression in the usual way, we get

```
:- not((department(D), not((tutor(T), employs(D,T))))
```

If we now interpret these negation symbols as *negation by failure*, the standard Prolog interpretation strategy will give us the right answer. The prover will check, for the first instance of a department, whether it can find a tutor such that he is employed by the department, and if it does it will fail due to the inner negation, backtrack to the next instance of a department, repeat the test, and go on until it runs out of entries for departments, and then succeed due to the outermost negation. If it finds one case of a department that does not meet the conditions it will fail the condition, succeed due to the inner negation and finally fail due to the outer negation. This method also works for embedded universal quantifiers in antecedents of implications:

18) All departments whose students are (all) lazy will be closed

i.e.

```
all(D): ((department(D) ∧ (all(S): (student(S,D) → lazy(S)))) → closed(D))
```

will give

```
:- not((department(D), not((student(S,D), not(lazy(S)))), not(closed(D))))
```

This double negation is a time-honoured trick of Prolog tradition, and it works beautifully under the given restrictions. For other types of quantifiers, however, it is useless. More about this will be said below.

In the manner described we can convert simple natural language yes/no-questions into legal Prolog queries provable over an axiom system consisting of ground facts. What can not be done in this manner is, among many other things, the processing of wh-questions which must often retrieve entire *sets* of entries. This is one of the subjects to be taken up in a later section. Before going into these more complex questions we must outline how we can actually *implement* the very simple interface just described.

Some Basic Techniques of Semantic Analysis in Logic Programming

Before we can translate natural language questions into Prolog queries in the manner described we must be able to map their syntactic structures into their logical correspondences. We had postponed this problem, and now we must approach ist.

The basic principle often used to solve it, is that of *compositionality*, and the standard technique to implement it is the *Lambda-calculus*.

The principle of compositionality says that the meaning of a complex (natural language) phrase should be derived solely from, first, the *meaning of its parts* and, second, the way these parts are combined, i.e. from the *syntax of the entire phrase*. What must *not* be used to compute the meaning, is any kind of *context* of the phrase. While the principle would, in theory, allow us to perform the interpretation of a natural language phrase *directly* it is, in practice, much easier to first translate it into a well-understood formalism like FOL, and then interpret the logical formula.

The principle of compositionality tends to sound self-evident to computer scientists but when natural language is concerned, it is anything but self-evident, and its implementation is very far from trivial. Although the principle itself can be said to be implicit in Frege's work already, it was made explicit only in the early seventies of this century by Richard *Montague*. In its most rigorous form (the one used by Montague himself) it requires that the translation of the whole phrase be computed exclusively by functional application of the translations of two partial phrases. There are two ways functional application can be used. Either the translation of *first* phrase becomes the *functor* applied to translation of the second phrase, or the other way round. In this view then, the only way how we can add some control to the computation of the meaning is in the choice of what becomes the functor. Normally, we can follow the order in the syntax structure.

The implementation of this principle requires the liberal use of lambda abstractions. Let us first consider proper names. We recall that they fulfil the same syntactic function as full noun phrases yet must result in totally different kinds of logical translations. From

```
'John' + 'works'
```

we must somehow get

```
works(john)
```

The problem is that the translation of one basic expression (a word form) now becomes the argument value of the translation of another basic expression while up until now all the argument positions have been taken up by variables, created by quantifiers. Apart from that, the surface order of the words must, loosely speaking, be inverted in order to get the right logical representation. This is possible if we translate the word "John" as

```
λ P. P(john)
```

If we now apply this expression to the translation into logic of "works", which we assume to be simply "works", we get

```
λ P. P(john)(works)
```

If we apply beta-reduction to this expression we get

```
works(john)
```

Thus the use of lambda abstraction allows us to get the correct representation in a purely compositional fashion. One of the effects of this, seemingly purely technical, move is that the denotation of a proper name like "John" is no longer the individual with this name but the property of being a property of John's or, in fully extensional terms, the set of all sets containing John as an element.

Now we turn to sentences with quantifiers which, as we said, should get a very different representation in logic, despite their syntactic similarity with sentences of the kind just mentioned. We know that the desired translation into logic of

19) Some student works

i.e.

```
19a) ∃ S: student(S) ∧ work(S)
```

would have to be composed exactly following its syntax structure

```
sent(np(det(some),
        n(student))
     vp(iv(work)))
```

This means that the translation would have to be composed following the pattern

```
some* + student*
```

```
"some student"* +  works*
```

```
"some student works"*
```

where 'N*' stands for the translation into logic of N, and '+' for functional application (in any direction). We start with suitable lambda-abstractions for the two quantifiers used in FOL. As the translation of "some" we use

```
λ P. (λ Q. ∃ X: (P(X) ∧ Q(X))
```

If we apply this to the translation of "student" in the order of the syntax structure (i.e. from left to right), we get

```
λ P. (λ Q. ∃ X: (P(X) ∧ Q(X))(student*)
```

We apply beta-reduction to the result, which gives as meaning representation for "some student"

```
λ Q. ∃ X: (student*(X) ∧ Q(X))
```

which will now be applied to "works*", and we get

```
λ Q. ∃ X: (student*(X) ∧ Q(X))(works*)
```

which gives, after reduction,

`∃ X: (student*(X) ∧ works*(X))`

which is the desired end product.

In completely analogous fashion, we assume the translation for the determiners "all"/"every" and for "the" to be

`every* = λ P. (λ Q. ∀ X: P(X) → Q(X))`
`the* =   λ P. (λ Q. ∃ Y: ¬ (∃ X: P(X) ∧ ¬ X=Y) ∧ Q(Y)))`

which gives, for "every student"

`λ P. (λ Q. ∀ X: P(X) → Q(X))(student*)`

and reduced

`λ Q. ∀ X: student*(X) → Q(X)`

For "Every student works" this will then give

`λ Q. ∀ X: student*(X) → Q(X)(works*)`

which reduces to

`∀ X: student*(X) → works*(X)`

Now we can in fact get the different logical representations from syntactically similar sentences in a perfectly regular way. The differences that we need in the final product have all been packed into the basic building blocks, i.e. into the lambda abstractions used as translations of the individual word categories.

The situation gets considerably more complicated when we turn to *transitive* verbs. In an extensional version of the theory, they require a translation of the form

`λ P. λ Q. P(λ Y. beat*(Y)(Q))`

where we use the notation `beat*(X)(Y)` instead of `beat*(Y,X)`. If we translate

10) Peter beats John

we start with the translation for "John", viz.

`λ R. R(john)`

take the translation for "beats"

`λ P. λ Q. P(λ Y. beat*(Y)(Q))`

functionally apply the first expression to the second

`λ P. λ Q. P(λ Y. beat*(Y)(Q))(λ R. R(john))`

and reduce the resulting expression in three steps

```
λ Q. λ R. R(john)(λ Y. beat*(Y)(Q))
λ Q. λ Y. beat*(Y)(Q)(john)
λ Q. beat*(john)(Q)
```

where the last expression is a predicate which means "John-beater". Now we apply the translation of "Peter", i.e. λ R. R(peter) to this newly created predicate which gives

```
λ R. R(peter)( λ Q. beat*(john)(Q))
```

and this can be reduced in two steps

```
λ Q. beat*(john)(Q)(peter)
beat*(john)(peter)
```

where the last expression is the same thing as, in standard notation,

```
beat*(peter,john),
```

which is what we wanted to get in the first place.

As can be seen, this procedure maps syntax structure into logical structure in an almost perfectly deterministic way. The only respect where the syntax structure does not, in itself, determine how to proceed, is the direction in which functional application works (in the examples above it was, however, always from left to right). It is therefore exceedingly simple to design a program that maps syntax into logic in this way. Logic Programming is particularly suited to this task as beta-reduction is the type of structure transformation that can be implemented very efficiently through unification. With comparatively few unification steps we can descend recursively through the highly complex source structures of lambda abstracted terms and simultaneously map them, step by step, into the desired target structures of FOL. Since the end result, i.e. expressions to be either proved or asserted in a relational data base, can itself be implemented in logic, we can also simulate the entire data base part in the same piece of software as the interface proper, which is very advantageous for developing purposes.

The parallel between syntax and logic is now so close that we can build up the logical translation exactly in tandem with the syntax structure. In fact, we can dispense with the latter altogether and create the logical translation directly. This is what is done in the following program, for illustrative purposes (after Warren 1983). In practice it is, however, advisable to put the syntax analysis and the translation into logic in separate modules of a program. In order to show how these concepts can be implemented we need to introduce a few notational conventions, mainly to accommodate the syntactic restrictions of Prolog:

1. the asterisk to denote the logical translation is dropped for basic terms

2. functional application is not written as `P(john)` which is illegal in most implementations of Prolog but as `@(P,john)`

3. As the notation `beat(john)(peter)` is not allowed in Prolog, either, we write it as `@(@(beat,john),peter)`

4. the lambda abstraction λ X. `student(X)` ∧ `lazy(X)` is written as `lambda(X, student(X) & lazy(X))`

5. We use the following symbols for operators: ∧ becomes &, ∨ becomes #, ¬ becomes ˜, → becomes ->, ↔ becomes <->

In the program below, the generation of the logical structures is done in absolutely regular fashion by always applying the translation of the first constituent to that of the second, and beta-reducing the result at once to get the translation of the whole phrase. All this is done in the expressions of the form

```
reduce(@(First,Second),WholePhrase)
```

The resulting overall translation is then communicated to the left hand side of the rule where it will be available for the computation of yet larger components.

```
decl(IDecl)              -->     np(Num,INp),
                                 vp(Num,IVp),
                                 {reduce(@(INp,IVp),IDecl)}.


np(sing,IPn)             -->     proper_name(IPn).
np(Num,INp)              -->     det(Num,IDet),
                                 n(Num,IN),
                                 {reduce(@(IDet,IN),INp)}.


vp(Num,IVp)              -->     verb(trans,Num,IV),
                                 np(Num1,INp),
                                 {reduce(@(IV,INp),IVp)}.
vp(Num,IV)               -->     verb(intrans,Num,IV).


/* ******* Translation of Basic Expressions *********** */

det(sing,lambda(P,lambda(Q,all(Z):@(P,Z) -> @(Q,Z)))) -->       [every].
det(plur,lambda(P,lambda(Q,exists(Z):@(P,Z) & @(Q,Z)))) --> []; [some].
det(sing,lambda(P,lambda(Q,exists(Z):@(P,Z) & @(Q,Z))))
                                        --> [a]; [an]; [some].
det(Nbr,lambda(P,lambda(Q,exists(Y):all(X):@(P,X) <-> X=Y & @(Q,Y))))
                                        --> [the].
verb(trans,Num,lambda(P,lambda(Q,@(P,lambda(Y,@(@(Root,Y),Q))))))
                        -->     [V], {is_verb_form(V,Root,Num),
                                verb_type(Root,main+trans)}.
verb(intrans,Num,Root)        -->       [V], {is_verb_form(V,Root,Num),
                                verb_type(Root,main+intrans)}.
verb(_,Num,lambda(P,lambda(X,@(P,lambda(Y,X=Y)))))
                        -->             [V], {is_verb_form(V,Root,Num),
                                verb_type(Root,aux+be)}.
```

```
proper_name(lambda(P,@(P,Pn)))   -->       [Pn], {is_proper_name(Pn)}.
n(Num,IN)                         --> [N], {is_noun_form(N,IN,Num)}.
```

Although it is really very straight-forward to design such a program, it is decidedly *not* simple to find the correct translation into lambda abstracted logical expressions of the basic categories (consider, for instance, the translation of transitive verbs). The reason is that functional application is a strictyl *sequential* operation, so that for every variable that must be bound, a separate level of lambda abstraction is needed. This makes some of these expressions very complex indeed.

Nevertheless, for relatively simple types of sentences this approach is very elegant, and since no unnecessary structures are computed it is also very efficient. Two test runs with commented extracts from the trace show compositionality in full action:

```
Some department employs a tutor

'some' + 'department':

reduce(@(lambda(Q,lambda(P,exists(D):@(Q,D) & @(P,D))),department),C).
==>
C = lambda(P,exists(D):@(department,D) & @(P,D)))


'a tutor':
lambda(R,exists(T):@(tutor,T) & @(R,T))

'employs':
lambda(X,lambda(Y,@(X,lambda(Z,@(@(employ,Z),Y)))))

'employs' + 'a tutor':
reduce(@(lambda(X,lambda(Y,@(X,lambda(Z,@(@(employ,Z),Y))))),
        lambda(R,exists(T):@(tutor,T) & @(R,T))),A).
==>
A = lambda(Y,exists(T):@(tutor,T) & @(@(employ,T),Y))

'some department' + 'employs a tutor':
reduce(@(lambda(P,exists(D):@(department,D) & @(P,D)),
        lambda(Y,exists(T):@(tutor,T) & @(@(employ,T),Y))),B).
==>
B = exists(D):@(department,D) & exists(T):@(tutor,T) & @(@(employ,T),D)


John employs a tutor

'a' + 'tutor':
reduce(@(lambda(P,lambda(Q,exists(T):@(P,T) & @(Q,T))),tutor),A)
==>
```

```
A = lambda(Q,exists(T):@(tutor,T) & @(Q,T))


'employs':
lambda(R,lambda(W,@(R,lambda(T,@(@(employ,T),W)))))


'employs' + 'a tutor':
reduce(@(lambda(R,lambda(W,@(R,lambda(T,@(@(employ,T),W))))),
         lambda(Q,exists(T):@(tutor,T) & @(Q,T))),B).
==>
B = lambda(W,exists(T):@(tutor,T) & @(@(employ,T),W))


'john':
lambda(S,@(S,john))


'john' + 'employs a tutor':

reduce(@(lambda(S,@(S,john))),
         lambda(W,exists(T):@(tutor,T) & @(@(employ,T),W))) ,C).
==>
C = exists(T):@(tutor,T) & @(@(employ,T),john)
```

While the approach just outlined manages to solve one of the basic problems of the semantic analysis of natural language, viz. to get two very different types of structures into correspondence, some other problems are still open. One of them is the structural ambiguity of many sentences. If we claim, as the compositionality principle in its strictest form does, that syntax and semantics are absolutely parallel, we have to claim that sentences such as 13 from above, i.e.

13) Everyone is a student in some class

not only have two *semantic* readings but also two different *syntactic* structures. This is, at the very least, not the most intuitively appealing explanation one can think of. Second, and possibly more important, the basic assumptions on which this approach is founded are too narrow for the purposes of NLP. These assumptions are still largely inspired by the goals of theoretical research in formal semantics. A generalisation of the interface example introduced above will show to what extent semantic analysis in NLP must go beyond these assumptions, even for a very simple application. It will become clear that, in the field of NLP, "semantic analysis" means, in many respects, something fairly different from what the formal semanticist understand by this term.

Special Techniques for the Analysis of Generalized Quantifiers

The first respect in which the interface example used above had been artificially made to fit some of these basic assumptions concerns the idea that the really fundamental question of semantics is whether a sentence is true or false. We stated that we merely wanted to *query* the data base in natural language. Then it will in fact be sufficient for the interface program to determine the truth value of the input sentences over the data base and, possibly, inform us of the variable bindings established in the process. However, when our task is to design a fully-fledged natural language interface to a DBMS, we will also want to *input new data* in natural language. But then declarative sentences must be dealt with, too, and it is *not* their truth value that will be of interest. Such sentences are meant to convey data to the system that is *new* to it, and trying to have the system establish their truth value over the existing data base would, by definition, give "false", in all interesting cases. An interface system will therefore have to translate declarative sentences into terms that are added to the data base, and interrogative sentences into terms that are proved over it. The distinction between these two types of sentences is therefore absolutely crucial in any even remotely application-oriented approach, and the strictly truth value oriented approach of standard semantic theory will not do any longer.

There is a second observation that leads us to extend the basic methods of semantic processing used in the initial interface example. It is the fact that only relatively few natural language determiners can get a first order quantificational representation at all. Determiners such as "most", "few", "many", (stressed) plural "some", (stressed) plural "several", "exactly seven" etc. all require a higher-order analysis. The standard quantifiers traditionally treated by formal semantics are in actual fact highly unusual in this respect. This was the finding that led, relatively recently, to the creation of Generalized Quantifier Theory (GQT; Barwise 1981, Gaerdenfors 1987). In a practical application we can, of course, not restrict ourselves to a few "well-behaved" exceptions. For this reason, generalized quantifiers of some kind or other have been used in NLP for a long time, under a variety of names (e.g. "three-branch quantifiers"). In many cases, researchers in NLP undoubtedly did not realise that the same concept had been developed, concurrently, in theoretical linguistics, too (e.g. Colmerauer 1982, Pereira 1983). There is a *second* reason why generalized quantifiers are favoured in NLP. For some properties of natural language it might be possible, in principle, to find a first order analysis but not in an intuitively transparent manner. It is often much easier to apply a GQT analysis to certain phenomena which are, strictly speaking, first order. In many respects the situation is fairly reminiscent of the way how context sensitive concepts are used in syntax for the analysis of certain, strictly speaking, context free constructions. The *third* reason why generalized quantifiers are popular in NLP is that many properties of natural language can be modelled much more efficiently if we use generalized quantifiers as starting point but do not interpret them according to their set theoretic definition but rather model certain of their properties directly. The resulting computational methods may themselves not always be complete but this is no limitation as we have, after all, the model theoretic definitions taking care of the foundations. We can thus often combine the best of both worlds: The sound theoretical basis of fully general model theory, and the efficiency of specific computational solutions.

A suitable representation of generalized quantifiers in a Logic Programming framework is as follows:

```
quant(Quantifier, Variable , Scope, Predication)
```

'Quantifier' is one of a (potentially infinite) list of generalized quantifiers, 'Variable' the variable over which the quantifier ranges, and both 'Scope' and 'Predication' either a generalized quantifier expression, a basic expression, or a Boolean combination of either type of expression. Such generalized quantifiers define a higher order relationship between the two *sets* of objects described by the terms in 'Scope' and 'Predication'. For the sentence "Most departments are closed" the quantifier structure would be

```
quant(most, D, department(D), closed(D))
```

with the meaning that the two sets stand in the relationship "is a majority of". Formally, the cardinality of the intersection of the set of departments and of the set of closed things is defined to be greater than, say, half of the cardinality of the first set, i.e.

```
|{D | department(D)} ∩ {X | closed(X)}| > |{D | department(D)}| /2
```

A sentence like "Most departments employ a tutor", which could be paraphrased as "Most departments are tutor-employing", would be represented as

```
quant(most, D, department(D), quant(exists,T,tutor(T), employ(D,T)))
```

meaning that the cardinality of the intersection of the set of departments and of the set of tutor employing entities is greater than half of the cardinality of the first set. The two classical quantifiers, "all" and "exists", can obviously be interpreted in this framework also. The interpretation of the generalized quantifiers

```
quant(exists, D, department(D), closed(D))
quant(all, D, department(D), closed(D))
```

is that the intersection of the two sets are non-empty in the existential case, and that it is a subset of the second set in the universal case.

The number of concrete quantifiers used in the tiny fragment of English for the example system here is very limited. The determiners "all" and "every" are both translated as quantifier "all", the determiners "some" and "a" as quantifier "exists", the interrogative pronouns "who" and "what" as quantifier "set", and the singular definite article as quantifier "the". The "non-classical" determiners which could not, in principle, be analyzed with first order quantifiers, have not been taken into account.

We now put these general notions to work. First, consider *interrogative* sentences. They are, in most respects, simpler to treat than declarative ones. *Yes-no-questions* are the easiest type of question, and existential questions are the easiest to analyze among them. The following question

```
Does some department employ a tutor?
```

would, under the approach outlined earlier, have to be translated into an existentially quantified logical sentence (due to the determiner "some"), which will give the generalized quantifier

```
?- quant(exists,D,department(D),quant(exists,T,tutor(T),employ(D,T)))
```

Its interpretation according to the definition would require us to compute, first, the full set of all departments in the data base, then the full set of all tutors employed by anybody or anything, then form their intersection, and finally determine its cardinality. It is clear that this is a hopelessly inefficient procedure. However, we can design a special-purpose interpretation for this type of generalized quantifier, without losing the sound model theoretic foundation. It amounts to reducing the generalized quantifier to its first order correlate. Both scope and predication are proved as straight-forward Prolog theorems, in sequence. The implementation is trivial:

```
prove_it(quant(exists,X,Scope,Prdcn))      :-
       prove_it(Scope),
       prove_it(Prdcn).
```

More interesting are universally quantified questions, such as
```
Does every department employ a tutor?
```

which results in the creation of the generalized quantifier

```
?- all(D): department(D), quant(exists,T,tutor(T), employ(D,T)))
```

Again, the interpretation according to its set theoretical definition would be unnecessarily time-consuming and, again, we can design a specific interpretation which reduces this generalized quantifier to its first order counterpart. The quantifier expression is interpreted as a proof of *every* expression in the predication for *every* instantiation of the scope. For the example above this means that, for every instance of 'department(D)', a proof of

```
?- quant(exists,T,tutor(T), employ(D,T))
```

is attempted (with the variable bindings established by the proof of the scope terms). The most efficient implementation uses the double negation we encountered earlier (ignore, for the time being, the predicate 'presup/2'):

```
prove_it(quant(all,X,Scope,Prdcn))     :-
       presupp(weak,Scope),
       not (( prove_it(Scope),
          not (prove_it(Prdcn))))).
```

The double negation, together with the machinery of backtracking, will make the interpreter step through the entire data base and check, for every single department *in isolation*, whether it employs a tutor. Once it encounters a department that does not meet the condition, it can break off the search at once and return a negative answer.

For natural language quantifiers that are strictly higher order this will not be possible, though. Consider a proportional determiner such as "most". If we ask a system "Do *most* departments employ tutors?" the system *cannot* go through its data base and look at things in isolation, and if it encounters a negative case it can, of course, not simply break off the search. It has to accumulate and compare cardinalities of entire sets, exactly according to the definition of generalized quantifiers. *In principle* this is easy to implement. We could certainly write

```
prove_it(quant(most,X,Scope,Prdcn))    :-
        presupp(weak,Scope),
        setof(X,Scope,Base),
        setof(X,Prdcn,Denotation),
        intersection(Base,Denotation,Inter),
        majority(Inter,Base).
```

with a suitable definition for the predicate 'majority/2' but this implementation will be appallingly inefficient.

Rather than going any deeper into this fairly thorny problem we want to show how the natural language interface system outlined above was overly simplified in yet another way. The interpretation of a universally quantified statement as a series of elementary proofs was certainly correct for the example sentence "Does *every* department employ a tutor?". However, it is far from clear that the same thing would hold for the sentence "Do *all* departments employ a tutor?" or, even clearer, "Do departments employ a tutor?". This is a general question, too, but it seems to be not about a number of *individual instances* of departments, tutors, and employment relationships but rather about one *general rule*. The first type of general question inquires about the *present* state of the data base only, i.e. about the departments known to the system right now. The second type of question inquires about the existence in the data base of a "standing rule", a rule which is also valid for any department introduced to the system at any point in the *future*. So far we have assumed that our data base would contain only basic facts, i.e. that it was a purely *extensional* data base, but this is a very limiting assumption. If we want to build up a data base containing rules as "first class citizens", i.e. if we want to have an *intensional* data base, we must also allow intensional questions.

If we continue to use the Prolog data base itself, we know what the rule must look like that we ought to retrieve when precessing an intensional question such as "Do *all* students work?". In such simple cases the rule would be

```
work(S)   :-    student(S).
```

In order to retrieve a rule directly we would need higher-order predicates (to find a match between our query and an entire rule). Prolog provides us, in principle, with such predicates, e.g. 'clause/2', so this question could be translated into

```
clause(works(sk-1), student(sk-1))
```

However, a slightly more complex question such as "Do all *foreign* students work?" will result in a 'clause'-predicate with more than one right hand side term. However, the order of the right hand side terms in the rule sought is (logically) arbitrary, and so a direct (higher-order) direct match is more difficult to achieve. We could impose a canonical order on the right hand side terms in rules, based on purely formal properties (some kind of alphabetic order) but this would work only for pure Prolog where order is irrelevant.

A more general and (slightly) cleaner solution is based on the following observations. Since a statement to be proved must be negated, we convert the negated FOL version of the rule above, viz.

¬ ∀ S: student(S) → work(S)

into HCL, and get

```
    student(sk-1).
:- work(sk-1).
```

As noticed earlier, the second term is a perfectly normal theorem, while the first term is not. However, if we interpret this first term as a *temporary assertion* (to be retracted once the entire proof has been performed) we get the desired behaviour. Always assuming that Skolem constants are unique, the resulting query

```
?- assert(student(sk-1)), prove(work(sk-1)), retract(student(sk-1)).
```

can succeed only if there is a *rule*

```
work(S)   :-      student(S).
```

in the data base. This also works if there are several right hand side terms.

Only marginally more complex are questions like "Do all departments employ a tutor?". They ask for the existence in the data base of *two* rules, viz.

```
tutor(sk-1(D))             :-  department(D).
employ(D,sk-1(D))          :-  department(D).
```

For the interrogative version of this sentence, the HCL form of the negated universal statement is

```
department(sk-2).
 :- tutor(T), employs(sk-2,T).
```

and the query derived from it

```
assert(department(sk-2)), tutor(T), employs(sk-2,T), retract(department(sk-2)).
```

This complex query can be derived *directly* from the generalized quantifier

```
quant(all,D,department(D),quant(exists,T,tutor(T),employ(T,D)))
```

by first skolemizing the variables in the terms in the scope part, asserting these terms temporarily, proving the terms in the predication part, and retracting the temporarily asserted terms.

Wh-questions require the collection of a *set of solutions*:

```
Who employs a tutor?
```

```
result set: john
```

This is another case where a first order semantics approach is at a loss. It cannot cope with sets, by definition. The generalized quantifier used here for this type of sentence is

```
quant(set,V, X, P(V))
```

meaning that *X* is identical with the set of instances satisfying *P(V)*. The completely unspecific "What employs a tutor?", which sounds pretty odd, would become

```
?- quant(set,P, X, quant(exists,T,tutor(T),employ(P,T)))
```

with a straight-forward direct implementation in Prolog

```
?- setof(P, T^(tutor(T),employ(P,T)), X)
```

In most cases we want some restrictions on the elements found. Thus "*Who* employs a tutor?" becomes

```
?- quant(set,P, p_name(P,N), quant(exists,T,tutor(T),employ(P,T)))
```

The proper names assigned to the objects found (first known only by the Skolem constants used to identify them) must be extracted from the solution set in an additional step (not shown here).

Wh-questions, too, come in an intensional and an extensional variety. The two questions

21) Which departments employ tutors?

22) What departments employ tutors?

are not quite synonymous. The first question is extensional. It asks for a list of *instances* of departments meeting the condition that they employ tutors. The second question, however, is *intensional*. It asks for a *general rule* defining what *sort* of department employs tutors. This may be all departments, present of future, where particularly large numbers of students are enrolled, or departments with a lot of equipment to operate, or some other such definition. If the data base does not contain a suitable rule it would be desirable to make the system *infer* one from the individual facts. However, this is far from trivial.

We can now turn to *declarative* sentences. The simplest declarative sentences are, again, of the kind of "Some department employs a tutor". Their FOL representation would be the same as for the interrogative sentences, only without a negation, viz.

$$\exists\ D: department(D) \wedge \exists\ T: tutor(T) \wedge employ(D,T)$$

But now we want the system to *add* facts to its data base, in this case one fact each for a department, a student, and an employment relationship between them. While, in the case of questions, it would have been possible in principle, if not in practice, to use the full definition of the generalized quantifier to perform a proof, it is difficult to see how this could be done for assertions. It is on the basis of GQT quantifier structures themselves that the system *directly* asserts individual entries to the data base.

As is to be expected in a Horn Clause Logic based system, Skolem constants must first be created for all existentially quantified variables. In a second step, the individual expressions in "Scope" and "Predication" are then all asserted, one after another. This way of interpreting existential quantifiers is very easy to implement, and the following predicate (with self-explanatory predicates) will do it:

```
assert_it(quant(exists,X,Scope,Prdcn))    :-
              skolemize_struct((Scope,Prdcn)),
              assert_it(Scope),
              assert_it(Prdcn).
```

For the example from above, viz.

```
quant(exists, D, department(D), quant(exists,T,tutor(T), employ(D,T)))
```

this would give the following assertions:

```
asserted: department(sk-2)
asserted: tutor(sk-3)
asserted: employ(sk-2,sk-3)
```

If a fact is already known it is pointless to assert it again. This would merely fill up the data base with "dead wood", slowing down any future proof. In such a case, nothing new should be added to the data base but a suitable message should be issued:

```
known: employ(sk-2,sk-3)
```

This may be a minor detail but in any application it is such details which make a system either acceptable or unacceptable to users. It also goes to show that a lot of useful distinctions must be made outside the logic proper of the language.

Proper names and *definite noun phrases* are special in the sense that they carry a (strong) presupposition of both existence and uniqueness with respect to the objects they name or describe. In a sentence "The department employs the tutor" we clearly assume that one department and one tutor are known to exist. If we input such a sentence we want the system to first find the pre-existing "hook" on which to "hang" the new information. The interpretation of the quantifier "the" must be defined accordingly. It must make sure that the expressions in the scope are tested for existence and uniqueness rather than asserted straight away, as one would otherwise expect in an expression derived from a declarative sentence:

```
assert_it(quant(the,X,Scope,Prdcn))    :-
       presupp(strong,Scope),
       assert_it(Prdcn).
```

This switch from proof mode to assert mode, in the middle of a quantified statement, is something the standard model theoretic semantics is particularly ill suited to cope with. Direct interpretation of generalized quantifier structures allows to do so without any difficulties. Note, however, that this has nothing to do directly with the fact that generalized quantifiers are potentially higher-order.

Definite quantifiers (as all others) can be embedded to any depth. In the following example

```
The department employs the tutor
```

the quantifier takes the form

```
quant(the,D,department(D),quant(the,T,tutor(T),employ(D,T)))
```

the predication is itself a definite quantifier structure. Again, the scope term ('tutor(T)') is proved, not asserted. Only the predication of this, inner, quantifier ('employ(D,T)') is now treated as a fact to be asserted. If a more complex syntax with relative clauses were admitted there would also be embedded scope terms. "The department that employs the tutor is closed" would give

```
quant(the,D, (department(D) & quant(the,T,tutor(T),employ(D,T))),
             closed(D))
```

Of great importance are situations when a presupposition is *violated*. Questions and declarative sentences behave basically the same way in this respect. In the case of definite noun phrases a presupposition failure can occur either when there is *no* object of the kind described, or if there are *several* of them. Even today semantic theory tends to follow the basic rules of Russell's "Theory of Descriptions", outlined above. However, experience in NLP has shown convincingly that this behaviour is quite useless, whether we consider declarative or interrogative sentences. Take first interrogative sentences. Someone may be asking a system

Did all of the students taking CS 101 pass the final exam?

If class CS 101 had not been offered the system would, under Russell's approach, nevertheless answer with a simple "no". If the user found out later that this class had not been offered in the first place he would feel duped, and correctly so. In the case of a *declarative* sentence such a system would have to simply refuse to accept the statement, a behaviour which is almost as unacceptable.

It is quite clear that, at the very least, an error message should be issued in this situation. It is normally also a good idea to assert the missing piece of information, *in addition* to complaining to the user. This step is called *"accommodation"*:

```
4: John employs a tutor

no "p_name(_4024,john)" known

accommodated: p_name(sk-5,john)
asserted: tutor(sk-6)
asserted: employ(sk-5,sk-6)
```

Again, this kind of reaction is very important for a smooth flow of interaction between system and user. Any exchange of information (whether it be between two humans, or between a human and a machine) would become unbearably boring if all the information that was new to one of the partners would have to be introduced as such *explicitly*. Accommodation even seems to be one of the *main* vehicles for the transfer of new information in real human communication.

Note that, for a variety of reasons irrelevant in the given context, proper names are not treated as identifiers in their own right. Instead, the existence of people (the only entities with proper names in this system) is recorded in the form of an entry about their name. Above, this means that the existence of an object "sk-5" with certain properties is asserted, and it is also asserted that this object has the proper name "john".

Let us now turn to universally quantified declarative sentences such as "Every department employs a tutor". The determiner *"every"* tells us that we must assert something about every single department known. For each of them, a separate new entry for a tutor and another one for an "employ"-relationship must added. This is achieved by treating a universally quantified assertive statement like

```
quant(all,D,department(D),quant(exists,T,tutor(T),employ(T,D)))
```

as a series of individual assertions of the expressions in the predication (here: itself an existentially quantified statement), for each instantiation of the scope expressions. Crucially, the proof of the predication must be performed with the variable bindings established by the proof of the scope terms. This, too, can be implemented efficiently as double negation:

```
assert_it(quant(all,X,Scope,Prdcn))      :-
       presupp(weak,Scope),
       ( not (( prove_it(Scope),
              not (assert_it(Prdcn)) )) ; true ).
```

Now note the term 'presupp(weak,Scope)'. It is meant to give a practicable solution to the age-old question of the so-called "existential import" of universal quantification. It is fairly clear that the determiner "every" presupposes that the domain over which it ranges (above: the domain of departments) is not empty. If it is empty, we want the system to tell us so. On the other hand, it would *not* be sensible for the system to accommodate the missing entries. This kind of "weak" existential presupposition, implemented in the definition of 'presupp(weak,Scope)', seems to hold for all those quantifiers that do not have a "strong" presupposition, like definite articles.

Again, the extensional interpretation just given to the universal declarative sentences is not the only possible one. In fact, for declarative sentences there are even clearer cases of intensionality than for interrogative sentences. The sentence "*All* departments employ a tutor" and, in particular, "*Any* department *will* employ a tutor" both assert the existence of a "standing rule", valid now and in the future. Asserting rules to the data base is quite simple as long as we restrict ourselves to assertions about singular entities (i.e. to those expressible by first order quantifiers). When the generalized quantifier

```
quant(all,D,department(D),quant(exists,T,tutor(T),employ(T,D)))
```

is reduced to its first order equivalent, i.e.

$$\forall \text{ D: department(D)} \rightarrow \exists \text{ T: tutor(T)} \wedge \text{employ(T,D)}$$

and this formula is converted into HCL, we get

```
tutor(sk-1(D))            :-   department(D).
employ(D,sk-1(D))         :-   department(D).
```

which could now be asserted to the data base.

These were some of the properties of natural language that must be modelled if one wants to design even a small natural language interface system. Many of these properties are outside the limits of FOL, and theories such as GQT are often used in NLP as a more powerful basis for better systems. Starting from this sound theoretical basis, more efficient implementations that work for specific cases are then often designed directly.

4. Literature

Barwise 1981: Barwise, J. and Cooper, R., "Generalized Quantifiers and Natural Language," *Linguistics and Philosophy*, no. 4, pp. 159-219, 1981.

Clocksin 1984: Clocksin, W.F. and Mellish, C.S., *Programming in Prolog,* Springer, Berlin etc., 1984.

Colmerauer 1982: Colmerauer, Alain, "An interesting subset of natural language," in: *Logic Programming*, ed. Keith L. Clark and Sten-Ake Tarnlund, pp. 45-66, Academic Press, London, 1982.

Gaerdenfors 1987: Gaerdenfors, P. ed., *Generalized Quantifiers,* Studies in Linguistics and Philosophy, 31, Reidel, Dordrecht/Boston/Lancaster/Tokyo, 1987.

Gal 1991: Gal, Annie, Lapalme, Guy, Saint-Dizier, Patrick, and Somers, Harry, *Prolog for Natural Language Processing,* Wiley, Chichester etc., 1991.

Gazdar 1989: Gazdar, G. and Mellish, Ch., *Natural Language Processing in Prolog*, Addison-Wesley, Wokingham etc., 1989.

McTear 1987: McTear, Michael, *The Articulate Computer*, Blackwell, Oxford, 1987.

Pereira 1983: Pereira, F.C.N., "Logic for Natural Language Analysis," SRI International Technical Note 275 , January 1983.

Pereira 1987: Pereira, F.C.N. and Shieber, S.M., *Prolog and Natural Language Analysis*, CSLI Lecture Notes, 10, Center for the Study of Language and Information, Menlo Park/Stanford/Palo Alto, 1987.

Shapiro 1987: Shapiro, Stuart C., *Encyclopedia of Artificial Intelligence*, John Wiley, 1987.

Smith 1991: Smith, George W., *Computers and Human Language*, Oxford University Press, New York, 1991.

Warren 1983: Warren, David S., "Using lambda-calculus to represent meanings in logic grammars," *ACL Proceedings, 21st Annual Meeting*, pp. 51-56, 1983.

PUNDIT -- Natural Language Interfaces

Deborah A. Dahl
Paramax Systems Corporation
(A Unisys Company)
P.O. Box 517, Paoli, PA 19301

Abstract. The PUNDIT natural language understanding system is a modular system implemented in Prolog which consists of distinct modules for syntactic,semantic, and pragmatic analysis. A central goal underlying PUNDIT's design is that the basic natural language processing functions should be independent of the system's domain of application and application. This approach to the design of natural language processing systems is motivated by the fact that in order to be practical, natural language systems must not require reimplementation of most of the system as they are ported to different domains. Thus, our goal in the design of PUNDIT is to reduce as much as possible the amount of effort that must be done to move the system to different applications.

1 PUNDIT

PUNDIT is a large, domain-independent natural language processing system which has been under development at Unisys since 1984. PUNDIT designed to be as domain and application independent as possible, with a central set of procedural language understanding components supplemented by data files providing specific information for particular applications, as shown in Figure 1.

In PUNDIT, the domain-specific information is supplied by data files describing the domain specific vocabulary, special purpose grammar rules, and the knowledge base. These data files are used by the domain independent procedural modules which perform linguistic analysis. The output of PUNDIT is an application independent data structure, the Integrated Discourse Representation, or IDR. In order to develop a specific application, for example, a database interface, an application module is added which maps the IDR into the format required by the application software. This may be, for example, a well-known format such as SQL, or a special purpose format required by a particular application.

PUNDIT has been used for both text-processing and interface applications. The differences are mainly in the application module. For example, the IDR for a text processing application could be passed to an application module which creates database entries from the input texts. How the input is received can also vary across applications. For example, interface applications include a dialog manager and tools for interfacing to speech recognizers.

PUNDIT runs on Sun 3's and Sun 4's and requires approximately 16M of memory. It consists of about 40K lines of domain-independent source code. An application might include around 50K lines of domain-dependent data.

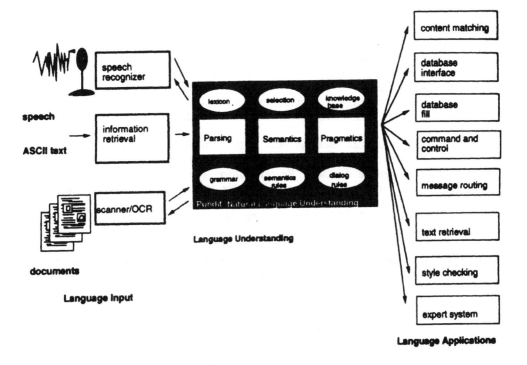

Figure 1 Relationship between PUNDIT, inputs, and applications.

Two interactive applications have recently been developed. The Voyager application includes 20,000 additional lines of source code defining the domain-specific customization [7]. The Air Travel Planning application includes 45,000 lines of domain-specific source code [18,19].

2 Processing in PUNDIT

Processing in PUNDIT consists of syntactic, semantic, and pragmatic processing.

2.1 Syntactic Processing

Overview. PUNDIT's syntactic component uses a grammar which is written in the Restriction Grammar framework [9,13]. Restriction Grammar is a logic grammar with facilities for writing and maintaining large grammars. Restriction Grammar is a descendent of Sager's string grammar [23]. The grammar consists of a set of context-free BNF definitions (currently numbering approximately 300), augmented by context-sensitive restrictions (approximately 140). The restrictions enforce context-sensitive well-formedness constraints and, in some cases, apply optimization strategies to prevent unnecessary structure-building. PUNDIT uses a top-down left-to-right parsing strategy, augmented by dynamic rule pruning for efficient parsing [9]. In addition, it uses a meta-grammatical approach to generate definitions for a full range of co-ordinate conjunction structures [11] and for wh-structures [12]. The current grammar covers declarative sentences, questions, sentence fragments, sentence adjuncts, conjunction, relative clauses, complex complement structures, and a wide variety of nominal structures, including compound nouns, nominalized verbs and embedded clauses.

The syntactic processor uses the rules of the grammar to produce a detailed surface structure parse for each sentence. This surface structure is converted into an *Intermediate Syntactic Representation* (ISR) which regularizes the syntactic parse and provides a more suitable interface between syntax and semantics. The parser makes use of selectional co-occurence patterns to limit its search space; PUNDIT's selection component is described in Section 2.1.4, and in more detail in [14].

Domain-Specific Syntactic Development. The porting of our system to different domains involves the augmentation of PUNDIT's syntactic component for the following reasons:

o To handle constructions specific to these applications, as in *Show me flights Dallas to Boston*, which does not occur in English generally.

o To fill in gaps in coverage or bugs in the domain-independent grammar.

o To re-order or delete options to reflect syntactic probabilities.

In one recent application (Air Travel Planning (ATIS)) 75 context-free grammar rules and 80 restrictions were added to the system.

Selection. PUNDIT makes use of some semantic information while parsing, using a component called Selectional Pattern Queries and Responses, or SPQR [14]. SPQR applies selectional co-occurrence patterns to filter subtrees of the parse tree as they are generated by the parser. The patterns are stored in a database, and may represent either allowable (good) or unwanted (bad) co-occurrences. Furthermore, SPQR makes use of a knowledge base of IS-A relationships in applying these patterns, so that generalizations can be captured in the database. For example, after parsing the noun phrase *flight to Boston*, SPQR consults a database of patterns and finds that *flight to city* is a good pattern, from which it is able to conclude that the noun phrase is semantically plausible. By contrast, if the parser attempted to create the noun phrase *airport to Boston* from input such as *I want to go from the airport to downtown Boston*, SPQR would find that *airport to city* is a bad pattern, and the noun phrase would be rejected, thus pruning that branch of the parse tree.

2.2 Semantic Processing

Overview. Semantic analysis in PUNDIT is based on the interpreter described in [20]. Verbs and other predicates are interpreted using declarative rules and are represented in the output of the system with a case-frame representation as described in [10]. Interaction between the semantics and pragmatics components provides a basis for PUNDIT to make much implicit information explicit ([12,15,21]). Pragmatic processing in PUNDIT includes reference resolution ([2,3,4,5], temporal processing ([38]) and dialog management ([1,17]. The basic approach to reference resolution is the focusing/centering paradigm described in [40, 16, 4, 17].

Arguments of verbs are assigned thematic role labels such as agent, patient, source, and so forth. Semantic interpretation is a search for suitable grammatical role/thematic role correspondences, using syntax-semantics mapping rules, which specify what syntactic constituents may fill a particular role; and semantic class constraints, which specify the semantic properties required of potential fillers. The syntactic constraints on potential role fillers are of two types: categorial constraints, which require that the potential filler be of a certain grammatical type such as subject, object, or prepositional phrase; and attachment constraints, which require that the potential filler occur within the phrase headed by the predicate of which it is to be an argument. categorial constraints are stated explicitly in the syntax-semantics mapping rules; the latter are implicit in the functioning of the semantic interpreter. For example, the source role of flight_C, the domain model predicate associated with the noun flight, can, in accordance with the syntax-semantics mapping rules, be filled by the entity associated with the object of a *from* prepositional phrase occurring within the same noun phrase as *flight* (e.g. *The flight from Boston takes three hours*).

Recent developments in the semantics component include a number of techniques for increasing the robustness of semantics, both with respect to less than optimal parses, and with respect to missing information. These are briefly reviewed here and described in more detail in [18].

Dynamic alternation of case frames. Semantic interpretation in PUNDIT is a process of case frame instantiation. Domain-specific files associate verbs and certain other lexical items with predicate-argument structures, and provide semantic constraints on role-fillers as well as linking rules which express allowable syntactic expressions of particular arguments. Role filling is driven by the decomposition associated with a given predicate. One difficulty for this approach was presented by adjuncts expressing location, time, manner, and other peripheral roles. In contrast to roles such as agent, patient, theme, actor, goal, source, and so forth, these expressions are not true arguments of the predicate they modify. These adjuncts were problematic for several reasons. First, unlike true arguments, they frequently allow multiple fillers, as in: *On Monday, I want to leave Dallas for Boston by 11 o'clock* contains two syntactically distinct time expressions; *I'm parked in Cambridge near Central Square, right in front of MIT* contains three locative expressions. A second difficulty presented by these roles is that unlike roles such as agent or patient, they comprise a set of sub-roles: in the sentence above, we cannot simply say that Cambridge, MIT and Central Square are location arguments; rather, they express what we term place_where, vicinity, and in_front_of.

In order to handle these adjunct roles, we have changed the functioning of the semantic interpreter to allow decompositions to be altered in accordance with what is in the constituent list. In PUNDIT, virtually all predicates are associated with generic_locative and generic_time roles. The semantic interpreter examines the list of syntactic constituents and replaces these generic location or time roles with more specific roles depending upon the actual preposition or adverb found in the constituent list.

Redistribution of Constituents.   Constituents that provide the semantic arguments of a   predicate are not always syntactically associated with it. There are several causes for such a mismatch between the parse and the intended interpretation. They include

(1) a variety of syntactic deformations which we refer to as extraposition (for example, *What flights do you have to Boston*, where the *to* prepositional phrase belongs in the subject noun phrase.

(2) metonymy (for example, *I want the $50.00 flight*, where the speaker means that s/he wants the flight whose FARE is $50.00), and

(3) suboptimal parses (e.g., parses with incorrect   prepositional phrase attachment).

Recent changes to the semantic interpreter allow it to fill roles correctly in cases such as the above, utilizing its existing knowledge of syntax-semantics correspondences, but relaxing certain expectations about the syntactic attachment of role-filling constituents. Thus the categorial constraints remain in force, but the attachment constraints have been loosened somewhat. Note that we continue to take syntax into account by observing the categorial requirements on potential role fillers while relaxing attachment expectations. If we ignored the syntax of leftover constituents and filled roles indiscriminately on the basis of semantic properties, we would lose an important source of constraint. Thus we achieve much more flexible and robust processing without sacrificing the constraints provided by a full syntactic analsysis.

Inter-Case Frame Inference Rules. While working with the ATIS domain, we noticed that PUNDIT was producing different representations (IDR's) for queries with different syntactic/lexical content but identical (or nearly identical) semantic content. For these queries, PUNDIT's semantic interpreter was correctly representing the meaning of a sentence but in an irregular way. For example, the instantiated decomposition produced for *flights from Boston* is:
flight_C(flight1, source(boston), ...)

while *flights leaving Boston* resulted in:

flight_C(flight1, source('), ...) leaveP(leave1, flight(flight1), source(boston), ...)

Clearly it would be preferable for the flight_C decomposition to be the same in both cases, but in the second case the source role of the decomposition associated with flight1 was unfilled, although it could be inferred from the leaveP decomposition that the flight's source was Boston. In other words, PUNDIT had not captured a basic synonymy relation between these noun phrases. Our response to this was to augment the semantic interpreter with a routine which can perform inferences involving more than one decomposition. The actual inferences are expressed in the form of rules which are domain-dependent;   the   inference-performing   mechanism   is domain-independent. For the above example, we have written a rule which, paraphrased in English, says that if a verb is one of a class of motion verbs used to express flying (e.g., *leave*), and if the source role of this verb is filled, propagate that filler to the source role of the flight involved. Thus the flight_C decomposition becomes the same for both inputs.

Automatic training. A unique feature of PUNDIT is its ability to automatically collect semantics training data and to use this training data to make informed guesses about the case frame structure of new predicates. This is important for speeding up the development process and for making the system more robust [5]. Previously, semantic processing would fail altogether if there was no case frame for a verb or predicate adjective, thus the system was not very robust when confronted with new words.

We have developed a guessing mechanism using the frequency of syntax-semantics mapping rules in a training corpus. This data is used to infer likely case roles, given a set of syntactic arguments. For example, the most frequent syntax/semantics mapping in the ATIS application is a mapping from the syntactic direct object to the theme of the predicate. Consequently, given an unknown predicate with a syntactic direct object, the system will guess that the predicate has a theme and that the direct object maps to the theme role. Other common mapping rules map the syntactic subject to the actor role, *from* prepositional phrases to the source role, and so on. So for example the presence of a *from* prepositional phrase in the parse will justify positing a 'source' role in the case frame.

The system can guess case frames in either of two modes. In the supervised mode the guessed case frames are presented to the user in an order reflecting their frequency in the training data. If the user rejects a proposed case frame a less frequent mapping for one of the roles will be selected and new case frames will be generated sequentially until the user accepts one of them. In the unsupervised mode the first guess is assumed to be correct and is used in the current analysis and is output to a file.

One interesting feature of this approach is that the newly guessed case frame is not assumed to represent the complete correct semantics of the verb. Since many verbs have optional arguments as well as several ways of expressing their arguments syntactically, it would be incorrect to simply assume that all the necessary information for the semantics of a verb is given by one instance. In the current system this results in a new guess for each instance of a verb in a corpus.

2.3 Pragmatic Processing

3.1 VFE

VFE is a dialog manager. It oversees the interaction of the system with the user, using PUNDIT and qtip as resources to interpret and respond to the user's requests. VFE administers the turn-taking structure, and maintains a higher-level model of the discourse than that available to PUNDIT. This level of knowledge enables it to provide PUNDIT and qtip with the context of an entire session, not just the current input utterance. VFE also keeps track of the current speaker and hearer, so that PUNDIT's Reference Resolution component can correctly interpret first and second person pronouns.

3.2 Deixis

Presenting tabular responses encourages the user to refer to entries in those responses. This introduces the need for a different kind of contextual reference. Whenever an answer table is returned, a discourse entity representing it is added to the discourse context, and a semantics for this entity is provided. Roughly speaking, if the query leading to the answer table is a request for X, the semantics can be thought of as being "the table of X" ([17]). For example, if the query was a request for flights from City1 to City2, the semantics assigned to the discourse entity rep resenting the answer is "the table of flights from City1 to City2".

4 Knowledge Base

In addition to the processing components of PUNDIT, the system makes also makes use of two auxiliary components; the knowledge base and the application module. The knowledge base is used in numerous ways during the processing of an utterance. SPQR, the module which checks selectional co-occurrence constraints, utilizes the KB to implement generalized selectional 9patterns, so that a pattern like "flight to city" can eliminate the need for individual patterns "flight to Boston", "flight to Denver", etc. The semantic interpreter makes frequent reference to the KB to check semantic class constraints, which limit the possible fillers which may be assigned to a slot of a decomposition (only cities and airports can fill the source and goal slots of the decomposition for the "flight" concept, for example). Finally, after PUNDIT has produced its output, the application module also makes heavy use of the KB to determine the relationships between the entities in the IDR.

The knowledge base is maintained in a semantic network supporting both IS-A and PART-OF hierarchies, and multiple inheritance in the IS-A hierarchy [16]. Contained in the knowledge base are concepts corresponding to most words in the lexicon, plus higher-level concepts useful for organizing the hierarchical network. As an example of the size of a typical knowledge base, the ATIS KB contains over 1200 concepts.

5 Application Module

The procedures used in PUNDIT's semantic and pragmatic analysis are domain and application independent, although when executing they make use of domain-specific information in a declarative form. The output of the analysis, is in a domain-independent format, the IDR (Integrated Discourse Representation). In order to use this output from PUNDIT, an additional module must be created which can take the IDR and use the information represented in it to perform whatever task or tasks a particular application calls for. This module is called qtip (Query Translation and Interface Program); a domain-dependent qtip needs to be created for each application [1,7].

6 Summary

The noteworthy features of PUNDIT are most apparent in the area of system design:

o Significant advances have been made in system design in the areas of domain and application independence and in modularity. These advances will help pave the way for practical spoken language systems which can be applied to real problems without prohibitive development cost.

o Domain independent dialog management: A domain- and application-independent dialog manager (VFE) has been designed and demonstrated in a number of distinct domains.

o Application independent architecture: An application-independent natural language understanding architecture has been developed which provides modular interaction with the application through a query translation module.

References

1. Catherine N. Ball, Deborah Dahl, Lewis M. Norton, Lynette Hirschman, Carl Weir, and Marcia Linebarger. Answers and questions: Processing messages and queries. In Proceedings of the DARPA Speech and Natural Language Workshop, Cape Cod, MA, October 1989.

2. Deborah A. Dahl. Focusing and reference resolution in PUNDIT. In Proceedings of the 5th National Conference on Artificial Intelligence, Philadelphia, PA, August 1986.

3. Deborah A. Dahl. Determiners, entities, and contexts. In Proceedings of TINLAP-3, Las Cruces, NM, January 1987.

4. Deborah A. Dahl. Evaluation of pragmatics processing in a direction finding domain. In Proceedings of the Fifth Rocky Mountain Conference on Artificial Intelligence, Las Cruces, New Mexico, 1990.

5. Deborah A. Dahl. Applications of training data in semantic processing, March 1991. AAAI Spring Symposium.

6. Deborah A. Dahl and Catherine N. Ball. Reference resolution in PUNDIT. In P. Saint- Dizier and S. Szpakowicz, editors, Logic and logic grammars for language processing. Ellis Horwood Limited, 1990.

7. Deborah A. Dahl, Lynette Hirschman, Lewis M. Norton, Marcia C. Linebarger, David Magerman, and Catherine N. Ball. Training and evaluation of a spoken language understanding system. In Proceedings of the DARPA Speech and Language Workshop, Hidden Valley, PA, June 1990.

8. Deborah A. Dahl, Martha S. Palmer, and Rebecca J. Passonneau. Nominalizations in PUNDIT. In Proceedings of the 25th Annual Meeting of the Association for Computational Linguistics, Stanford University, Stanford, CA, July 1987.

9. John Dowding and Lynette Hirschman. Dynamic translation for rule

pruning in restriction grammar. In Proceedings of the 2nd International Workshop On Natural Language Understanding and Logic Programming, Vancouver, B.C., Canada, 1987.

10. C. Fillmore. The case for case. In Bach and Harms, editors, Universals in Linguistic Theory, pages 1-88. Holt, Reinhart, and Winston, New York,

11. Lynette Hirschman. Conjunction in meta-restriction grammar. Journal of Logic Programming, 4:299-328, 1986.

12. Lynette Hirschman. A meta-rule treatment for English wh-constructions. In Proceedings of META88 Workshop on Meta-Programming in Logic Programming. University of Bristol, Bristol, 1988.

13. Lynette Hirschman and John Dowding. Restriction grammar: A logic grammar. In P. Saint-Dizier and S. Szpakowicz, editors, Logic and Logic Grammars for Language Processing, pages 141-167. Ellis Horwood, 1990.

14. Fran,cois-Michel Lang and Lynette Hirschman. Improved portability and parsing through interactive acquisition of semantic information. In Proceedings of the Second Conference on Applied Natural Language Processing, Austin, TX, February 1988.

15. Marcia C. Linebarger, Deborah A. Dahl, Lynette Hirschman, and Rebecca J. Passonneau. Sentence fragments regular structures. In Proceedings of the 26th Annual Meeting of the Association for Computational Linguistics, Buffalo, NY, June 1988.

16. David L. Matuszek. K-Pack: A programmer's interface to KNET. Technical Memo 61, Unisys Corporation, P.O. Box 517, Paoli, PA 19301, October 1987.

17. Lewis M. Norton, Deborah A. Dahl, Donald P. McKay, Lynette Hirschman, Marcia C. Linebarger, David Magerman, and Catherine N. Ball. Management and evaluation of interactive dialog in the air travel domain. In Proceedings of the DARPA Speech and Language Workshop, Hidden Valley, PA, June 1990.

18. Lewis M. Norton, Marcia C. Linebarger, Deborah A. Dahl, and Nghi Nguyen. Augmented role filling capabilities for semantic interpretation of natural language. In Proceedings of the DARPA Speech and Language Workshop, Pacific Grove, CA, February, 1991.

19. David S. Pallett. DARPA resource management and ATIS benchmark poster session. In Proceedings of the DARPA Speech and Language Workshop, Pacific Grove, CA, February, 1991.

20. Martha Palmer. Semantic Processing for Finite Domains. Cambridge University Press, Cambridge, England, 1990.

21. Martha S. Palmer, Deborah A. Dahl, Rebecca J. Schiffman. Passonneau, Lynette Hirschman, Marcia Linebarger, and John Dowding. Recovering implicit information. In Proceedings of the 24th Annual Meeting of the

Association for Computational Linguistics, Columbia University, New York, August 1986.

22. Rebecca J. Passonneau. A computational model of the semantics of tense and aspect. Computational Linguistics, 14(2):44-60, 1988.

23. Naomi Sager. Natural Language Information Processing: A Computer Grammar of English and Its Applications. Addison-Wesley, 1981.

The ESTEAM-316 Dialogue Manager

Thomas Grossi, Didier Bronisz and François Jean-Marie

Cap Gemini Innovation, Grenoble, France

Abstract. Advising someone seems easy to do but, like any human behavior, advice-giving requires numerous and elaborate mechanisms. The advice-giver must have a good understanding of the needs of the user, be able to solve the problem or answer the request of the user, and finally to explain the presented solution. In ESPRIT project ESTEAM-316[1], we developed an advice-giving system for financial investment. In this paper we give an overview of the entire project (which ended in 1989) and a more detailed presentation of the dialogue management component. We put particular emphasis on the data structures used and on the use of planning mechanisms in conducting the dialogue.

1 The Project ESTEAM-316 in General

Automated advice-giving (i.e., providing "expert" solutions to human "novices" by computer) is a complex task requiring the integration of knowledge and data from a variety of sources. ESTEAM-316's objective was to study the design and construction of mixed-initiative, *advice-giving expert systems* (henceforth referred to as AGES) which are general, distributed and cooperative:

- *mixed-initiative* in the sense that, although the system has a general plan for conducting the dialogue, it can accommodate requests for clarifications, change of subject, etc., from the user;
- *advice-giving* in the sense that it provides "expert" solutions to the "novice" user's problems;
- *general* in the sense that the system architecture and design are essentially domain independent, allowing the use of self-contained modules or *agents* that encapsulate knowledge or expertise;
- *distributed* in the sense that the system may be composed of multiple, logically independent software components which work together to respond to the user's request; and
- *cooperative* in the sense that the system actively helps the user to formulate his request and provides responses adapted to the user's expectations.

This objective presents a number of interesting and difficult research problems in the fields of user-system dialogue management, knowledge representation, database access and overall system architecture. To address these issues, a consortium was formed comprising four partners and one sub-contractor: Cap

---

[1] Supported in part by the Commission of the European Communities.

Sesa Innovation (Prime Contractor) in Grenoble, France; ONERA–CERT in Toulouse, France; Philips Research Lab in Brussels, Belgium; CSELT in Turin, Italy; and the University of Turin on contract from CSELT. The project began in January, 1985, and ended in December 1989. It had a budgeted manpower of 106 person/years.

The project was divided into five major tasks: an architecture for cooperation between agents; knowledge representation and acquisition; dialogue management; data and knowledge bases; and a demonstrator incorporating much of the research results of the other tasks. This demonstrator and its individual components will be discussed in detail in the next section.

1.1 The Demonstrator

The final phase of the project was oriented towards the development of a demonstrator or prototype, incorporating ESTEAM-316 research results and showing the feasibility of an AGES. The important modules consist of a *Problem Solver* specialized in the development of financial portfolios satisfying a number of parameters and constraints; a *Cooperative Answering Module* which acts as a front-end to a data base and provides more helpful answers to queries; a *Dialogue Manager* which interacts with the user and directs queries to the Problem Solver or to the Cooperative Answering Module as appropriate, taking into account the vagaries of human conversation; all tied together by a *Cooperation Architecture*. We will examine each of these modules briefly.

The Problem Solver. The Problem Solver is the module which solves the user's problems in a financial domain and provides proof trees that the Dialogue Manager can use to generate explanations. This Problem Solver module is based upon a knowledge representation formalism which integrates an object-oriented approach and logic. It is implemented in Prolog. From a Basic Investment Situation which contains the parameters describing the user's problem, the Problem Solver gives a plan which is a set of investment recommendations. At this state the user's desiderata (restrictions on financial institutions or countries...) can be taken into account to give the final solution which is a list of optimed portfolios, each portfolio including a set of pairs (security, amount) with the return and the risk of the portfolio. At any point, the Problem Solver can produce a "proof tree" describing how it arrived at a particular conclusion. See also Bruffaerts and Henin (1988) and Bruffaerts et al. (1989).

The Cooperative Answering Module. In traditional applications devoted to company management, like payroll computation, people who access data in a database have a precise definition of the data they want. There are many other applications where people want to access data in order to make a decision, or to solve a problem whose solution cannot be found applying a simple algorithm. An important feature of this context, from the point of view of data retrieval, is that users don't have a precise idea of the data which can help them to solve a problem, or to make a decision. The objective of the Cooperative Answering module is to simulate the behavior of a person who wants to help as much as possible a user seeking information.

The Cooperative Answering operates by transforming the user's query to retrieve additional information. The additional information provided in the answers falls into the following categories:

- Additional entities having a different type than those requested in the query.
- Additional entities satisfying "neighborhood" conditions.
- Additional attribute values.

Examples:

- If a bond is in a currency different from that of the user, then return the currency.
- If a query about bonds has no answer, return a stock with very low risk.

See also Cuppens and Demolombe (1990).

The Dialogue Manager. The role of the Dialogue Manager is to provide a "friendly" user interface that makes available to the user the expertise of the various modules in a flexible and uniform fashion (Grossi 1988) (Bronisz et al. 1990). In the following section we will present the organization of the Dialogue Manager and we will examine its role in the demonstrator.

The cooperation architecture. To provide a design method and to support the efficient implementation of AGES, we built a special architecture. The intended "user" of the architecture is thus an AGES designer. Support is provided via a set of primitives for defining Agents, i.e. programs with given functionalities, and implementing their interactions. Architectural support can be seen as a "glue" for AGES designers who need to interface components of a complex system cooperatively. The architecture aims to keep separate the design and implementation of internal Agent features from external features handling cooperative interaction. An architectural approach at the level of cooperating Agents aims, in general, to facilitate the changing of domain for an AGES, the addition of new capabilities to a system, and experimentation with AGES functionality and design. The above can be achieved by providing communication and control mechanisms independent of the implementation of each component. The architectural approach to communication allows Agents to share tasks by problem decomposition and to share knowledge by using special Agents for Data and Knowledge Base Management. See also Govannini and Cortese (1987).

2 The Dialogue Manager Closer Up

The organization of a human dialogue is very complex. In a two-person dialogue, both participants build, according to their own intentions, the structure of the discussion. Some topics are more important than others for one, whereas they may be less so for the other. Ordinarily, the dialogue proceeds with many digressions like explanations or reminders of information. The topics discussed are ordered according to the preferences of each dialoguer, and the participants may often come back to a previous topic or begin another one before the current one is finished. Moreover, there is great deal of contextual information during the discussion. The typical Question/Answer pair refers to some previously given piece of information. During a dialogue the consultant builds and maintains a model of the whole plan of the client. In our view, the role of a consultant is to guide the consultation dialogue. However, this must be done in a flexible way in order to allow the client to express himself.

In the context of the project ESTEAM-316, we limited our study to a specific type of advice giving situations: dialogue for the consultation of the ESTEAM-316 financial problem solver and cooperatively-answering data base. In such a dialogue, the Dialogue Manager (DM) and the user cooperate in building an appropriate description of the user's financial situation and his investment needs. More specifically, an investment problem definition is broken down into a number of subparts (emergency funds, fixed-term investments and long-term investments) and the amounts placed in each subpart are constrained to be equal to the total amount to be invested. An intermediate solution contains a list of types of portfolios that would satisfy the user's needs. The session may be stopped at this point if desired. However, if the user wishes to continue, the system will offer a final solution consisting of a detailed plan with concrete means (savings account, securities) for achieving the initial goals.

2.1 A Sample Dialogue

Before describing the ESTEAM-316 Dialogue Manager in detail, we will motivate the presentation with a sample dialogue illustrating its various possibilities. This is the actual scenario typically used in demonstrations, and we will refer to it in later discussions.

1	*System*:	Let's talk about the investment plan.
2	*User*:	I want to buy a car in 3 years.
3	*System*:	Yes.
		An investment plan has a amount. An investment plan has an emergency fund and a fixed-term investment and a long-term investment.
		What is the amount of the investment plan?
4	*User*:	100000 Belgian francs.
5	*System*:	Let's talk about the emergency fund. An emergency fund has a amount and a delay.
		What is the amount of the emergency fund?
6	*User*:	Let's talk about about the fixed-term investment.
7	*System*:	Yes. A fixed-term investment has a amount and a date.
		Is the date of the fixed-term investment in 3 years?
8	*User*:	Yes.

9	*System*:	What is the amount of the fixed-term investment?
10	*User*:	70000 Belgian francs
11	*System*:	What is the management style of the fixed-term investment?
12	*User*:	What are the possible values for the management style?
13	*System*:	The management style of the fixed-term investment is a management style. The possible value is active or passive. What is the management style of the fixed-term investment?
14	*User*:	active.
15	*System*:	I recommend a liquid portfolio whose delay is 1 months , and whose amount is 70000 Belgian francs.
		I recommend a reconsideration when interest rates fall.
		Do you want me to refine the fixed-term investment?
16	*User*:	Could you justify the recommendation?
17	*System*:	If the investment situation is a fixed-term investment and it is expected that interest rates rise and the management style is active then I recommend a liquid portfolio whose amount is the amount of the fixed-term investment , and whose delay is 1 months and I recommend a reconsideration when it is expected that interest rates fall.
		I recommend a liquid portfolio whose delay is 1 months , and whose amount is 70000 Belgian francs.
		I recommend a reconsideration when interest rates fall.
		Do you want me to refine the fixed-term investment?
18	*User*:	Yes.
19	*System*:	Do you want to add a restriction to the fixed-term investment?
20	*User*:	No.
21	*System*:	I recommend a liquid portfolio. The portfolio contains 70000 Belgian francs of a fixed-term deposit at Paribas whose term is 1 month. The return of the liquid portfolio is 6.75.
		Do you accept the liquid portfolio?
22	*User*:	I want only non Belgium securities
23	*System*:	I recommend a liquid portfolio whose delay is 1 months , and whose amount is 70000 Belgian francs.
		I recommend a reconsideration when interest rates fall.
		You insist on only non Belgium securities.
		Do you want to modify the restriction of the fixed-term investment?
24	*User*:	No.
25	*System*:	I recommend a liquid portfolio. The portfolio contains 70000 Belgian francs of a fixed-term deposit at gen-bq whose term is 1 months. The return of the liquid portfolio is 6.5.
		Do you accept the liquid portfolio?
26	*User*:	What is/are the issuer country of the fixed term deposit whose issuer is Gen-bq?
27	*System*:	The issuer country of the fixed-term deposit at gen-bq whose term is 1 , and whose currency is Belgian francs is Luxemburg.
		The issuer country of the fixed-term deposit at gen-bq whose term is 3

, and whose currency is Belgian francs is Luxemburg.

The issuer country of the fixed-term deposit at gen-bq whose term is 6
, and whose currency is Belgian francs is Luxemburg.

I recommend a liquid portfolio whose delay is 1 months , and whose
amount is 70000 Belgian francs.

I recommend a reconsideration when interest rates fall.

You insist on only non Belgium securities.

I recommend a liquid portfolio. The portfolio contains 70000 Belgian
francs of a fixed-term deposit at gen-bq whose term is 1 months. The
return of the liquid portfolio is 6.5.

Do you accept the liquid portfolio?

28	*User:*	Let's suppose the management style of the fixed-term investment is passive.
29	*System:*	Is the amount of the fixed-term investment 70000 Belgian francs?
30	*User:*	Yes.
31	*System:*	Is the date of the fixed-term investment in 3 years?
32	*User:*	Yes.
33	*System:*	I recommend a risk free portfolio whose term is 36 months , and whose delay is 1 months , and whose amount is 70000 Belgian francs.

You insist on only non Belgium securities.

Do you want to modify the restriction of the fixed-term investment?

34	*User:*	No.
35	*System:*	I relaxed the following restrictions.

You insist on only non Belgium securities.

I recommend a risk free portfolio.

The portfolio contains 70000 Belgian francs of a savings account at sefb.

The return of the risk free portfolio is 6.0.

Do you accept the risk free portfolio?

36	*User:*	I want the 2th fixed-term investment.
37	*System:*	I recommend a liquid portfolio whose delay is 1 months , and whose amount is 70000 Belgian francs

I recommend a reconsideration when interest rates fall.

You insist on only non Belgium securities.

I recommend a liquid portfolio. The portfolio contains 70000 Belgian
francs of a fixed-term deposit at gen-bq whose term is 1 months. The
return of the liquid portfolio is 6.5.

Do you accept the liquid portfolio?

38	*User:*	Yes
39	*System:*	Let's talk about the emergency fund again.

What is the amount of the emergency fund?

40	*User:*	Negation of: I want the emergency fund
41	*System:*	The emergency fund is obligatory.

What is the amount of the emergency fund?

42	*User:*	20000 Belgian francs.
43	*System:*	What is the delay of the emergency fund?

44	*User:*	3 months.
45	*System:*	I recommend a liquid portfolio whose delay is 3 months , and whose amount is 20000 Belgian francs.
		Do you want me to refine the emergency fund?
46	*User:*	No.
47	*System:*	Let's talk about the long-term investment. A long-term investment has a amount. Do you want a long-term investment?
48	*User:*	No
49	*System:*	The constraint *equal_amount applies to the amount of the investment plan and the amount of the emergency fund and the amount of the fixed-term investment and the amount of the long-term investment.
		The amount of the long-term investment does not satisfy the constraint *equal_amount.
		Please resolve the constraint.
50	*User:*	What is the amount of the investment plan?
51	*System:*	The amount of the investment plan is 100000 Belgian francs.
		The amount of the long-term investment does not satisfy the constraint *equal_amount.
		Please resolve the constraint.
52	*User:*	The amount of the emergency fund is 30000 Belgian francs.
53	*System:*	You have resolved the constraint *equal_amount.
		Let's talk about the emergency fund again.
		Is the amount of the emergency fund 30000 Belgian francs?
54	*User:*	Yes
55	*System:*	Is the delay of the emergency fund 3 months?
56	*User:*	Yes
57	*System:*	I recommend a liquid portfolio whose delay is 3 months, and whose amount is 30000 Belgian francs.
		Do you accept the investment plan?
58	*User:*	Yes
59	*System:*	Goodbye

2.2 Analysis

From this dialogue, we can illustrate the different points we mentioned above.

- In sentence 6 the user prefers to begin the discussion with the fixed-term investment, while the system began the discussion with the emergency fund. The system changes the order of the different topics according to the will of the user and the feasibility of such an exchange.
- In sentence 10, the user does not say "The amount of the emergency fund is 70,000 BEF" but only "70,000 BEF". The system must understand such an elliptical response. Normally, the two participants hide a lot of information. They just give sufficient information to have an unambiguous dialogue. But having a partial knowledge necessitates being able to confirm sometimes what has been understood.

- In sentence 7, the system asks the user for a confirmation of the value of the fixed-term investment (2 years) which was previously given by him in the sentence 2. The system has understood that buying a car is equivalent to having a fixed-term investment, but it asks him to verify this supposition.
- In sentence 13 the user begins a digression in order to have an explanation about a new term introduced by the system in asking for a value. The system has to recognize this new user's intention, cope with it (it may be longer than a two-turn dialogue (User, system) as in the discussion of a given solution) and come back to the previous dialogue.
- In sentence 18 the user accepts the system's offer to come up with a detailed investment plan, while in sentence 26 he decides to interrogate the data base; the Cooperative Answering module provides information on the currency even though it was not asked for. In response to another query to the data base — sentence 36 — the Cooperative Answering module informs the user that it was obliged to relax a constraint in order to come up with an answer.
- In sentences 28 – 36 the system and the user explore the implications of a modification of one of the parameters.
- In sentence 52 the system points out that a certain domain constraint — namely that the total amount invested be equal to the sum of the amounts invested in each of the subparts — has been violated, and asks the user to remedy the situation.

2.3 Theoretical Foundations

The discourse theory of Grosz and Sidner (1985, 1986), distinguishes among three aspects of dialogue structures: linguistic, attentional and intentional structures. Briefly summarized, linguistic structure concerns the literal utterances of the participants; attentional structure deals with the focus of attention at each point in the dialogue; and intentional structures represent the goals of each participant which underlie and inform their utterances.

Within the Dialogue Manager, the intentional and attentional aspects are the concern of the Planner. The links with the linguistic structure are handled by the Recognizer and the Expression Specialists. The process of recognition of user intentions results in the update of the *Structure of System Intentions* (SSI), which represents the system's view of the user's intentions and the system's own intended reactions to them (Litman 1984). Interchange of literal meanings (Appelt 1985) with the natural-language *Front End* (Lancel et al., 1988) uses a knowledge representation formalism known as *functional descriptions* and was the inspiration for our own internal representation.

Furthermore, we explicitly distinguish between the *task level* and the *communicative level*. In an advice giving dialogue, the "real" problem of the user (e.g., the investment problem) is represented by the task level plan; the system has only communicative goals for constructing, refining, and explaining the task level plan to the user. This distinction is crucial and must be kept in mind to understand what follows.

2.4 The Structure of the Dialogue Manager

The general organisation of the ESTEAM-316 Dialogue Manager is depicted in fig. 1. The *Recognizer* relates the observed surface act to the expectations of the system by trying to recognize the intentions underlying the user's input. The *Planner* conducts the dialogue and chooses the reactions to the user's intentions. The Planner also interacts with the *Problem Solver* via the *Cooperation Architecture* for solving the user's problems and with the *Cooperative Answering Module* to obtain specific data from the data base. It sends the communicative acts it intends to perform to the *Expression Specialists* for an appropriate literal meaning. The Expression Specialists interact with the user via a pseudo-natural-language *Front End*.

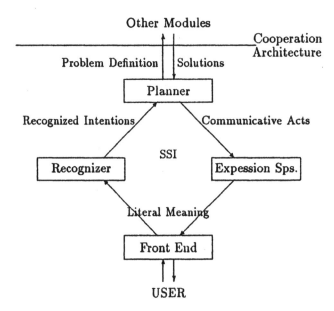

Fig. 1. Architecture of the Dialogue Manager

2.5 Data and Knowledge Representation

The Functional Description (FD) formalism (Kay 1981, Lancel et al. 1988) is our common knowledge representation formalism. In the FD formalism, as in many other knowledge representation schemes, knowledge is represented as objects having slots to indicate their key properties. These slots are type restricted, where the type indicates some portion of the object abstraction hierarchy. From a purely representational point of view, all slots are equivalent — they consist of a slot name, a modifier, and a value (a modifier is used for attaching auxiliary information such as whether the User or the System has referred to a certain slot, or refused it outright). Different slots are used for specific purposes by the various modules of the system; the meaning of a slot can only be understood by examining how it is used.

The Dialogue Manager uses several conceptual knowledge bases, which contain models of objects, actions, and plans. In these knowledge bases we find domain actions, such as invest_cash_need or buy; communicative acts, such as a classic inform or request; higher-level communicative plans, such as ask_parameter or start_discourse_segment, for each of which there exists an Expression Specialist; and dialogue plans, such as advise, which control the overall unfolding of the advice-giving session. Several relations, expressed as slots in these plans, appear in many different action models. The most important of these is the decomposition slot, which points to a group or sequence of possible actions which can realize this plan. In many cases we also find a restriction slot which limits the applicability of a plan or an action. These restrictions take the form of conditions expressed in first order logic predicates on parameters or on elements of the decomposition.

Now let us consider the main structures used by the DM and based on this representation.

The Task Plan. This is used both to structure and to store the task level problem of the user. It consists of instances of domain concepts defined in the dictionary of concepts. The following example will illustrate the structure used:

```
:- create_model('*invest_plan','*invest_action',
   [ amount = Total_Sum:[ type = '*sum' ],
     agent = '*user' ,
     decomposition =
       [ type = '*dec_obj',
         action = EF: [ type    = '*emergency_fund',
         amount = EF_Sum:[ type = '*sum' ] ],
         action = [ type    = '*cash_need',
            amount = CN_Sum:[ type = '*sum' ] ],
         action = [ type    = '*long_term',
            amount = LT_Sum:[ type = '*sum' ] ],
         constraint =
           equal( :Total_Sum, :EF_Sum, :CN_Sum, :LT_Sum ),
         mandatory  = :EF,
         discuss_order = indifferent ],
     in_parameter = amount,
     precondition = [type    = '*possess',
             agent  = '*user',
             object = :Total_Sum]
   ]).
```

This means that the type "investment plan" is a sub-type of "investment actions". An investment plan is decomposed into three sub-actions, namely an emergency fund [investment], a cash need [investment] and a long-term investment. There exists a constraint within the investment plan: that the total amount be equal to the sum of the amounts in the sub-investments. The colon (:) notation in the above structure allows the object filling a given slot to be referenced elsewhere within the same structure.

The Task Plan was designed with two different views: the Recognizer's view for collecting exactly what has been recognized in the user's speech acts and the Planner's view for maintaining an approved version of the Task Plan. Each object in the Task Plan exists in both of these views.

Dialogue Plans and Communicative Goals. Stereotypical knowledge about planning the communicative goals of the system is expressed by a set of models: the dialogue plans for the abstract communicative goals and the communicative goals models for the atomic goals. The dialogue plans express the structure of the overall advice-giving session. The purpose of these plans is to capture procedural knowledge for an "ideal"[2] advice-giving session. Dialogue plans are designed to allow the Structure of System Intentions (described below) to be developed by progressive refinements of the main goal of an advice-giving dialogue session. Each model of a dialogue plan is associated with one abstract communicative goal of the system, and basically describes how it can be decomposed into sub-goals, and how it relates to expectations toward the user intentions. A model of a dialogue plan is for instance:

```
[(name, discuss_plan),
 (type,[dialogue]),
 (restriction, mandatory_plan),
 (decomposition, [seq,
                  open_discourse_segment,
                  ask_information,
                  accept_user_initiative,
                  treat_in_detail,
                  close_discourse_segment)
         ]),
 (expect,[id_parm_plan,
          req_expl_plan,
          refuse_plan
          ])
].
```

Here we see the several steps that make up the version of plan discuss_plan which applies to mandatory plans, along with the system's expectations for *normal* user reactions.

---

[2] i.e., without any digression from the user

The Structure of System Intentions (SSI). This is a network of nodes representing the system's communicative goals. The nodes are instantiations of the dialogue plan and communicative goals described previously. Each node contains a description of the communicative goal and its associated subject in the Task Plan. Dominance and ordering among nodes in the SSI are represented by additional nodes that do not directly correspond to communicative goals, i.e. their role is purely structural. Ordering is expressed in a quite simplistic way: the SSI can only specify that a list of goals must be achieved in a given order (the SEQ construct) or that they can be achieved in any order (the BAG construct).

Input and Output with the Dialogue Module. In a domain as complex as ours, we felt that interactions with the user should take place in natural language. In fact, in the earlier stages of the project we had a natural language front-end, which partially motivated our choice of Functional Descriptions as a representation language. Development of this front end did not keep pace with the development of ESTEAM-316, however, and we were obliged to drop it. On the other hand, since we still felt that natural language was necessary, we kept the FDs and the "real" input to the system consists of speech acts represented as FDs. But since discussion with the system via FDs is extremely laborious, we needed another way of communicating. The result was a module which uses template grammars which allow the system and the user both to express themselves in a very restricted natural language. The user composes sentences by selecting templates containing the desired phrases, and the system builds a sentence by selecting templates based on their FD structure. A portion of the top-level template is shown below. In each group, the pattern selected by the user is shown first, followed by the actual surface string produced (containing calls to other templates), after which we find the FD structure used by the system.

```
menu( starts, [
    [ 'Let us talk about the <object>.',
      [ 'Let''s talk about the', (Object, objects), '.' ],
      [ instance_of = '*s_inform',
        speaker = '*user',
        hearer = '*system',
        proposition =
          [ instance_of = '*want',
            actor = '*user',
            proposition =
              [ instance_of = '*discuss',
                object = Object ] ] ] ],

    [ 'I have <object>.',
      [ 'I have a(n)', (Object, objects), '.' ],
      [ instance_of = '*s_inform',
        speaker = '*user',
```

```
        hearer = '*system',
        proposition =
          [ instance_of = '*possess',
            actor = '*user',
            object = Object ] ] ],

  [ 'I have <number> <currency>.',
    [ 'I have', (Number, number), (Currency, currencies), '.' ],
    [ instance_of = '*s_inform',
      speaker = '*user',
      hearer = '*system',
      proposition =
        [ instance_of = '*possess',
          actor = '*user',
          object =
            [ instance_of = '*sum',
              amount = Number,
              currency = Currency ] ] ] ],

  ...
] ).
```

2.6 The Execution Cycle

To start the advice-giving session, the Planner initializes the SSI with the single high-level goal advise(investment_plan). The nodes of the task plan that dominate the current node are called the *active* nodes. The path between the current node and the root of the SSI is called the *active branch*. One step in the control cycle corresponds to the processing of the current node. The processing depends on the type of the node.

Abstract (non-atomic) nodes are expanded according to their model. This expansion consists of three steps:

- selection of a model (use of name slot and restriction slot),
- instantiation of parameters,
- construction of child nodes according to the decomposition slot.

When atomic communicative goals are reached, the Planner collects them into a buffer called the Agenda and calls the Expression Specialists to achieve them. The Expression Specialists are a group of highly specialized plans that know how to achieve specific communicative goals aimed at the user. There is at least one specialist for each type of goal. The result of these plans is the construction of Functional Descriptions which describe a speech act, and which are then handed over to the Front End for conversion into a natural-language sentence. These speech acts can be of two types: *one-turn games*, meaning that the system has something to say to the user but expects no reaction, and *two-turn games* which means that the next "move" is up to the user.

(The purpose of the buffering is to allow the Expression Specialists to combine the atomic communicative goals when possible. For example, a goal to confirm an input before changing a subject might be combined with the change of subject, since we can suppose that if the system doesn't object the information is valid.)

When the Expression Specialists indicate that a two-turn game has been initiated, the Planner generates the Expectation Stack corresponding to the current state of the SSI and calls the Recognizer. The skeleton of the Expectation Stack is the active branch of the SSI. The top level corresponds to the current two-turn goal, and hence contains the most probable reactions of the user. The bottom level corresponds to the overall goal of the session. The Planner puts into each layer the list of expectations found in the model of the corresponding goal.

The Recognizer processes the surface speech act of the user. Recognition of the user's intention normally results in selecting one level in the Expectation Stack and one of the expected communicative acts within that level. The Recognizer passes to the Planner a list of the changes it has made in its view of the Task Plan. These changes are called the *side effects* of a speech act. The Planner must verify that they are compatible with the previous state of the Task Plan and the coherence constraints before executing them in the planner's view. One of the major roles of the Planner is to repair disparities between the two views by conducting clarification sub-dialogues with the user. If the Recognizer fails to match the user's input with anything in the Expectation Stack (meaning that it was unable to recognize any intention at all), the Planner considers that the user's input is irrelevant to the current dialogue. In this case, the planner inserts the atomic goal irrelevant (to express misunderstanding) to the left of the unanswered request and updates the current goal to this new goal. Revision deals both with interaction among user's goals inside the Task Plan and interaction between system's and user's goals.

The Planner first analyses the side effects in order to verify that they keep the Task Plan coherent. This process is called *constraint verification*. If the modifications implied by the side effects would result in an incoherent Task Plan, the side effects are not executed in the Planner's view, and a transformation associated with constraint violation is applied to the SSI. The transformation results in the insertion of a clarification sub-dialogue just before the current goal in the SSI. The root of the clarification sub-dialogue becomes the current goal and control returns to the top cycle. If no constraint violation was detected, the Planner executes the side effects in the Planner's view of the Task Plan and then considers the impact of user's intentions on the current state of the SSI.

If the recognized intention corresponds to the immediate expectation of the two-turn game (the "normal" case), all the goals in the Agenda are marked as achieved, and the current goal is updated to the next goal in the SSI. Control returns to the top cycle. Otherwise, the SSI and thus the current goal are modified according to the transformation associated with the recognized communicative act.

The revision cases that are currently handled by the Planner concern changes of subject among different entities in the Task Plan, requests for explanations of concepts in the task domain, and refusal of the task plan or sub-plans. A transformation corresponds to the conjunction of a communicative act of the user and a level in the Expectation Stack. The node of the SSI attached to that level is the main parameter of a transformation. It is the root of the sub-dialogue containing the current node that is affected by the transformation.

3   Remarks on the Implementation

We have often been asked to evaluate our choices for the implementation of the Demonstrator, and here we will take the opportunity to express our opinion.

Quintus Prolog was chosen first for the Problem Solver and the Dialogue Manager, and later also for the Cooperative Answering module. This was no coincidence – Prolog turns out to be a very easy language to use for developing prototypes. Its trace mechanism allows for easy debugging and it is easy to modify a single function without recompiling an entire prototype. Also, its backtracking mechanism is ideal for a program that has to try many alternatives before finding a correct solution.

On the other hand, we sorely felt the lack of structured data types. A significant part of the activity in task C (Dialogue Management) was dedicated to developing a representation heavily inspired by the object–oriented paradigm. It is possible also that the existance of complex data types would have facilitated work in task B (Knowledge Representation) as well. It should also be pointed out that the "module" mechanism in Quintus Prolog is cumbersome and sometimes difficult to use, especially if one likes to scatter one's source files in different directories.

In years past, Prolog's rather inefficient evaluation mechanism led to performance problems, but it has been our experience that on a reasonably rapid modern machine (e.g. a Sun-3/60) with a respectable amount of local memory (at least 8 Mb.), the Demonstrator functions at a perfectly acceptable speed. In fact, during a demonstration, the only point at which the speaker must consciously fill in the time while the machine is processing is when the user has asked for a justification. (Quite a lot of work goes into building these justifications: the Problem Solver returns only the most immediate justifications for a fact, and the obtain the justifications for the justifications, the Dialogue Manager must make repeated calls to the Problem Solver. Each call requires a resimulation of the entire chain of reasoning. This, added to the fact that for reasons of "information hiding" the Problem Solver is called to access the attributes of all the domain objects it generates, explains why the justifications take so long to build.)

As for "infrastructure", Sun workstations with Unix were the rule. Their widespread availability made it easy for partners to work at separate sites and come together with their code for integration. The popularity of Sun workstations has also made possible several demonstrations of the Demonstrator at conferences (Coling-90, ECAI-90, as well as the present LPSS). Finally, the use of *uucp* — which can be installed on any Unix machine — greatly facilitated communication between partners via electronic mail.

4 Conclusion

Natural language is, of course, the most "natural" way for humans to express themselves, but it is also notoriously difficult because humans do not say everything they mean. In this description of the EsTEAM-316Dialogue manager, we have tried to show the usefulness of the planning paradigm in human-machine interfaces. In the Dialogue Manager, an understanding of why a user might have said something allows the system to react in a much more helpful way than would be possible in a system that interpreted its input literally, and also permits much more flexibility than in a system that had a fixed agenda of actions to be carried out. As the set of plans and intentions represented by the system grows, we can hope that its usefulness and friendliness will approach that of "real" humans.

References

Allen, J.F. and C.R. Perrault: "Analyzing intentions in utterances". *Artificial Intelligence*, 3(15):143-178, 1980.

Appelt, Douglas E.: *Planning English Sentences.* Cambridge University Press, Reading (MA), 1985.

Bronisz, D., T. Grossi, and F. Jean-Marie: "Advice-Giving Dialogue: An Integrated System." *13th International Conference on Computational Linguistics*, Helsinki, August 16-18, 1990.

Bruffaerts, A. and E. Henin: "Some Claims about Effective Explanation Generation in Expert Systems." *Proc. of the AAAI'88 Workshop on Explanation*, M. Wick (ed.), Minneapolis, August 22-26, 1988, pp. 83-86.

Cuppens, F. and R. Demolombe: "Extending answers to neighbour entities in a Cooperative Answering context." *The International Journal of Decision Support Systems*, vol. 6, no. 1, 1990.

Bruffaerts, A., E. Henin and A. Pirotte: "A Sound Basis for the Generation of Explanations in Expert Systems", *Philips Technical Review*, v.44, no. 8/9/10, May 1989, pp. 287-295.

Giovannini F. and G. Cortese, "An expert system architecture supporting cooperation among loosely-coupled heterogeneous agents." *IASTED International Conference on Expert Systems — Theory & Applications*, Geneva, Switzerland, June 16–18, 1987.

Grossi, Thomas. "Planning." In M. McTear (ed.), *Understanding Cognitive Science*, Ellis Horwood Limited, Chichester, 1988.

Grosz, Barbara J.: "Discourse Stucture and the Proper Treatment of Interruptions." *Proceedings of the IXth IJCAI*, Los Angeles (USA), 1985.

Grosz, Barbara J. and Candace L. Sidner: "Attention, intentions and the structure of discourse." *Computational Linguistics*, 12(3):175-205, 1986.

Kay M., "Unification Grammars," *Xerox publication*, 1981.

Lancel, Jean-Marie, Miyo Otani and Nathalie Simonin, "Sentence parsing and generation with a semantic dictionary and a lexicon–grammar." *12th Int. Conf. on Computational Linguistics (Coling)*, Budapest, Hungary, August 22–27, 1988.

Litman, D.J. and J.F. Allen: "A Plan Recognition Model for Subdialogues in Conversations." *Technical Report TR 141, Computer Science Dpt.*, University of Rochester, November 84.

Legislation as Logic Programs

Robert A. Kowalski

Department of Computing
Imperial College of Science, Technology and Medicine
London SW7 2BZ, UK

Abstract. The linguistic style in which legislation is normally written has many similarities with the language of logic programming. However, examples of legal language taken from the British Nationality Act 1981, the University of Michigan lease termination clause, and the London Underground emergency notice suggest several ways in which the basic model of logic programming could usefully be extended. These extensions include the introduction of types, relative clauses, both ordinary negation and negation by failure, integrity constraints, metalevel reasoning and procedural notation.

In addition to the resemblance between legislation and programs, the law has other important similarities with computing. It needs for example to validate legislation against social and political specifications, and it needs to organise, develop, maintain and reuse large and complex bodies of legal codes and procedures. Such parallels between computing and law suggest that it might be possible to transfer useful results and techniques in both directions between these different fields. One possibility explored in this paper is that the linguistic structures of an appropriately extended logic programming language might indicate ways in which the language of legislation itself could be made simpler and clearer.

1   Introduction

The characteristic feature of the language of legislation is that it uses natural language to express general rules, in order to regulate human affairs. To be effective for this purpose, it needs to be more precise than ordinary language and, as much as possible, it needs to be understood by different people in the same way. In this respect legislation can be viewed as programs expressed in human language to be executed by humans rather than by computers.

Thus the language of legislation might also serve as a model for computing, suggesting ways in which programming languages might be made more like human languages, while still remaining machine executable. In this paper I shall focus on a comparison between the language of legislation and the language of logic programming. I shall argue that, although logic programming fares well in this comparison, it needs to be improved by incorporating such extensions as types, relative clauses, both ordinary negation and negation by failure, integrity constraints, metalevel reasoning, and procedural notation. I shall also argue that in some cases legislation itself can be improved by re-expressing it in a style more closely resembling such an extended logic programming form.

I shall investigate three examples. The first consists of several sections from the British Nationality Act 1981; the second is the University of Michigan lease termination clause; and the third is the London underground emergency notice. The first example was investigated earlier by the author and his colleagues [24] as an illustration of the use of logic programming for representing legislation. The second was investigated by Allen and Saxon [1] as an example of the use of logic to eliminate ambiguities in the formulation of a legal contract. The third was identified by the author [13] as an example of a public notice which is meant not only to be precise but also to be as clear as possible to ordinary people.

In our earlier investigation of the British Nationality Act 1981 [10] we emphasized both the prospects of using logic programming to build legal applications as well as the problems of attempting to use logic programming for knowledge representation. In this paper I am concerned only with the second of these matters, but more specifically with investigating linguistic similarities and differences between logic programming and legislation, and more generally with exploring other parallels between computing and the law.

2 The British Nationality Act 1981

The following four examples from the British Nationality Act illustrate some of the complexity and precision of legal language. They also illustrate the treatment of time, default reasoning, negative conclusions and reasoning about belief.

2.1 Acquisition by Birth

The first subsection of the British Nationality Act deals with the case of acquisition of citizenship by virtue of birth in the United Kingdom after commencement (1 January 1983, the date on which the Act took affect).

> 1.-(1) A person born in the United Kingdom after commencement shall be a British citizen if at the time of the birth his father or mother is -
> (a)   a British citizen; or
> (b)   settled in the United Kingdom.

The English of this clause is already close to logic programming form, even to the extent of expressing the conclusion before (most of) the conditions. Using infix notation for predicates and upper case letters for variables, 1.1 can be paraphrased in logic programming form by:

> X acquires british citizenship by section 1.1
> > if   X is born in the uk at T
> > and T is after commencement
> > and Y is parent if X
> > and Y is a british citizen at T or
> > > Y is settled in the uk at T

This has the propositional form

A if [B and C and D and [E or F]]

which is equivalent to two rules

A if B and C and D and E
A if B and C and D and F

in normal logic programming form.

In this paper I shall use the term **logic program** to refer to any set of sentences which are equivalent to a set of universally quantified implications in the **normal logic programming form**

A if B_1 and ... and B_n

where A is an atomic formula, B_i for $0 \leq i \leq n$ is an atomic formula or the negation of an atomic formula, and all variables, e.g. X_1, ..., X_m occurring in the implication are assumed to be universally quantified, i.e.

for all X_1, ..., X_m [A if B_1 and ... and B_n].

The logic programming representation of 1.1 can be made more like the English, while remaining formal, by introducing types and various forms of syntactic sugar. For example:

a person who is born in the uk at a time
          which is after commencement
          acquires british citizenship by section 1.1
          if a parent of the person is a british citizen at the time,
          or a parent of the person is settled in the uk at the time.

Here "person" and "time" are type identifiers; "a person" is the first occurrence of a variable of type "person"; "a time" is the first occurrence of a variable of type "time"; "the person" and "the time" stand for later occurrences of the same variables. The relative pronouns "who" and "which" also stand for additional occurrences of the variables they follow. "who" stands for an occurrence of type "person", whereas "which" stands for an occurrence of any type of variable. Relative clauses in expressions of the form

... V which P ***

for example, are syntactic sugar for

... V *** if V P

where "V" is a variable, and "P" is a predicate which applies to "V".

Similarly an expression of the form

    ... a R of T P ∗∗∗

is syntactic sugar for

    ... V P ∗∗∗ if V R of T

where "R of" is a binary predicate, "T" is a term, and "V" is any variable not occurring elsewhere in the sentence.

Notice that the two transformations described above need to be combined with the simplication of formulae of the form

    (A if B) if C

to the form

    A if B and C

This kind of typing and syntactic sugar can be defined more precisely and can be extended to deal with several variables of the same type, pronouns, and more flexible kinds of relative clauses. In this way English can serve as a model to improve the naturalness of logic programming without sacrificing its precision

I shall argue elsewhere in this paper that, conversely, the use of conclusion-conditions form, which characterises the syntax of logic programming, can sometimes improve the clarity of natural languages such as English.

2.2 Representation of Time

In the representation of 1.1 time has been represented by an explicit parameter of type "time". The expression

    ... after ∗∗∗

is interpreted as short-hand for

    ... at a time which is after ∗∗∗
i.e.

    ... at T if T is after ∗∗∗.

This explicit representation of time contrasts with modal representations, where temporal relationships are represented by modal operators, and time itself is implicit rather than explicit.

As mentioned for example in [11], to reason about time, an explicit axiom of persistence can be formulated to express that

> a property holds at a time which is after another time
>> if   an event occurs at the other time
>> and the event initiates the property
>> and it is not the case that
>>> another event occurs at yet another time
>>> which is between the time and the other time
>>> and the other event terminates the property.

> "a person acquires british citizenship by section 1.1" initiates
> "the person is a british citizen".

Perhaps this is an example where symbolic notation with explicit representation of variables might be easier for some people to follow. Here "a time", "another time", and "yet another time" introduce different variables of the same type "time". Notice that the English suggests that the variables refer to distinct individuals, whereas the usual logical convention is that different variables of the same type can refer to the same individual. This is one of several discrepancies which would need to be attended to in a more systematic study of the correspondence between logic and a precise style of English.

Notice also in the two axioms above how events and properties are treated metalogically as names of sentences.

2.3   Abandoned Children and Default Reasoning

The second subsection of the British Nationality Act is conceptually one of the most complex sentences of the Act.

> 1.-(2) A new-born infant who, after commencement, is found
>   abandoned in the United Kingdom shall, unless the
>   contrary is shown, be deemed for the purposes of
>   subsection (1)-
>> (a)   to have been born in the United Kingdom after
>>   commencement; and
>> (b)   to have been born to a parent who at the time of
>>   the birth was a British citizen or settled in the United
>>   Kingdom.

Under the procedural interpretation of logic programs, conclusions of sentences are interpreted as goals and conditions as subgoals. According to this interpretation, the conclusion of a sentence identifies its purpose. Thus we can interpret the phrase "the purposes of subsection (1)" as a metalevel reference to the logical conclusion of 1.1, namely to acquire British citizenship. Moreover the object level phrases 1.2.a and 1.2.b are exactly the logical conditions of 1.1. Thus we can regard the entire sentence 1.2 as a mixed object level and metalevel sentence which expresses that

> the conditions of 1.1 shall be assumed to hold for a person
>> if   the person is found newborn abandoned in the uk
>>   at a time which is after commencement.
>> and the contrary of the conditions of 1.1 are not shown

This can be reformulated at the object level alone by replacing the metalevel descriptions by their object level counterparts:

```
a person   who is found newborn abandoned in the uk at a time
           which is after commencement
           acquires british citizenship by section 1.2
           if    it is not shown that it is not the case that
                 the person is born in the uk at a time
                 which is after commencement
           and either it is not shown that it is not the case that
                      a parent of the person is a british citizen at the time of birth
               or     it is not shown that it is not the case that
                      a parent of the person is settled in the uk
                      at the time of birth
```

This seems to be a case where the mixed object-level meta-level expression may be easier to understand than the purely object level representation.

Conditions of the form

> it is not shown that it is not the case that P

in the object level sentence above, can be interpreted as combining negation as failure "not" and ordinary negation "$\neg$", i.e.

> not $\neg P$.

Thus, for another example, the statements

> A bird flies if it is not shown that it is not the case that the bird flies.
> It is not the case that an ostrich flies.

can be formalised by

> a bird flies if not $\neg$ the bird flies
> $\neg$ an ostrich flies

Just such an extension of logic programming to include both negation by failure and ordinary negation has been investigated by Gelfond and Lifschitz [8] and by Kowalski and Sadri [14].

Negation by failure is a form of default reasoning and is non-monotonic. Thus a person who acquires citizenship by 1.2 might non-monotonically have citizenship withdrawn in the light of new information. It is unlikely, however, that parliament intended that citizenship be withdrawn in this way. Both such an intention and the opposite intention can be catered for by introducing an extra layer of time concerned with the time for which beliefs are held in addition to the historical time for which properties hold true in the world. A logic programming approach to such a joint representation of belief time and historical time has been developed by Sripada [25].

It is important to emphasize that when formalising legislation as (extended) logic programs we do not attempt to define concepts which occur in conditions of the legislation but are not defined in the legislation itself. Thus, for example, we do not attempt to define the concept "new born infant" which occurs in the conditions of 1.2. This means, as a consequence, that a formalisation of the British Nationality Act has only limited applicability by itself. To be used in a particular case it would need to be supplemented, if not by a set of definitions of such vague terms, at least by a set of facts or assumptions which express judgements about whether or not such terms apply to the case in hand.

2.4 Deprivation of Citizenship and Negative Conclusions

Except for its occurrence in conditions of the form

$$\text{not } \neg P$$

ordinary negation $\neg$ seems to be needed only in the conclusions of rules. In such cases, a negative conclusion typically expresses an exception to a general rule, as in the example

It is not the case that an ostrich flies.

which expresses an exception to the general rule that all birds fly.

Exceptions, expressed by sentences with negative conclusions, are common in legislation [12]. The provisions for depriving British citizens of their citizenship exemplify this use of negation:

40.-(1) Subject to the provisions of this section, the Secretary of State may by order deprive any British citizen to whom this subsection applies of his British citizenship if the Secretary of State is satisfied that the registration or certificate of naturalisation by virtue of which he is such a citizen was obtained by means of fraud, false representation or the concealment of any material fact.

40.-(5) The Secretary of State -
(a) shall not deprive a person of British citizenship under this section unless he is satisfied that it is not conducive to the public good that that person should continue to be a British citizen; ...

40.1 has the logical form

$$P \text{ if } Q$$

whereas 40.5 has the form

$$\neg P \text{ if not } R$$

If both conditions Q and not R hold, then by ordinary logic it would be possible to deduce a contradiction

P and ¬P.

But this is not the intention of the legislation, which is rather that the exception should override the rule, or equivalently that the rule should be understood as having an extra, implicit condition.

P if Q and not ¬P.

In fact, the metalevel phrase "subject to the provisions of this section" at the beginning of 40.1 can be regarded as a caution that the meaning of 40.1 cannot be understood in isolation of the rest of the section as a whole.

The extension of logic programming to allow negative conclusions, for the purpose of representing exceptions, has been investigated by Kowalski and Sadri [14]. They also show that such extended logic programs can be transformed into normal logic programs. In particular a rule with a single exception

P if Q
¬P if not R

can be transformed into the simpler rule

P if Q and R.

Both representations can be useful for different purposes. A representation in terms of rules and exceptions is often easier to develop and to maintain. However, the simpler representation as normal logic programs is usually clearer and easier to understand. The first representation, accordingly, might be preferred by a draftsman, who codifies the law; the second might be preferred by an administrator who executes the law.

In this discussion of the provisions for deprivation of citizenship we have considered only the propositional structure of the English sentences. We have not considered the meaning of such conditions as

"he is satisfied that it is not conducive
to the public good that that person
should continue to be a British citizen".

This is partly because it would be very difficult to do so; but also because we have restricted our attention to representing formally only what is defined explicitly in the legislation itself. Nonetheless, reasoning about reasoning can, at least to some extent, be formalised by metalogic or by logics of knowledge and belief.

2.5 Naturalisation and the Representation of Belief

Like the provisions for deprivation of citizenship, the provisions for naturalisation contain conditions concerning the Secretary of State's beliefs. In addition, however, they also contain rules governing the subject matter of those beliefs. This leads us to consider whether we can establish a logical connection between the two.

Section 6.1 contains the main provision for naturalisation:

> 6.-(1) If, on an application for naturalisation as a British citizen made by a person of full age and capacity, the Secretary of State is satisfied that the applicant fulfills the requirements of Schedule 1 for naturalisation as such a citizen under this sub-section, he may, if he thinks fit, grant to him a certificate of naturalisation as such a citizen.

At the propositional level this is equivalent to a sentence in conclusion-conditions form:

> the secretary of state may grant a certificate of
> naturalisation to a person by section 6.1
> > if   the person applies for naturalisation
> > and   the person is of full age and capacity
> > and   the secretary of state is satisfied that
> >   the person fulfills the requirements of
> >   schedule 1 for naturalisation by 6.1
> > and   the secretary of state thinks fit
> >   to grant the person a certificate of naturalisation.

The last two conditions vest considerable powers of discretion in the Secretary of State. The last condition is totally inscrutable and can only be given as part of the input for a given case. But the meaning of the penultimate condition ought at least to be constrained by the meaning of Schedule 1. This schedule is quite long, and it is convenient therefore to summarise its contents:

> a person fulfills the requirements of
> schedule 1 for naturalisation by 6.1
> > if   either the person fulfills residency
> >   requirements specified in subparagraph 1.1.2
> >   or the person fulfills crown service
> >   requirements specified in subparagraph 1.1.3
> > and the person is of good character
> > and the person has sufficient knowledge of
> >   english, welsh, or scottish gaelic
> > and either the person intends to reside in the uk
> >   in the event of being granted naturalisation
> >   or the person intends to enter or continue in crown service
> >   in the event of being granted naturalisation.

To understand the connection between 6.1 and Schedule 1, it is necessary to understand the connection between meeting the requirements for naturalisation specified in Schedule 1 and satisfying the Secretary of State that those requirements are met. Fortunately, this can be done, at least in part, by regarding satisfaction as a kind of belief. The appropriate rules of belief can be formalised in both modal logic and metalogic. The following formalisation in metalogic has the form of a metainterpreter.

> a person is satisfied that P
>> if   the person is satisfied that P ← Q
>> and the person is satisfied that Q

> a person is satisfied that P ∧ Q
>> if   the person is satisfied that P
>> and the person is satisfied that Q

> a person is satisfied that P ∨ Q
>> if   the person is satisfied that P
>> or   the person is satisfied that Q

Here "←", "∧", and "∨" are infix function symbols naming implication, conjunction, and disjunction respectively.

We may safely assume that

> the secretary of state is satisfied that P
>> if P is a representation of the meaning
>>> of a provision of the british nationality act 1981

Thus the Secretary of State is satisfied in particular that the implication which represents the meaning of Schedule 1 holds. This assumption and the metainterpreters above are all we need to establish a logical connection between 6.1 and Schedule 1. This connection can be made more explicit, however, if we transform the metainterpreter using the technique of partial evaluation [7, 26]:

> the secretary of state is satisfied that a person fulfills
> the requirements for naturalisation by 6.1
> if   either the secretary of state is satisfied that
>> the person fulfills residency requirements specified in paragraph 1.1.2
> or the secretary of state is satisfied that
>> the person fulfills crown service requirements specified in paragraph 1.1.3
> and the secretary of state is satisfied that
>> the person is of good character
> and the secretary of state is satisfied that
>> the person has sufficient knowledge of english, welsh, or scottish gaelic
> and either the secretary of state is satisfied that
>> the person intends to reside in the uk in
>> the event of being granted naturalisation
> or the secretary of state is satisfied that
>> the person intends to enter or continue in
>> crown service in the event of being granted naturalisation.

The result is an explicit, though somewhat tedious, statement of what it means to satisfy the Secretary of State concerning the requirements for naturalisation. Clearly the statement could be made a little less tedious if we used a pronoun, "he" or "she" for all references to the Secretary of State after the first.

The language of the British Nationality Act 1981 is for the most part extraordinarily precise. It is also very complex. Most of this complexity is inherent in the meaning of the Act. However, some of the complexity can be reduced by the explicit use of conclusion-conditions form and by the use of meaning-preserving transformations of the kind illustrated in the last two examples.

By comparison with ordinary language and even with legal language in general, the Act is also surprisingly unambiguous. However, as we have already seen, it does contain vague terms and undefined concepts. Such vagueness is often confused with ambiguity. Although, like genuine ambiguity, vagueness causes problems of interpretation, it is also useful, because it allows the law to evolve and adapt to changing circumstances.

Genuine ambiguity, on the other hand, generally serves no useful purpose. Moreover, whereas logic can easily accommodate vagueness, it cannot tolerate ambiguity.

The University of Michigan lease termination clause, presented in the next section, was originally investigated by Allen and Saxon [1] to illustrate the use of propositional logic to formulate a precise interpretation of an ambiguous legal text. I shall argue that the use of logic programming conclusion-conditions form has the further advantage of rendering many of the possible interpretations logically implausible.

3 The University of Michigan Lease Termination Clause

The clause consists of a single, long sentence which has the underlying, logically ambiguous form

> A if A1 and A2 or A3 or A4 or A5 or A6 or A7
> unless B1 or B2 or B3 or B4 or B5 in which cases B.

Different ways of introducing parentheses produce different interpretations. Some of these are logically equivalent because of the associativity of "or", for example. After accounting for these equivalences, Allen and Saxon identify approximately 80 questions that might need to be asked in order to distinguish between the different parenthesizations. As a result of this analysis they identify one intended interpretation which has the form

> ((A if (A1 and(A2 or A3)) or A4 or A5 or A6 or A7)
> if not (B1 or B2 or B3 or B4 or B5)) and
> (if (B1 or B2 or B3 or B4 or B5) then B)

where "unless" has been translated as "if not". It is interesting that this interpretation has a logic programming form.

The logic programming representation can be simplified if, as Allen and Saxon maintain, conditions B1-B5 are the only ones under which conclusion B holds. In that case the conditions not(B1 or B2 or B3 or B4 or B5) can be replaced by not B. Thus the intended interpretation can be represented by the simplified, normal logic program:

A if A1 and A2 and not B
A if A1 and A3 and not B
A if A4 and not B
A if A5 and not B
A if A6 and not B
A if A7 and not B
B if B1
B if B2
B if B3
B if B4
B if B5

This logical analysis of the propositional structure of the sentence should be compared with the English text of the sentence:

"The University may terminate this lease when the Lessee, having made application and executed this lease in advance of enrollment, is not eligible to enroll or fails to enroll in the University or leaves the University at any time prior to the expiration of this lease, or for violation of any provisions of this lease, or for violation of any University regulation relative to Resident Halls, or for health reasons, by providing the student with written notice of this termination 30 days prior to the effective time of termination; unless life, limb, or property would be jeapordized, the Lessee engages in the sales or purchase of controlled substances in violation of federal, state or local law, or the Lessee is no longer enrolled as a student, or the Lessee engages in the use or possession of firearms, explosives, inflammable liquids, fireworks, or other dangerous weapons within the building, or turns in a false alarm, in which cases a maximum of 24 hours notice would be sufficient".

Notice how the conclusion A of the first half of the sentence is split into two parts by the insertion of the conditions A1-A7. Notice also that the language of the sentence is so complicated and so confused that the drafters mistakenly wrote "maximum of 24 hours" when they must have meant "minimum of 24 hours".

In fact I have slightly misrepresented Allen and Saxon's analysis of the sentence. In addition to identifying the intended placement of parentheses, they analyse for each of the three occurrences of "if" in the apparent meaning of the sentence whether or not "if and only if" is really intended. They conclude that in the first two cases (of the words "when" and "unless") it is not intended, whereas, in the third case (of the words "in which cases") it is. Thus their real analysis of the intended interpretation has the form

((A if (A1 and (A2 or A3)) or A4 or A5 or A6 or A7)
if not (B1 or B2 or B3 or B4 or B5)) and
(if (B1 or B2 or B3 or B4 or B5) then B) and
(if not (B1 or B2 or B3 or B4 or B5) then not B).

In contrast, with this change of representation using ordinary logic, the logic programming representation is not affected by this change of interpretation. In the logic program there is no difference between the representation of "if" and the representation of "if and only if". The difference between the two interpretations depends upon whether or not the "closed world assumption" [6] is applied. The closed world assumption for a predicate P is the assumption that all the implications

P if Q1
P if Q2
:
P if Qn

with conclusion P in a program represent all the conditions under which conclusion P holds. It is this assumption that justifies the negation as failure rule:

not P holds if P fails to hold, i.e.
not P holds if all ways of trying to show P result in failure.

Thus, in the example of the lease termination clause, in the case of the word "when", the interpretation "if and only if" is not intended because there are other situations referred to elsewhere in the lease under which the University may terminate the lease with 30 days written notice. But in the case of the words "in which case", the interpretation "if and only if" is intended because there are no other cases under which the University may terminate the lease with 24 hours notice. In the case of the word "unless", the question is not relevant because in the context in which it occurs the closed world assumption is not applicable.

Allen and Saxon argue that the logical representation of the lease termination clause does not express what the drafters must have actually intended. After all the ambiguities have been resolved, the English text expresses that for the University to be able to terminate the lease with 30 days written notice, not only must one of the conditions

(A1 and (A2 or A3)) or A4 or A5 or A6 or A7

hold but none of the conditions

B1 or B2 or B3 or B4 or B5,

under which it may terminate the lease with 24 hours notice, may hold. But these extra negative conditions play no useful role. They serve only to make the conditions under which conclusion holds exclusive of the conditions under which conclusion B holds.

The simpler rules

A if ((A1 and (A2 or A3) or A4 or A5 or A6 or A7)
B if (B1 or B2 or B3 or B4 or B5)

are more flexible. Compared with the original rules they give the university the extra option of giving students 30 days notice in cases where they would otherwise be forced to give 24 hour notice.

Using indentation, and the expressions "both ... and", and "either ... or" in place of parentheses, this new interpretation can be written in a form which arguably has both the precision and simplicity of logic programming and the naturalness of English:

> The university may terminate this lease by providing the lessee with written notice of the termination 30 days prior to the effective time of termination
> > if   both the lessee has applied for and executed
> > > this lease in advance of enrollment
> > > and   either the lessee is not eligible to enroll
> > > or the lessee fails to enroll
> > or   the lessee leaves the university at any
> > > time prior to the expiration of this lease
> > or   the lessee violates any provisions of this lease
> > or   the lessee violates any university regulations
> > > relative to residence halls
> > or   there are health reasons for terminating this lease.

> The university may terminate this lease by providing the lessee with notice of the termination a minimum of 24 hours prior to the effective time of termination
> > if   life, limb or property would be jeopardized by
> > > continuation of the lease
> > or   the lessee engages in the sale or purchase of
> > > controlled substances in violation of federal, state or local law
> > or   the lessee is no longer enrolled as a student
> > or   the lessee engages in the use or possession of
> > > firearms, explosives, inflammable liquids,
> > > fireworks, or other dangerous weapons within the building
> > or   the lessee turns in a false fire alarm.

The University of Michigan lease termination clause is not a good illustration of our thesis that legal language can be a good guide for improving computer languages. If anything, it seems to suggest the converse, that some computer languages might be a useful guide for improving the language of the law.

In fact, few legal documents are written to the standards of precision found in the acts of parliament; and hardly any legal documents at all are written not only to be precise but also to be clear and easy to understand. However, public notices, which are meant to be understood by ordinary people, are for the most part an important exception to this rule. The London underground emergency notice is a good example of such an exception.

4   The London Underground Emergency Notice

The notice has many characteristics of a logic program, but with some interesting differences:

EMERGENCIES

Press the alarm signal button
to alert the driver.

The driver will stop immediately
if any part of the train is in a station.

If not, the train will continue to the next station,
where help can more easily be given.

There is a £50 penalty
for improper use.

From a knowledge representation point of view, the first sentence is probably the most interesting. Expressed in a procedural style, it shows that a procedural form of expression can sometimes be more appropriate than an "equivalent" statement in declarative style:

You alert the driver
if You press the alarm signal button.

Notice, however, that the procedural form can be regarded as a compiled version of the procedural interpretation of the declarative form. Like most compiled representations of knowledge, it requires less work on the part of the recipient to put the knowledge into effect.

This example and others like it suggest that logic programming could be made more like natural language if it allowed both declarative and procedural syntax. Under the procedural interpretation of logic programming, both the declarative syntax

A if B and C

and the procedural syntax

to do A do B and do C

would be equivalent. In fact both styles of expression would have the same declarative meaning

A if B and C

and the same procedural meaning

to do A do B and do C.

A procedural syntax for logic programs would not, however, include arbitrary imperative programming language constructs. It would not, for example, without further extension, include such purely imperative statements as

> press the alarm signal button.

All imperative statements in a logic programming language would have to be imbedded in a procedure, which contains an expression of its purpose. I shall discuss the possible extension of logic programs to include purposeless procedures, viewed as integrity constraints, in sections 5.1 and 5.2.

To simplify the discussion of the emergency notice, I have ignored and, for the most part, will continue to ignore the temporal relationships between the different actions and situations referred to in the notice. We should note however, that to be accurate the title of the notice should be incorporated into the conclusion of the sentence:

> press the alarm signal button,
> to alert the driver to an emergency.

The second sentence of the notice is explicitly expressed in a logic programming form. However, even allowing for the fact that the phrase

> the driver will stop immediately

is shorthand for

> the driver will stop the train immediately,

the sentence fails to express its intended meaning, because it is missing an entire condition. The meaning of the sentence can be made explicit, by supplying the missing condition from the conclusion of the previous sentence:

> the driver will stop the train immediately
> if You alert the driver to an emergency
> and any part of the train is in a station.

Certainly this precise expression of the meaning of the sentence is more cumbersome that the English. However, it is hard to see how the logic programming representation could be simplified so that it more closely resembles the English, without loosing its precision.

The third sentence begins with an allusion to the explicitly stated condition of the previous sentence. Ignoring for the moment, the comment at the end, the sentence with all its conditions made fully explicit has the logical form:

> the train will continue to the next station
> if You alert the driver to an emergency
> and not any part of the train is in a station.

But this alone cannot be all that it is intended by the English, because the train will generally continue to the next station whether or not the driver is alerted to an emergency. Surely, what is meant is that the train will stop at the next station and that help will be given there. This is part of the meaning of the phrase

> where help can more easily be given.

Moreover, presumably help will be given at a station whether it is the next station or not. Thus we can obtain a better approximation to the intended meaning of the third sentence with the two sentences:

> the train will stop at the next station
> if You alert the driver to an emergency
> and not any part of the train is in a station.

> help will be given in an emergency
> if You alert the driver to the emergency
> and the train is stopped in a station.

This second sentence of the revised formulation of the sentence captures part of the meaning of the comment at the end of the sentence. Presumably the rest of its meaning could be expressed by the meta statement that this procedure for getting help is better than the alternative procedure of stopping the train when it is not in a station.

The last sentence of the notice has a simple formulation in conclusion-conditions form:

> there is a £50 penalty
> if You use the alarm signal button improperly.

This contrasts with a purely imperative statement, which expresses a prohibition without expressing a purpose:

> do not use the alarm signal button improperly.

In contrast with the purely imperative statement of prohibition, the procedural interpretation of the English sentence contains a clear expression of purpose:

> if You want a £50 penalty,
> then press the alarm signal button improperly!

Notice, by the way, how different the procedural syntax of a sentence can be from its declarative meaning. The English procedural sentence

> if You want A, then do B

actually has the underlying declarative meaning

> A if B.

Although the English of the London underground notice can be improved, it is undoubtably clear and easy to understand. I believe its clarity is due to at least three characteristics

- the explicit use of conclusion-conditions form
- the appropriate use of procedural form, and
- the use of ellipsis to avoid unnecessarily stating the obvious.

The first two characteristics can usefully be applied to the design and improvement of computer languages today. The third characteristic is harder to achieve, although some progress along these lines might be possible in the future.

5 Other Computing Paradigms

The preceding examples illustrate some of the typical characteristics of legal language and its relationship to logic programming form. It is also possible, however, to find indications of other computing paradigms.

5.1 Condition-Action Production Rules

Condition-action rules were developed by Newell and Simon [19] as a model of human psychology and have been used to implement expert systems [27]. They can also be found in the language of public notices. For example, the following notice is displayed in the carriages of the London underground

Please give up this seat
if an elderly or handicapped person needs it

This is a distinct improvement over the earlier, ambiguous, and potentially disturbing notice

please give up this seat
to an elderly or handicapped person.

But even with the explicit use of the word "if", the sentence falls short of logic programming form, because the apparent conclusion

please give up this seat

is imperative rather than declarative. Moreover the sentence does not express a purpose.

The condition-action form in which the rule is expressed can be converted into logic programming form by making the purpose, e.g.

to do a good deed

explicit rather than implicit. The resulting statement can be expressed procedurally

to do a good deed
give up this seat
if an elderly or handicapped person needs it

or declaratively

You do a good deed
if You give up Your seat to a person
who needs Your seat and
who is elderly or handicapped.

The claim that every command has an explicit or implicit purpose is an important theory in legal philosophy. The use of logic programming form, which forces purposes to be made explicit, is in the spirit of this theory. Associating explicit purposes with commands makes it possible to reason about the relative merits of conflicting commands and even to reason whether a command is appropriate in a given context.

Nonetheless, natural language does allow the expression of commands without purpose, and there even seems to be a logic programming analogue of this in the form of integrity constraints.

5.3 Integrity Constraints

For many years the London underground displayed the following notice above the automatic doors of its carriages

Obstructing the doors causes
delay and can be dangerous.

In other words

there will be a delay
if You obstruct the doors.

there can be danger
if You obstruct the doors.

As long as delay and danger are regarded as undesirable, a thinking person will conclude that obstructing the doors is undesirable too.

But the London underground authorities have recently changed the wording of the notice on some of its trains. The new sign reads

Do not obstruct the doors.

A sad reflection of our changing times. Either delay and danger are no longer regarded as undesirable, or the public cannot be relied upon to reason about the consequences of its behaviour.

But for a logic programmer the new notice is worrying, not only because it indicates the possibly deteriorating state of British underground society, but also because it represents a move away from a logic programming style of communication to a more imperative style. But on closer consideration, the change of wording is reminiscent of recent efforts to extend logic programming by the inclusion of integrity constraints.

This extension is motivated by database applications of logic programming. For these applications, a number of studies [5, 17, 21, 22] have investigated the nature of integrity constraints in logic programming and the development of efficient integrity checking methods. In all of these approaches integrity constraints are viewed as properties which a database or program must satisfy as it changes over the course of time. To the extent that the contents of a database describe states of affairs in the world, commands, which impose obligations or prohibitions on states of the world, can be interpreted as integrity constraints on states of the database.

An integrity constraint can be expressed in the form of any sentence of first-order logic including a denial. Thus the command

> do not obstruct the doors

might be represented by a denial

> ¬You obstruct the doors

which expresses an integrity constraint on descriptions of events which take place in the world.

Similarly the condition-action rule

> please give up this seat
> if an elderly or handicapped person needs it

might be interpreted as an integrity constraint which has the form of an implication

> You give up a seat to a person
> if You are sitting in the seat
> and the person needs Your seat
> and the person is elderly or handicapped.

Thus, given a database that records events that take place in the world, the integrity of the database will be violated if the database records that a person is sitting in a seat which an elderly or handicapped person needs and the database does not contain a record of that person giving up the seat to the elderly or handicapped person. It is another problem, if integrity has been violated, to decide how integrity should be restored. Perhaps this is where "purpose" or "sanctions" might play a useful role.

Thus commands without purpose seem to be compatible with logic programs extended by the inclusion of integrity constraints. Moreover, there is even a transformation between integrity constraints and logic program rules, which is analogous to a transformation between commands without purpose and procedures with purpose:

Given an integrity constraint expressed as a first-order sentence

   C

introduce a new predicate S and convert the constraint to the rule

   S if not C

together with the new constraint

   ¬S.

The new predicate S can be interpreted as a "sanction" which applies if the original constraint is violated. This transformation has been used in the literature on integrity constraints in deductive databases to convert arbitrary first-order integrity constraints into denial form.

The analogy between this transformation and the legal doctrine of sanctions suggest the possibility of adapting legal techniques for dealing with violations of commands to the problem of restoring integrity in deductive databases. This is an intriguing possibility that merits closer investigation.

5.3 Object-Oriented Programming

The paradigm of object-oriented programming has become increasingly important in computing in recent years. It is interesting to investigate, therefore, to what extent it has analogues in natural language and in legislative language more particularly.

We have already seen some characteristics of object-orientation in English when we saw the use of common nouns such as "person", "time" and "lessee" as a kind of object-oriented typing of variables. Other manifestations of object-orientation seem to be more difficult to find in the actual language of legislation, but easier to find both in descriptions of individual cases and in the organisation of law as a whole.

In natural language descriptions, it is common to group sentences together around a single topic placed at the beginning of each of the sentences. Such topics help to organise communication similar to the way in which objects can be used to organise knowledge in computing.

Compare, for example, the pair of sentences

   The Prime Minister stepped out of the plane.
   Journalists immediately surrounded her.

with the pair

> The Prime Minister stepped out of the plane
> She was immediately surrounded by journalists.

Psycho-linguists have found that the second pair of sentences is easier to understand than the first, despite the fact that the second pair uses the passive rather than the active voice. The two sentences in the more comprehensible pair have the same topic, whereas the two sentences in the other pair have different topics. Such examples suggest that organising knowledge around objects makes the knowledge more coherent and easier for humans to understand.

In the domain of law, it is common to organise the different areas of law into hierarchies, which are similar to hierarchies of objects. Thus a country might have one statute governing criminal law in general, another statute covering behaviour in public places, and yet another dealing with behaviour in public buildings. Assault and battery, for example, might be prohibited everywhere, whether in public places or not. Going about naked, however, might be prohibited only in public places, but be allowed in the privacy of one's own home. Smoking, on the other hand, might be prohibited only in public buildings but be allowed everywhere else.

Thus natural language seems to support two notions of objects: objects in the small, which are used like types and topics to organise descriptions of individuals; and objects in the large, which are used in hierarchies to organise whole areas of knowledge. From this point of view, logic programming and object-orientation correspond to different aspects of natural language and are complementary.

However, the notion of object in computing has other characteristics, such as change of state, which do not have such obvious counterparts in natural language. These characteristics seem to be more closely associated with simulating the behaviour of objects in the world than with describing their behaviour.

There have been several attempts to apply object-orientation to legal reasoning. Some of these, like Gordon's Oblog [9], are based on a view of objects as types and topics, which is entirely compatible both with logic programming and with the representation of natural language meanings. Others, like the treatment of patent law by Nitta et al [20] are based on the use of objects to simulate behaviour.

The use of objects for simulation in the patent law example is especially interesting because of the way in which patent procedures, obligations and prohibitions are used to generate and filter changing states of the simulation of a patent application. It seems possible that, if the changing states of the simulation are viewed as database states, then the obligations and prohibitions expressed in the patent law might be viewed as integrity constraints. This possibility would establish an interesting link between imperative statements in object-oriented programming and integrity constraints in deductive databases and logic programming.

No matter what the outcome of a more detailed investigation of these possibilities, there can be little doubt that legislation provides a rich domain outside computing science itself within which relationships between different computing paradigms can be studied. These studies need not be confined to programming languages alone, but could usefully be extended to many other aspects of computing.

6 Other Relationships Between Computing and Law

To the extent that we can truly regard legislation as programs to be executed by people, we can also expect to find analogues in the law of such other computing matters as program specification and software management.

6.1 An Analogy Between Specifications and Policies

In the same way that programs are written to meet specifications, laws are drafted to achieve policies, which are social or political objectives. The purpose of the British Nationality Act 1981, for example, was "to make fresh provisions about citizenship and nationality, and to amend the Immigration Act 1971 as regards the right of abode in the United Kingdom", and in particular to restrict immigration to the United Kingdom by residents of the former British colonies. The purposes of the University of Michigan lease termination clause presumably include such goals as discouraging unsociable behaviour in the halls of residence, restricting residency to legitimate students, and not causing undue hardship for individuals who are obliged to terminate their residence. The rules for dealing with London Underground emergencies, on the other hand, are designed to facilitate the provision of help as effectively and quickly as possible in the case of genuine emergencies and to avoid inconvenience and unnecessary trouble in the case of false alarms.

Program specifications have many characteristics in common with the policies of legal documents. In the same way, for example, that the primary obligation of a program might be to meet its specification, the primary duty of a legal document should be to achieve its social and political objectives. In both cases, moreover, .specifications and policies are often ill-defined, inconsistent, or the result of compromise between conflicting demands.

The formal methods developed in computing to verify that programs meet their specifications are much more advanced than any corresponding methods developed for the law. A pilot study of the possibility of adapting formal methods of logic-based software verification to the problem of verifying social security regulations has been made by Bench-Capon [2].

Thus the transfer of techniques for program verification is one area in which the field of law might be able to benefit from its similarities with computing. In other areas, such as software management, the benefits might apply more equally to both fields.

6.2 An Analogy Between Software Maintenance and Maintenance of the Law

In the same way that programs need to be modified to meet changing specifications, legislation needs to be modified to meet changing social and political needs. But programs are both difficult to construct and difficult to change. So much so in fact, that programs are often still in use long after they have outlived their specifications.

The situation is not much better in the law, where legislation often lags far behind social and political changes. Obsolete and incorrect legislation is enforced simply for the sake of "law and order".

But the drafters of legislation have developed some ingenious devices for adapting, modifying and revising old legislation. The liberal use of vague and undefined terms such as "good character", "life, limb or property would be jeopardized" and "improper use" greatly contribute to the flexibility of legislation and to its ability to adapt to change. Such use of vague terms is reminiscent of the use of data abstraction and encapsulation in computer programming, which allow the lower levels of a program to change, while leaving the higher levels intact.

Much legislation is explicitly concerned with the repeal or amendment of earlier legislation. The British Nationality Act 1981, for example, repeals the British Nationality Acts 1948 to 1965 and amends the Immigration Act 1971. Amendments in particular are typically expressed by metalevel statements which describe how an old piece of text should be edited to create a new text. Metalevel statements are also used to create a new provision from a similar provision in the same act.

Section 6.2 of the British Nationality Act 1981, for example, makes special provision for naturalisation of people who are married to British citizens. The requirements are similar to those for people who apply under section 6.1, but include shorter residency requirements, omit the requirement of having sufficient knowledge of English, Welsh, or Scottish Gaelic, and include

"the requirement specified in paragraph 1(1)(b)".

This metalevel reference to 1(1)(b) is in fact a reference to the requirement

"that he is of good character".

This particular use of metalanguage is rather unusual in that the English expression of the metalinguistic form is actually longer than the equivalent object level expression. Usually the metalinguistic formulation is more concise than the object level formulation.

Thus the source code of legislation often mixes object level statements about the domain of discourse with metalevel statements about the text of other legislation or other provisions in the same legislation. The principle objective of using such metalevel statements in preference to equivalent object level statements is to make explicit the relationship between different but similar texts.

The language of legislation also employs remarkable techniques for reusing previous legislation. In the British Nationality Act 1981, for example, it states that one of the conditions for being a British citizen by descent under the 1981 Act is to be a person who

under any provision of the British Nationality Acts 1948 to 1965,
was deemed for the purposes of the proviso to section 5(1) of the 1948 Act
to be a citizen of the United Kingdom and Colonies by descent only,
or would have been so deemed if male.

The last phrase is an example of a counterfactual condition. A metalogical interpretation of such counterfactuals has been proposed by Bench-Capon [3]. It is possible to imagine how metaprogramming might be used to implement such counterfactual reuse of software in a logic programming environment.

6.3 The Relationship Between Case-Based and Rule-Based Reasoning

In artificial intelligence a contrast is sometimes made between case-based and rule-based reasoning, and a conflict is often held to exist between these two kinds of reasoning [23]. People, it is argued, reason by means of analogies between different cases rather than by means of the deductive application of rules.

The distinction between these two kinds of reasoning also lies at the heart of law. To some extent it is even reflected among the distinguishing features of the two main western legal traditions. Common law systems, such as those in England and the United States, place greater emphasis on reasoning by means of cases. Civil law systems, such as those on the continent of Europe, place greater emphasis on reasoning by means of codified rules. In fact, in both systems of law the two kinds of reasoning interact and complement one another.

In rule-based legislation, for example, case-based reasoning plays a fundamental role in determining the meaning of vague concepts. Previous cases of a concept serve as precedents for new cases.

On the other hand, in case-based legal argumentation, the justification for a decision in a precedent setting case is often expressed in general terms and appeals to general principles. Moreover, authorative restatements of case law effectively reformulate the precedents set in individual cases into general, rule-based form, even though such case-based rules do not have the same binding force as rules in legislation. Indeed it can be argued that there is a natural evolution in the law from reasoning by means of cases to reasoning by means of rules.

7. Conclusion

The similarities between computing and the law seem to cover all areas of computing software. Moreover, the linguistic style in which legislation is drafted combines in one language the expressive power of computer languages for such diverse areas as programming, program specification, database description and query, integrity constraints, and knowledge representation in artificial intelligence. This linguistic style might be a good guide therefore to how these different areas of computing might be unified in the future.

The similarities between computing and law go beyond those of linguistic style. They extend also to the problems that the two fields share of developing, maintaining and reusing large and complex bodies of linguistic texts. Here too, it may be possible to transfer useful techniques between the two fields.

In this paper I have concentrated on similarities between logic programming and legislation. I have indicated several ways in which the language of legislation suggests that the basic model of logic programming can usefully be extended, to include types, relative clauses, both ordinary negation and negation by failure, integrity constraints, metalevel reasoning, and procedural notation. I believe that with the aid of such extensions logic programming can provide the foundations for a future, single computer language that will be suitable for all areas of computing in the same way that natural language is suitable for all areas of law.

Acknowledgement

This work was supported initially by the Science and Engineering Research Council and more recently by the ESPRIT Basic Research Action, "Computational Logic". I am especially indebted to my colleagues, Trevor Bench-Capon, Fariba Sadri and Marek Sergot, whose work on legislation and logic programming has provided much of the background for this paper.

References

[1] Allen, L. E., and Saxon, C.S. [1984] "Computer Aided Normalizing and Unpacking: Some Interesting Machine-Processable Transformation of Legal Rules", Computing Power and Legal Reasoning (C. Walter, ed.) West Publishing Company, pp. 495-572.

[2] Bench-Capon, T.J.M. [1987]: "Support for policy makers: formulating legislation with the aid of logical models", Proc. of the First International Conference on AI and Law, ACM Press, pp. 181-189.

[3] Bench-Capon, T. [1989] "Representing Counterfactual Conditionals". Proceedings of Artificial Intelligence and the Simulation of Behaviour (A. Cohn, Ed.) Pitman Publishing Co.

[4] Bowen, K. A. and Kowalski, R. A. [1982]: "Amalgamating Language and Metalanguage in Logic Programming", in Logic Programming (Clark, K.L. and Tärnlund, S.-Å., editors), Academic Press, pp. 153-173.

[5] Bry, F., Decker, H., and Manthey, R. [1988] "A uniform approach to constraint satisfaction and constraint satisfiability in deductive databases", Proceedings of Extending Database Technology, pp. 488-505.

[6] Clark, K. L. [1978]: "negation by failure", in "Logic and databases", Gallaire, H. and Minker, J. [eds], Plenum Press, pp. 293-322.

[7] Gallagher, J. [1986] "Transforming Logic Programs by Specializing Interpreters", Proc. of 7th European Conference on Artificial Intelligence, pp. 109-122.

[8] Gelfond, M. and Lifschitz, V. [1990]: "Logic programs with classical negation", Proceedings of the Seventh International Conference on Logic Programming, MIT Press, pp. 579-597.

[9] Gordon, T. F. [1987] "Oblog-2 a Hybrid Knowledge Representation System for Defeasible Reasoning" Proc. First International Conference on Artificial Intelligence and Law. ACM Press, pp. 231-239.

[10] H.M.S.O. [1981]: "British Nationality Act 1981", Her Majesty's Stationery Office, London.

[11] Kowalski, R. A. and Sergot, M. J. [1986]: "A logic-based calculus of events", New Generation Computing, Vol. 4, No. 1, pp. 67-95.

[12] Kowalski, R. A. [1989]: "The treatment of negation in logic programs for representing legislation", Proceedings of the Second International Conference on Artificial Intelligence and Law, pp. 11-15.

[13] Kowalski [1990] "English as a Logic Programming Language", New Generation Computing, Volume 8, pp. 91-93.

[14] Kowalski, R. A. and Sadri, F. [1990], "Logic programs with exceptions", Proceedings of the Seventh International Conference on Logic Programming, MIT Press, pp. 598-613.

[15] Kowalski, R. A., Sergot, M. J. [1990]: "The use of logical models in legal problem solving", Ratio Juris, Vol. 3, No. 2, pp. 201-218.

[16] Lloyd, J. W. and Topor, R. W. [1984]: "Making Prolog more expressive", Journal of Logic Programming, Vol. 3, No. 1, pp. 225-240.

[17] Lloyd, J. W. and Topor, R. W. [1985] "A Basis for Deductive Database Systems", J. Logic Programming, Volume 2, Number 2, pp. 93-109.

[18] Mitchell, T. M., Keller, R. M. and Kedar-Cabelli [1986] "Explanation-based Generalization: A Unifying View" Machine Learning, Volume 1, pp. 47-80.

[19] Newell, A. and Simon, H. A. [1972] "Human problem solving", Prentice-Hall.

[20] Nitta, K., Nagao, J., and Mizutori, T., [1988] "A Knowledge Representation and Inference System for Procedural Law", New Generation Computing, pp. 319-359.

[21] Reiter, R. [1990]: "On asking what a database knows", Proc. Symposium on Computational Logic, Springer-Verlag.

[22] Sadri, F. and Kowalski, R. A. [1987]: "A theorem proving approach to database integrity", In Foundations of deductive databases and logic programming (J. Minker, editor), Morgan Kaufmann, pp. 313-362.

[23] Schank, R. C. [1983] "The current state of AI: One man's opinion", AI Magazine, Volume 4, No. 1, pp. 1-8.

[24] Sergot, M. J., Sadri, F., Kowalski, R. A., Kriwaczek, F., Hammond, P. and Cory, H. T. [1986]: " The British Nationality Act as a logic program", CACM, Vol. 29, No. 5, pp. 370-386.

[25] Sripada, S. M. [1991] "Temporal Reasoning in Deductive Databases". Department of Computing, Imperial College, London.

[26] Takeuchi, A. and Furukawa, K. [1986] "Partial evaluation of PROLOG programs and its application to metaprogramming", Proc. of IFIP 86, North-Holland, pp. 415-420.

[27] Waterman, D. A. and Hayes-Roth [1978] "Pattern-directed Inference Systems", Academic Press, New York.

Knowledge Representation
for
Natural Language Processing

Udo Pletat

Software Architectures and Technologies
IBM Germany
P. O. Box 80 08 80
D-7000 Stuttgart 80
Germany
email : PLETAT at DS0LILOG.BITNET

Abstract

We give an overview of the typed predicate logic L_{LILOG} which serves as the target language for translating the information provided in German texts into machine processible form. Being part of the natural language understanding system LEU/2, the knowledge representation system built around L_{LILOG} serves different purposes. Its knowledge engineering environment has been used for modeling the semantical backgound knowledge for the application domain of LEU/2. The inference engine implementing L_{LILOG} is a flexible theorem prover for processing the information extracted from natural language texts.

1   Introduction

Our discussion of knowledge representation for natural language processing reports on experiences from the LILOG project (see [25]) on natural language understanding. The objective of the LILOG project was to develop technologies for extracting the semantic contents of German texts in the sense of translating them into a logical representation over which information processing algorithms can compute.

The extraction of information from texts requires a certain background knowledge with respect to which the new information can be acquired. All this knowledge needs to be represented within a natural language understanding system and thus respective representation languages have to be available.

In the LILOG project three major representation aspects have been identified:

* The linguistic background knowledge given in terms of the grammar and the lexicon for analysing the syntactic structure of natural language texts.

The syntax approach of the LEU/2[1] system has been HSPG ([28]) while for the first LILOG prototype we used the CUG paradigm, c.f. [21]. The grammar as well as the lexicon have been defined in STUF ([17], and [5]), a formalism for defining complex attribute value structures standing in the tradition of [18] or [30].

- The construction of the semantical information requires several activities such as the resolution of anaphoric references and the disambiguation of multiple readings of a sentence. For this purpose the discourse structure of a text has to be represented.

  These intermediate representations of a natural language text have been provided in extended discourse representation structures containing already parts of the final logical representation together with linguistic information. The respective representation formalism is based on Kamp's dicourse representation theory, [15].

- The result of understanding a natural language text is a collection of logical formulas formulated in the typed predicate logic L_{LILOG}.

  This is the target language of the translation process from natural language to logic and serves as the computational basis for processing the semantical contents of natural language texts. A formal definition of L_{LILOG} the fundamentals of the inference calculus for processing the language, and the implementation of the corresponding theorem prover have been documented in [27], [6], [8], and [7], respectively.

Our major concern here will be representation languages for the final output of the process of understanding a natural language text. We will argue that a sophistiated typed predicate logic such as L_{LILOG} can satisfy a number of requirements one is likely to set up for a representation formalism serving as the target language for translating information given in natural langue into machine processible form.

2 Representing Semantic Knowledge

To make the knowledge contained in a natural language text available for information processing systems it needs to be compiled into a computer processible form. This intended use of information originally given in natural language sets up several requirements for the target language of the transformation process:

- The language has to be expressive enough to capture the complex information structures that can be formulated in natural language. This expresiveness is a prerequesite for making the translation process feasible.

- The language has to be computationally adequate in the sense of allowing for the intended handling of the information with reasonable efficiency.

---

[1]LILOG Experimentier Umgebung Version 2 or LILOG Experimenting Environment Version 2

While using (pure) predicate logic as the formal representation language for the semantic contents of natural language texts originates from philosophy and traditional linguistics , efforts in building natural language understanding systems for realistics application domains are confronted with non-trivial knowledge engineering problems. The systems have to be provided with huge amounts of background knowledge in order to draw expected conclusions from information acquired from texts. To model this backgorund knowledge is comparable to classical software engineering problems and therefore languages of high expressiveness - including type systems - are very welcome. A natural step forward following the tradition to use logic as the basic representation formalism is to consider typed predicate logic ([26], [34], [10], [12], or [32]).

Various nice properties offered by typed predicate logic make it an adequate formalism to be used as the target language of the translation process making information given in natural language available for information processing systems:

- The typedness allows for structuring the universe of objects within an application subject into subdomains.

- The relational view of the world which is inherent in logic allows for describing relationships between the objects of the universe of discourse.

- The various logical connectives and quantifiers allow for complex descriptions of the relationships mentioned above.

While the above arguments for typed predicate logic address the *static* aspects of the representation formalism, let us also mention some *dynamic* aspects of computing with typed logic:

- PROLOG has become an accepted programming language demonstrating that predicate logic and the principle of logical deduction form a competitive computational framework for information processsing, see [20] and [11].

- Introducing type systems into predicate logic improves the computational efficiency of logic. This has been verified for several extensions of PROLOG ([24], [1], [14], or [3]) as well as for theorem proving techniques exploiting type information to reduce the computational complexity of logical deduction, c.f. [33], [12], [13].

2.1 Representing Knowledge in L_{LILOG}

The knowledge representation language L_{LILOG} we have developed within the LILOG project has been designed in the spirit of order-sorted predicate logic and integrates KL-ONE like constructs for describing complex sorts, c.f. [27].

The data domains of an application may be modeled as atomic (*sort names*) or complex sorts (*sort expressions* formed with respect to a number of type constructors as known from th KL-ONE family of languages). Moreover, sort names and sort expressions may be used as unary predicates in order to formulate arbitrary conditions under which an object belongs to some data domain. L_{LILOG} has the expressiveness of first order predicate logic and therefore it has to be compared with

KRYPTON ([9]) and the LLAMA logic ([12]) - two logics for knowledge representation where types can be decsribed in a highly sophisticated way.

Let us give some modeling examples employing L_{LILOG} for describing a taxonomy of data domains (called *sorts* in the L_{LILOG} jargon) together with some axioms relating objects of the respective sorts.

> **sort** *person* $<$ *top;*
>> **features** *sex : sexes;*
>>> *age : integer.*
>
> **sort** *sexes* $<$ *top;*
>> **atoms** *female, male.*
>
> **sort** *woman* $=$ *person* **and** *sex : { female }*
> **sort** *female-twen* $<$ *woman.*
> **sort** *man* $<$ *person;*
>> **disjoint** *woman.*
>
> **sort** *car* $<$ *top;*
>> **features** *doors : integer;*
>> **roles** *passenger :: person.*
>
> **sort** *coupe* $<$ *car* **and** *doors : { 2 }.*

The above declarations describe a sort hierarchy by introducing sorts and some functional attributes, i.e., features like the *sex* or the *age* of a *person*. Relational attributes like the role *passenger* asssociated to a *vehicle* indicate that there may be an arbitrary number of *persons* using a vehicle. The atoms *female* and *male* of the sort *sexes* are objects of that sort and we make a unique names assumption for these atoms, i.e., *female* and *male* stand for different objects.

Next we introduce some constants standing for objects of the respective sorts, e.g., *mary* is a *25*-year-old *person*. Due to the formula *{ female }(sex(mary))* we will be able to derive that *mary* cannot be a *man*.

> **constant** *mary : person* **and** *age : { 25 }.*
> **axiom** *{ female }(sex(mary)).*

Finally we describe the age of female twens by

> **axiom** *forall W : woman;*
>> *age : [20..29](W)*
>> $\Rightarrow$
>> *female-twen(W).*

Besides its sophisticated type system, L_{LILOG} offers the posibility to formulate default rules and thus allows for employing both full explicit negation as well as negation by failure for modeling purposes. The L_{LILOG} approach to non-monotonic reasoning is described further in [23] and [29].

Another aspect of knowledge modeling supported by L_{LILOG} is to control inferences. This can be done in two ways: (1) certain deductions can be delegated to external reasoning devices. Technically this is achieved by labeling certain literals of a logical formula to be processed outside the general theorem prover for L_{LILOG}. In the LILOG project we attached the so-called depiction module (see [19]) - a special reasoner for processing spatial information represented in an analog style by means of cell matrices - to the inference engine.

Finally, we are able to select axioms of a knowledge base to be employed for forward or backward chaining. This feature has been motivated by the two fundamental modes of operation of our LEU/2 text understanding system: when understanding a text the acquired new information has to be combined with existing knowledge. This is a typical task for a forward chainer. On the other hand, existing knowledge can be queried in natural language. Therefore, the logical formulas representing the original questions have to be run as goal queries against the knowledge base of LEU/2: a typical backward chaining task.

3 Inferential Computations

Choosing (typed) predicate logic as the target language for the process of understanding natural language, the information originally given in textual form can now be processed according to the principle of logical deduction seen as a computational model.

The basic mechanism underlying this computational model is that of deriving a logical formula - the *goal query* $G(x)$ where x represents the query variables for the solutions to be computed - from a knowledge base **KB**. I.e., the computation tries to verify whether the implication

$$\mathbf{KB} \implies G(x)$$

holds. Since the query variables x are existentially quantified, the above implication can be interpreted as

Compute an x *such that* $G(x)$ *follows from* **KB***!*

Technically, the computation is performed for the reformulation

$$\mathbf{KB} \land \neg\ G(x)$$

of the problem and sets up the following task

Assume that **KB** *and* $\neg\ G(x)$ *for any* x *hold.*
Find an x *contradicting this assumption!*

The above idea of computing on the basis of logical deduction, requires a so-called *inference calculus* for performing the necessary inferences to calculate the solutions. The most prominent such inference calculus is the SLD resolution calculus on which PROLOG ([11]) implementations are based, see [2] or [22].

3.1 Sample Computations

Since Horn logic is a representation formalism with only one basic modeling construct - the Horn clause, the SLD calculus is rather simple and therefore very effcient to implement. However, complex knowledge representation languages like L_{LILOG} offer a large variety of modeling features each of which contributes to the complexity of an inference calculus for implementing the language.

Below we demonstrate that a number of additional inference rules is necessary for processing L_{LILOG} and we point out the role of types in these inference based computations.

Given the knowledge encoded above we now want to verify that *mary* is not a *man* by leading the assumption *man(mary)* to a contradiction.
Thus given the goal literal

    man(mary)

we perform a proof where a clause is the negation of the original goal clause or a resolvent, and *knowledge* is information taken from the knowledge base with respect to which the goal shall be solved.

    man(mary)

$$female\text{-}twen(W{:}woman) \lor \neg\ age\ :\ [20..29]\ (W)$$

$$\overline{\quad\neg\ age:[20..29](mary)\ \lor\ \neg\ woman(mary)\quad}$$

The above resolution between the two positive literals may be performed since the subsumption checker for sort expressions computes for us that the sorts *man* and *female-twen* are disjoint[2]. Compared to standard resolution where only pairs of positive and negative literals can be resolved, this inference step requires a new inference rule taking into account that also two positive literals can be contradictory. Knowing only that *mary* is a *person*, we have to generate a sort literal as a new subgoal in order to verify that *mary* is a *woman* due to taxonomic information encoded in the clauses of the knowledge base. This extension of the concept of unification has been described in [4] and is related to Stickel's partial theory resolution ([31]) which has been employed for implementing KRYPTON.

In order to solve the subgoal testing whether *mary* is a *woman*, we have to expand the definition of the sort *woman* and thereafter simplify it until the sort literals have the form *sortname(term)* and { *atom* }(*term*). The expansion of sort names and the simplification form a collection of inference rules that is necessary for coping with definable sorts and complex sort expressions.

---

[2]Subsumption checking ([16]) for sort expressions is a generalisation of testing the subsort relationship when computing according to the principle of order-sorted unification (see [35]). This effort has to be added to unification, if type information is attached to terms to be unified.

¬ age:[20..29](mary) ∨ ¬ woman(mary)

woman = person **and** *sex:{female}*

¬ age:[20..29](mary) ∨ ¬ (person and sex:{female})(mary)

¬ age:[20..29](mary) ∨ ¬ person(mary) ∨ ¬ sex:{female} (mary)

¬ age:[20..29](mary) ∨ ¬ person(mary) ∨ ¬ {female}(sex(mary))

The above transformation of the goal clause

    ¬ age:[20..29](mary) ∨ ¬ woman(mary)

to the goal clause

    ¬ age:[20..29](mary) ∨ ¬ person(mary) ∨ ¬ {female}(sex(mary))

is a sequence of inference steps that has to be part of the computation in order to make the information of the knowledge base applicable for solving the goal.

Now we can employ the sort information concerning the constant *mary*: Knowing that *mary* is a *person* of the *age* of *25*, we can apply the subsort elimination rule and cancel the literals ¬ person(mary) as well as ¬ age:[20..29](mary) from the goal clause. This is possible since, e.g., knowing that Mary is a person at the age of 25 and the assumption that she's not a person is contradictory information that may be resolved away as part of the proof. The final inference step is a standard resolution step and we can drop the last literal of the goal giving us the empty clause □ which indicates that our original goal has been proven.

    ¬ age:[20..29](mary) ∨ ¬ person(mary) ∨ ¬ {female}(sex(mary))

mary : person **and** *age:{25}*

¬ {female}(sex(mary))

{female}(sex(mary))

□

As we have seen from the examples above, upon each logical inference cycle one has to decide which calculus rules apply and which one shall be used for the next inference step. In contrast to this, SLD resolution employs only one inference rule. Therefore, the logical deductions required for realizing PROLOG can be implemented efficiently: e.g., deciding which inference rule can be applied is an administrational overhead that occurs for every logical inference step of the LILOG inference engine, while PROLOG systems are not burdened with this extra effort.

3.2 Configurating The Inferential Capabilities

The complexity of the inference mechanism required by knowledge representation languages like L_{LILOG} suggest to tailor the inference calculus and the search strategy for traversing the search space. The architecture of the LILOG inference for processing L_{LILOG} has been designed with the intention of realizing an *open* computational device, where inference calculi and search strategies can be exchanged easily. The flexibility we have achieved allows us to adapt the inferential capabilities of the knowledge processor to the contents of the knowledge bases to be handled.

In the current implementation these adaptions have to be made by hand. However, future versions of the inference engine should be intelligent enough to adapt themselves to the contents of the knowledge base to be processed. The result would be a reasoning system that always uses the least expensive inference calculus for the knowledge base it has to process.

To realize the 'intelligence' of such an adaptive knowledge processing system the following is necessary:

- A grammar for L_{LILOG} such that the parser can detect the complexity of the knowledge base being fed into the system. This complexity can range from an untyped Horn logic program to a knowledge base employing the full complexity of the type system, full first order logic with equality, and may be even default rules.

- Knowing how complex the logic of the current knowledge base is, the required inference calculus can be configured from a library of inference rules, unification algorithms, and an optionally available truth maintenance system in case default rules have occured in the knowledge base.
  The idea behind this automatic selection of the inference calculus is that the language L_{LILOG} as defined now represents the upper bound for the inferential complexity of a knowledge base. However, there may be situations where a less complex logic is used for modeling an application. If that is the case, we don't need the full inference calculus for processing L_{LILOG}; rather a more efficient sub-calculus can realize the required computations.

- Selecting the search strategy automatically is a more complex task.
  Deciding to use the depth-first left-to-right strategy when being confronted with a Horn logic knowledge base is not too difficult. However, inferring over arbitrary first oder logic knowledge bases requires a good analysis of the structure of the knowledge to be processed.
  If, for example there is a large amount of simple facts in the knowledge base, resolving these unit clauses first is a good heuristics. This idea stems also from apporaches to partial evaluation of logic programs . To find the appropriate search strategy for processing a particular knowledge base is also very much goal dependent. Thus, learning algorithms evaluating the inferential behaviour of different strategies for processing different goal queries may be necessary for determining the right search strategy. This effort will be of particular interest for knowledge bases with a presumably long lifetime.

When developing the LEU/2 natural language understanding system, the fine tuning of the inference engine has been an important effort that improved the performance of the inferential services considerably. However, all this has been manual work, but given the experience we gained it is worth to make an investment to automate these efforts.

4 Conclusion

To use a highly expressive knowledge representation language as the target for translating natural language text into a logical form appears as a must for two reasons: (1) it 'shortens' the translation process, since we don't have to compile 'down' to some low level executable formalism; (2) it allows for high level modeling of the semantic background knowledge which is required for the process of understanding the semantic contents of natural language texts.

Given these assumptions, a sohisticated typed predicate logic is an adaquate representation formalism. However, as the representation language is rather complex, it requires powerful inference mechanisms for processing it. Part of the computational complexity of such a typed logic can be compensated through flexible theorem provers interpreting the representation language. However, in the long run we will need compilation-based implementations for such languages: the amount of information to be handled by natural language understanding systems will grow with every improvement of the technology.

References

[1] H. Aït-Kaçi and R. Nasr. LOGIN: A logic Programming Language with Built-in Inheritance. *Journal of Logic Programming*, 3:185–215, 1986.

[2] K. R. Apt and M. H. van Emden. Contributions to the Logic of Programming. *Journal of The ACM*, 29:185–215, 1982.

[3] C. Beierle. Types, Modules and Databases in The Logic Programming Language PROTOS-L. In K. H. Bläsius and U. Hedtstück and C.-R. Rollinger, editor, *Sorts and Types for Artificial Intelligence*, volume 418 of *Lecture Notes in Artificial Intelligence*. Springer-Verlag, Berlin, Heidelberg, New York, 1990.

[4] C. Beierle, U. Hedtstück, U. Pletat, J. Siekmann, and P. H. Schmitt. An order-sorted predicate logic for knowledge representation systems. IWBS-Report 113, IBM Deutschland GmbH, Stuttgart, 1990. To appear in AI Journal 1992.

[5] C. Beierle, U. Pletat, and H. Uszkoreit. An algebraic characterization of STUF. In I. S. Batori, U. Hahn, M. Pinkal, and W. Wahlster, editors, *Computerlinguistik und ihre theoretischen Grundlagen. Informatik-Fachberichte 195*. Springer-Verlag, 1988.

[6] T. Bollinger. A Model Elimination Calculus for Generalized Clauses. In *Proceedings IJCAI 91*, 1991.

[7] T. Bollinger and U. Pletat. Knowledge in Operation. IWBS Report 165, IBM Deutschland, 1991.

[8] T. Bollinger and U. Pletat. An Order-Sorted Predicate Logic with Sophisticated Sort Hierarchies. IWBS Report, IBM Deutschland, 1992. to appear.

[9] R. J. Brachman, V. P. Gilbert, and H. J. Levesque. An Essential Hybrid Reasoning, System: Knowledge and Symbol Level Accounts of KRYPTON. In *Proceedings IJCAI-85*, pages 532–539, 1985.

[10] R. J. Brachman and J. G. Schmolze. An overview of the KL-ONE knowledge representation system. *Cognitive Science*, 9(2):171–216, April 1985.

[11] W. F. Clocksin and C. S. Mellish. *Programming in Prolog.* Springer Verlag, Berlin, Heidelberg, New York, 1981.

[12] A. G. Cohn. A More Expressive Formulation of Many Sorted Logic. *Journal of Automated Reasoning*, 3:113–200, 1987.

[13] A. M. Frisch. The substitutional framework for sorted deduction: fundamental results on hybrid reasoning. *Artificial Intelligence*, 49:161–198, 1991.

[14] J.A. Goguen and J. Meseguer. EQLOG: Equality, Types and Generic Modules for Logic Programming. In D. DeGroot and G. Lindstrom, editors, *Logic Programming, Functions, Relations and Equations*. Prentice Hall, 1986.

[15] H. Kamp. A theory of truth and semantic representation. In J. A. G. Groenendijk and T. M. V. Janssen and M. B. J. Stokhof, editor, *Formal Methods in The Study of Language*, volume 135. Mathematical Center Tracts, Amsterdam, 1981.

[16] B. Hollunder, W. Nutt, and M. Schmidt-Schauss. Subsumption Algorithms for Concept Description Languages. In *Proc. ECAI 90*, 1990.

[17] J. Dörre and R. Seiffert. A Formalism for NAtural Language - STUF. In O. Herzog and C.-R. Rollinger, editor, *Textunderstanding in LILOG*, volume 546 of *Lecture Notes in Artificial Intelligence*. Springer Verlag, Berlin, Heidelberg, New York, 1991.

[18] R. T. Kasper and W. C. Rounds. A logical semantics for feature structures. In *Proceedings of the 24th Annual Meeting of the Association for Computational Linguistics*, pages 257–265, Columbia University, New York, 1986.

[19] M. N. Khenkhar. DEPIC-2D: Eine Komponente zur depiktionalen Repräsentation und Verarbeitung räumlichen Wissens. In D. Metzing, editor, *GWAI 89*, pages 318–322, Berlin, Heidelberg, New York, 1989. Springer Verlag.

[20] R. A. Kowalski. Predicate Logic as a Programming Language. In *IFIP 74*, 1974.

[21] L. Karttunen. Radical lexicalism. In A. Broch and M. Baltin, editor, *Alternative Conceptions of Phrase Structure*. Chicago University Press, Chicago, 1989.

[22] J. W. Lloyd. *Foundations of Logic Programming.* Symbolic Computation. Springer-Verlag, Berlin, Heidelberg, New York, 1984.

[23] S. Lorenz. Nichtmonotones Schließen mit ordnungssortierten Defaults. IWBS-Report 100, IBM Deutschland, Scientific Center, January 1990.

[24] A. Mycroft and R. A. O'Keefe. A polymorphic type system for Prolog. *Artificial Intelligence,* 23:295–307, 1984.

[25] O. Herzog and C.-R. Rollinger, editor. *Textunderstanding in LILOG,* volume 546 of *Lecture Notes in Artificial Intelligence.* Springer Verlag, Berlin, Heidelberg, New York, 1991.

[26] A. Oberschelp. Untersuchungen zur mehrsortigen Quantorenlogik. *Mathematische Annalen,* 145:297–333, 1962.

[27] U. Pletat and K. v. Luck. Knowledge Representation in LILOG. In K. H. Bläsius and U. Hedtstück and C.-R. Rollinger, editor, *Sorts and Types for Artificial Intelligence,* volume 418 of *Lecture Notes in Artificial Intelligence.* Springer-Verlag, Berlin, Heidelberg, New York, 1990.

[28] C. Pollard and I. A. Sag. *Information Based Syntax and Semantics. Vol. I: Fundamentals.* Chicago University Press, Chicago, 1987.

[29] K. Schlechta. Defeasible Inheritance: Coherence Properties and Semantics. In Michael Morreau, editor, *SNS-Bericht 89-47.* Seminar für natürlich-sprachliche Systeme, Univertsität Tübingen, 1989.

[30] S. M. Shieber, H. Uszkoreit, F. C. N. Pereira, J. J. Robinson, and M. Tyson. The Formalism and Implementaion of PATR-II. In J. Bresnan, editor, *Research on Interactive Acquisition and Use of Knowledge.* Artificial Intelligence Center, SRI International, Menlo Park, CA, 1983.

[31] M. E. Stickel. Automated Deduction by Theory Resolution. *Journal of Automated Reasoning,* 1:333–355, 1985.

[32] W. A. Woods. Understanding subsumption and taxonomy: A framework for progress. TR-19-90, Harvard University, Center for Research in Computing Technology, Cambridge, MA, 1990.

[33] C. Walther. A Many-Sorted Calculus Based on Resolution and Paramodulation. In *Proceedings IJCAI 83,* 1983.

[34] C. Walther. A Mechanical Solution of Schubert's Steamroller by Many-Sorted Resolution. *Artificial Intelligence,* 26:217–224, 1985.

[35] C. Walther. Many-sorted unification. *Journal of the ACM,* 35(1):1–17, January 1988.

A Set of Tools for VHDL Design

Peter Reintjes

DAZIX/Intergraph Corporation

Abstract. Prolog is uniquely suited to be an implementation language for electronic circuit design tools. Its strength in language processing and basis in logic make it a perfect environment for manipulating the formal languages used to describe logical systems. Furthermore, *language-oriented* design, in which the hardware designer writes, compiles, and tests textual hardware descriptions has become increasingly important. Prolog meets this need for a powerful, *language-oriented*, software development environment.

1   Introduction

In the early 1980's, the United States Department of Defense began the Very High Speed Integrated Circuits (VHSIC) project to push forward integrated circuit and systems technology. The VHSIC Hardware Description Language (VHDL) was developed as a consistent representation for all Department of Defense projects. Today, nearly all electronic CAD vendors support the VHDL language, and complete systems to view, analyze, and simulate VHDL designs are approaching a million lines-of-code.

Designers create complex system descriptions in VHDL, using the behavioral descriptions to model the external environment, or components which have not yet been designed while the structural subset of VHDL is used to describe the actual data-paths and component connectivity of the circuit being designed.

While focusing on Prolog tools which support VHDL, this talk presents techniques for implementing the following components of a CAD system:

- *Parsers and Generators for Hardware Description Languages*
- *Query-Systems, Style- and Rule-Checkers*
- *Translators*
- *Compilers*
- *Verification and Simulation Tools*
- *Logic Synthesis*
- *Interactive Graphics*

2   Parsers and Generators

2.1 VHDL Parser Development

The following programs were developed with a VHDL parser/printer created by scanning the BNF for the VHDL grammar directly from the appendix of [Lipsett] and then editing it into DCG syntax. The original grammar, consisting of 129 productions, was translated into 140 Prolog rules. Testing this "executable specification" of the VHDL grammar on files from a VHDL benchmark suite then uncovered several errors in the textual specification [Lipsett].

The parser can be called with the goal `vhdl_read(+File,-VHDL)` and unifies `VHDL` with a list of `design_unit/2` terms. These terms contain lists of structures representing VHDL statements. Alternatively, `vhdl_read(+File)` will assert these terms into the Prolog database.

2.2 A Query/Rule-Checking System

The simplest Prolog program for VHDL is a question answering system that operates on a VHDL description loaded into the database. It can answer such questions as – *What components are contained in architecture X?, In what designs does component Y appear?*, or *Do all latches in the system use a consistent clocking scheme?*. To begin, we define the following predicate to identify the structural sub-components of an architecture.

```
substructure(Comp, Arch) :-
    design_unit(_, arch(Arch, _, _, Statements)),
    member(comp_instant(_, Comp, _, _), Statements).
substructure(Comp, Arch) :-
    design_unit(_, arch(Arch, _, _, Statements)),
    member(comp_instant(_, SubArch, _, _), Statements),
    substructure(Comp, SubArch).

contains(Arch, Comp, structure) :- substructure(Comp, Arch).
contains(Arch, Comp, Type)   :- subroutine(Comp, Arch, Type).
```

Examples of queries that can be answered by this system are:

```
English:   What functions are used by the entity "alu"?
Prolog:    ?- contains(alu,Function,function).

English: What entity is not used as part of any other design?
Prolog:  ?- design_unit(_,entity(Name,_,_)),
            not contains(_,Name,_).
```

3 Translators

3.1 A Pretty Printer

The next application uses a language generator to produce the textual form of VHDL from the internal data structure. Minimally, such a program can read erratically formatted VHDL source, perhaps with inconsistent use of upper and lower-case keywords, and produce uniformly indented text with consistent use of character cases. A program of this sort is sometimes called a "pretty printer".

Unfortunately, we cannot simply run the parser backwards to produce the VHDL text for two reasons: First, the white space (blanks, tab characters, and newlines) were eliminated deterministically by the tokenizer and we need a uniform method of reinserting them. Second, the clauses of rules which recognize optional constructs must be ordered one way for a parser and the reverse for a printer. For comparison, the parser and printer versions of the optional association list rules are given.

```
opt_association_list(A)  --> association_list(A).
```

```
opt_association_list([]) --> [].
```

```
write_opt_association_list([]) --> !, [].
write_opt_association_list(A)  --> write_association_list(A).
```

3.2   A Translator

Current CAD systems employ dozens of languages, many with advantages over VHDL. For this reason, a CAD system must provide extensive translation capabilities. These translators begin with parsers and generators for several languages and can be constructed by defining a predicate to transform the input language's parse-tree to one expected by the output language.

```
runtime_entry(start) :-
    unix(argv([In, Out])),
    vhdl_read(In, VHDL),            % VHDL Parser
    vhdl_to_newhdl(VHDL, NEWHDL),
    newhdl_write(Out, NEWHDL).      % NEW-HDL Printer
```

For each such translator, a programmer must write a specialized langX_to_langY/2 predicate to perform the translation. This task will be simplified if we have a discipline for building similar (if not identical) parse-trees from different languages to minimize the effort required for each translation.

3.3   Improved Translator

The most successful way to enforce a "discipline" is to hide that level of detail from the programmer. There is a discipline for writing lexical analyzers described in [OKeefe], but we have gone an extra step and defined PLEX (a version of LEX for Prolog). Furthermore, instead of writing a parser and using this as a model for a hand-written printer, we use a slightly higher level language called BNF to define both simultaneously. In order to automate the construction of a printer from the parser, it is necessary to formalize (and hide) the operations on the parse tree (just as DCGs hide the token list). This higher-level language consists of lexical and grammar rules from which all lexical analyzers, parsers, generators, and parse tree operations are synthesized.

Because all parsers and generators built by MULTI/PLEX operate on "compatible" data structures, a simple translator can be defined simply by giving two language specifications. This can be demonstrated with the following subsets of natural languages. By using the same English words (subject, cat, dog) to store the semantic information from German and Spanish texts, these specifications define a bi-directional translator.

Figure 1. German Specification

```
german lexicon
''[A-Za-z]+'' is Word  if name(Word,text).

german ::= file('name.german',german), subjekt, beliebige_subjekte.
```

```
beliebige_subjekte ::= subjekt, beliebige_subjekte.
beliebige_subjekte ::= [].

subjekt ::= artikel(Gender), hauptwort(N,Gender), update(subject, N).

hauptwort(dog, maennlich) ::= [ 'Hund' ].
hauptwort(cat, weiblich) ::= [ 'Katze' ].

artikel(mannlich)  ::= [ der ].
artikel(weiblich)  ::= [ die ].
artikel(saechlich) ::= [ das ].
```

Figure 2. Spanish Specification

```
spanish lexicon
''[a-z]+'' is Word if name(Word,text).

spanish ::= file('name.sp',spanish), sujeto, sujetivos_opcionales.

sujecto ::= articulo(Gender), nombre(N,Gender), update(subject, N).

sujetivos_opcionales ::= sujeto, sujetivos_opcionales.
sujetivos_opcionales ::= [].

nombre(dog, masculino) ::= [ perro ].
nombre(cat, masculino) ::= [ gato ].

articulo(masculino) ::= [ el ].
articulo(feminino)  ::= [ la ].
```

While the grammars for these languages deal explicitly with gender, no gender information is stored in the parse-tree. Gender is an example of information which belongs in the grammar, but which we do not require (or want) in the data-base. Since "cat" is feminine in German and masculine in Spanish, preserving the gender would result in an incorrect translation.

This simple model works best when the languages represent identical levels of detail and have corresponding statements and sub-elements. However, we cannot expect translation to consist of a one-for-one rewriting of statements. In general, a translation also requires computation. Since the specification language is Prolog, general predicates can be included in the language specification and executed as part of the parser or generator.

A more complex translator which performs feature mapping between structurally diverse languages, can be built by using MULTIPLEX as a *module* and specifying a transformation function which will turn the input parse-tree into the necessary output parse-tree. The *parse-tree* is simply a list containing either the primitive data elements ID:Type:Value or the hierarchical objects subcell(ID,ListofObjects).

```
transform(In, Out) -->
  compatible(hierarchy_model, In, Out),
  compatible(timing_model,    In, Out),
  compatible(naming_model,    In, Out).
```

Information about different language features and ways to reconcile them can be easily added to this translator. These facts and functions can be built into the translator or brought in as part of the language specification files. This organization is similar to the one developed earlier for the AUNT translator [Rein90a].

4 A VHDL Compiler

We next define a compiler to transform the high-level hardware description into a network of boolean gates. The term "compilation" is appropriate since relatively high-level description is transformed into one at a lower level, a process quite similar to the compilation of programming languages. In particular, all of the techniques and benefits described in [Clocksin] and [Warren] are applicable to this process.

The statements in the high-level language include assignment, if-then-else, and case statements. We first define a rule to compile the assignment statement, and then show how more complex statements can be decomposed into assignment statements. If we define complex transformations recursively in terms of slightly less complex ones, we will achieve maximum generality. With a minimum of effort, we create a compiler which handles complex statements such as case-statements of any size in which arbitrary statements are nested to any depth. This may seem trivial, but many HDL compilers impose arbitrary restrictions on depth of nesting and size of expressions.

4.1 Compiling Statements and Expressions

The heart of the compiler is compile_stmt(+VHDL,-Gates,☐) which constructs a list of boolean gates from a VHDL statement. We begin with assignment statements, which are the simplest form of statement. Below we have the original VHDL source, the Prolog term created by the parser, and the desired gate-list which the compiler should produce.

```
VHDL:  v = w AND ^( ^x  XOR  y ) ;

TERM:  assign(name(v),expr(and,[name(w),expr(not,expr(xor,
                      [vhdl_expr(not,name(x)),name(y)]))]))

GATES: [gate(not,[C,x]),gate(xor,[B,y,C]),
        gate(not,[A,B]),gate(and,[v,w,A])]
```

This compilation, including the creation of the internal variables (A,B,C) is accomplished simply by compiling the expression on the right-hand side of the statement with the output variable as the second argument.

```
compile_stmt(assign(Out, Expr)) --> compile_expr(Expr, Out).
```

To compile the assignment shown above, we need the following four clauses for compile_expr//2. After handling lists of expressions, we unify the base variables with the output expression, or create a boolean gate to compute the expression and recursively compile the gate's input expressions.

```
compile_expr([],           [])   --> [].
compile_expr([E|Es],[O|Os])      --> compile_expr(E,O),
                                     compile_expr(Es,Os).
compile_expr(name(Out), Out)  --> [].

compile_expr(expr(Op,In),Out) --> [ gate(Op, [Out|Expr]) ],
                                   compile_expr(In,Expr).
```

We define compilation for more complex statements in terms of simpler ones by establishing a partial ordering of statement complexity.

```
             assignment
             if-expression
   if-statement       case-expression
                      case-statement
```

With mutually recursive definitions of statement and expression compilers we can progressively define the rules for more complex statements.

Figure 3. Compiling If-Expressions and If-Statements

```
compile_expr(ifexpr(E, TStmts, EStmts), Out) -->
    [ gate( or, [Out, ThenC, ElseC]),
      gate(and, [ThenC, Then, Condition]),
      gate(not, [ ElseCondition, Condition]),
      gate(and, [ElseC, Else, ElseCondition]) ],
    compile_expr(E, Condition),
    compile_expr(TStmts, Then),
    compile_expr(EStmts, Else).

compile_stmt(if(E, TStmts, EStmts)) -->
    { extract_lhs([TStmts,EStmts], [TExpr,EExpr], LHS) },
    compile_expr(ifexpr(E, TExpr, EExpr), LHS).
```

5   Verification and Simulation

The predicate **extract_lhs/3** transforms structures containing arbitrary assignment statements into a functional form, an important step in preparing formal language descriptions for theorem provers.

Likewise, the simplest strategy for compiling HDLs for simulation (rather than synthesis) is to transform the internal form into one compatible with a MULTI/PLEX grammar for a standard programming language such as "C". We can produce efficient executable programs in an imperative language that corresponds to the circuit to be simulated (theorem to be proved, numerical algorithm to be executed). We therefore have an effective procedure for side-stepping the issue of Prolog's inefficiency in these areas.

6   A Logic Synthesis System

A logic synthesis system optimizes a network of gates for speed, power, or chip area based on the constraints of a specific technology. Note below that `synthesize/3` expects a tree data-structure, rather than the list of gates produced by the compiler.

```
runtime_entry(start) :-
        unix(argv([In, Out, Tech])),
        vhdl_read(In, VHDL1),              % VHDL Parser
        compile(VHDL1, VHDL2),            % Compiler
        list_to_tree(VHDL2, Tree0),
        synthesize(Tech, Tree0, Tree1),  % Logic Synthesis
        tree_to_list(Tree1, Netlist),
        edif_write(Out, Netlist).         % Netlist Language Printer
```

This program reads a VHDL description and writes an EDIF network suitable for fabrication in the technology specified by the third command-line argument (e.g. cmos). A complex system of transformations, possibly with other arguments in addition to the technology, can be easily defined as DCG rules.

```
synthesize(Tech) --> canonize,
                     min1,
                     min2(Tech),
                     implement(Tech).


min2(cmos)  -->  cmos_min.
min2(nmos)  -->  mos_min.
```

6.1   Two Data Structures for Logic Synthesis

Logic synthesis programs are re-writing systems. The re-write rules themselves are often very simple, but the circuits being re-written can contain tens- to hundreds-of-thousands of sub-terms (gates). The program architecture based upon DCG's might suggest that we are using lists to store the circuit description, but in synthesis, one must index on node identifiers for reasonable efficiency.

The first data organization proposed is a quad-tree with with update operations that require $O(M log_4(N))$ in time and space, where M is the number of pins on the circuit element being added, and N is the total number of circuit elements. If each node is identified by a unique integer, a quad-tree of depth D can hold the circuit if 4^D is greater than the largest node identifier. This identifier is the key we will use to locate a given leaf node in D steps. The predicate `quadtree(D,Tree)` defines an empty quad-tree of depth D.

```
quadtree(0, []).
quadtree(D, t(A,B,C,D)) :- D > 0, DN is D-1,
                           quadtree(DN,A),
                           quadtree(DN,B),
                           quadtree(DN,C),
                           quadtree(DN,D).
```

Gates are placed in the tree at the leaf nodes corresponding to the gate's nodes. For example, an AND gate which has input nodes 3 and 5 and output node 10 is written as gate(and,[10,3,5]) and will appear at locations 10, 3 and 5 in the tree. A nodes "location" is found by choosing a branch of the quad tree based on the lowest two bits in the index. As this branch is taken, these bits are shifted out of the index. The definition of the in_tree/3 predicate requires only 25 goals, while merge/4 requires 6, with calls to remove/3 and append/3. The following rule combines two similar N and M-input positive gates (AND or OR) which share a node into a single (N+M-1) input gate:

```
transform(I, TreeIn, TreeOut) :-
    in_tree([G1,G2]:I, TreeIn, _),  % G1,G2 are alone on Node I
    ( merge(I,G1,G2,G3)             %    G1 can merge into G2
    ; merge(I,G2,G1,G3)             % or G2 can merge into G1
    ),!,
    in_tree([G1,G2],TreeIn,Tree2),  % remove G1, G2
    in_tree(G3, TreeOut, Tree2).    % and insert G3.
```

6.2   A Logical version of Espresso

The second data organization follows the popular Espresso algorithm for minimization of logic functions ([Brayton]). Espresso is defined in terms of matrix operations on a small finite algebra. A row of 0,1,2 elements defines the positive, negative, don't-care variables of a single product term of a boolean expression, while a complete matrix represents the sum of these terms (e.g. a combinatorial function). Our sparse-matrix approach takes advantage of the fact that "don't care" values become frequent in large examples, and we only need one bit to to represent the [0,1] values. We combine this bit with the *shifted* variable index. Operationally, we encode an element in a sparse matrix by left-shifting its index sufficiently to pack the value into the lower bits. Thus a 4 represents a 0 at position 2 $(4 = (2 \ll 1) \vee 0)$.

The following predicates implement the description of the Espresso algorithm given in [Brayton]. The Prolog implementation is actually shorter than the pseudocode definition and eliminates the seven go-to's and four labeled statements.

```
lexpress -->  unwrap, complement,
              check_cost(1,_),
              expand,
              essential_primes,
              lexpress(1),
              add_to_care,
              sub_from_dont_care,
              make_sparse.

lexpress(Step) --> check_cost(Step, Next), process(Next).

process(1) --> irredundant, lexpress(2).
process(2) --> reduce,      lexpress(3).
process(3) --> expand,      lexpress(1).
process(4) --> last_gasp.
```

Figure 4. Complete Matrix, Sparse Encoding and Co-Factor Calculation

$$
\begin{array}{ccccc}
X(1) & X(2) & X(3) & X(4) & X(5)
\end{array}
\begin{pmatrix}
2 & 0 & 1 & 2 & 2 \\
0 & 0 & 2 & 2 & 2 \\
1 & 2 & 2 & 2 & 1 \\
2 & 2 & 2 & 2 & 2
\end{pmatrix}
=
\begin{array}{ccccc}
X(1) & X(2) & X(3) & X(4) & X(5)
\end{array}
\begin{pmatrix}
 & 4 & 7 & & \\
2 & 4 & & & \\
3 & & & & 13 \\
 & & & &
\end{pmatrix}
$$

$$
(c_p^i)_k = \begin{cases} \phi & \text{if } c^i \cap p = \phi \\ 2 & \text{if } p_k = 0 \text{ or } 1 \\ c_k^i & \text{otherwise} \end{cases}
$$

```
cofactor([],_,[]).
cofactor([C|Cs],F,Xs) :- ( C   =:=  F    -> Xs = Cs
                         ; C>>1 > F>>1 -> Xs = [C|Cs]
                         ; C>>1 < F>>1 -> Xs = [C|X1s],
                           cofactor(Cs,F,X1s)
                         ).
```

7   Interactive Graphics

Graphic-oriented design is experiencing a renaissance, with the increasing popularity and standardization of graphical and menu-driven interfaces. With Prolog, menus and other graphical objects are easily defined as compiled facts, and new objects can be easily defined and asserted in the same form (cf. [Rein90b]):

```
macro(buffer,_,
       [ port(a,in,-17,6),port(y,out,23,8),
         line(-12,8,-2,8),
         lines_rel(-2,-7,[0:30,20:-15,-20:-15)]],
         bbox(-22,-22,23,23)).
```

Once **draw/2** and **erase/2** predicates are defined to draw an object on a specified window, object (and list-of-objects) creation and destruction are implemented as:

```
create([],_)       :- !.
create([H|T],Win) :- !, create(H,Win),
                         create(T,Win).
create(Obj,Win)    :-    draw(Obj,Win),
                         assert(Obj).
```

By defining the inverses of these basic operations and maintaining a history of commands, a flexible UNDO capability can be implemented easily.

```
inverse(create(O,W),    destroy(O,W)).
inverse(destroy(O,W),   create(O,W)).
inverse(replace(O,N,W), replace(N,O,W)).
```

```
action(undo, Win) :-
    retract(history(Fs,[U|Us])),
    inverse(U,F),
    assert(history([F|Fs],Us)),
    call(U).
```

8   Conclusion and Results

The first version of an ECAD library developed for Quintus consisted of 11,000 goals and included hand-crafted parser/generators for VHDL, Verilog, EDIF and CIF. In the current version, we have replaced these parsers by MULTI/PLEX specifications. Including the additional code for MULTI/PLEX and two more languages (Synopsys and FPDL), the ECAD library has been reduced to 7,000 goals. The first graphical VLSI editor in Prolog seemed remarkably concise at 8000 goals (MCNC version ca.1986) when compared with a version in "C" (50K lines-of-code). Since then, the entire system has been re-written four times and the current implementation adds several functions while being completely defined in about 2000 goals.

When executed on a 3-MIP (VAX) workstation, The quad-tree logic synthesis algorithm transformed an 8,500-gate (16K nodes) circuit in 37 seconds and a 15K-gate circuit (26K nodes) in 60 seconds. The VHDL parser has been measured at 2000 lines-per-minute on a 3-MIP and 10K lines-per-minute on a 14-MIP workstation.

The programs described here demonstrate vividly that complex VLSI systems will not require millions of lines of code if they are described at an appropriate level of abstraction. In fact, the bulk of these programs are the language specifications (about 1200 goals for VHDL) and compare quite favorably with other implementations (the Intermetrics VHDL parser alone is 10,000 lines-of-code).

References

[Brayton], Robert Brayton, et. al. *Logic Minimization Algorithms for VLSI Synthesis*, Kluwer Academic Publishers, 1984

[Clocksin], W.F. Clocksin, "Logic Programming and Digital Circuit Analysis", *The Journal of Logic Programming, 1987:4:59,82*, March 1987.

[Lipsett] Lipsett et al., *VHDL: Hardware Description and Design*, Kluwer Academic Press, 1989

[OKeefe] Richard A. O'Keefe, *The Craft of Prolog*, 1990 MIT Press.

[Rein90a] "AUNT: A Universal Netlist Translator", *Journal of Logic Programming*, 1990:8:5-19 North Holland.

[Rein90b] *PREDITOR: A Prolog-based VLSI editor*, chapter three in "The Practice of Prolog", Leon Sterling, Editor. November 1990, MIT Press.

[Warren] D.H.D. Warren, "Logic Programming and Compiler Writing", *Software – Practice and Experience*, Vol 10, Number 2, pp 97-125, 1980, John Wiley and Sons, Ltd.

Tutorial Notes: Reasoning about Logic Programs*

Alan Bundy
bundy@edinburgh.ac.uk

Department of Artificial Intelligence
University of Edinburgh

1   Introduction

In this tutorial we will describe techniques for reasoning about logic programs.

Why should we want to reason about logic programs?

We will see that reasoning about programs is an important software engineering tool. It can be used to improve the efficiency and reliability of computer programs. Such reasoning is particularly well suited to logic programs. In fact, the ease of reasoning with logic programs is one of their main advantages over programs in other languages.

The kind of reasoning tasks we will consider are as follows.

Verification: to verify that a program meets a specification of its behaviour.
Synthesis: to synthesise a program that meets a specification.
Transformation: to transform a program into a more efficient program meeting the same specification.
Termination: to show that a run of a program will always terminate.
Abstraction: to abstract from the program information about the types of its input/output, its modes of use, *etc.*

In conventional, imperative, programming languages these four tasks are very different, but in logic programming some of them merge together. The reason is that specifications are usually written as logical formulae. As such, they can be interpreted as logic programs. They may be very inefficient and they may not always terminate, but they can often be used as a prototype of the desired program. This means that synthesis and transformation are not different in kind, but only in degree. Moreover, synthesis can be seen as verification of a partially specified program. All this imparts a simplicity and unity to reasoning with logic programs.

---

* I would like to thank the members of the mathematical reasoning group at Edinburgh and members of the CompuLog Network for helpful advice and feedback on these notes. In particular, Ina Kraan, Andrew Ireland, Helen Lowe, Danny De Schreye and Michael Maher were especially helpful. Special thanks are due to Ian Green and Andrew Ireland for getting the Springer LaTeX style file to work. Some of the research reported in this paper was supported by Esprit BRA grant 3012 (CompuLog).

1.1 Semantics of Logic Programs

To reason formally about a computer program we must have a method of turning a question about programs into a mathematical conjecture. The program reasoning task can then be converted into a theorem proving task. The usual way to do this is to associate a *semantics* with the programming language. By 'semantics' we mean an assignment of a mathematical expression to each program statement. The mathematical expression is often thought of as the *meaning* of the program statement.

To a first approximation it is unnecessary to give a semantics to a logic program; it is already a mathematical expression, namely a clause in first-order logic. Unfortunately, this is only true of *pure* logic programs. In practice, logic programs, *e.g.* Prolog programs, often contain impure features, *e.g.* negation as failure, **assert/retract**, **var**, the cut operator, the search strategy, *etc.* These cannot be *directly* interpreted in logical terms.

There are (at least) three solutions to this problem.

1. Provide a semantics for the particular logic programming language, *e.g.* Prolog, in which both the pure and impure features of the language are assigned a mathematical interpretation.
2. Work only within a pure subset of the language. Implement a logic programming language based on this pure subset.
3. Reason about specifications of logic programs. Introduce the impure features as necessary during a final compilation of the specification into your logic programming language of choice.

We will discuss the tradeoffs between these approaches in §2.

1.2 Specifications of Logic Programs

Specifications of programs are usually given as logical relations between input and output. For instance, a sort program takes in a list and outputs another list in which the same elements are put in order, *i.e.* the output is an ordered permutation of the input. This program can be specified by the relation[2]:

$$perm(IL, OL) \ \& \ ordered(OL)$$

where IL is the input list, OL is the output list and *perm* and *ordered* are defined appropriately.

Suppose **sort1** is a logic program for sorting lists, then to verify **sort1** we must prove as a theorem:

$$\forall IL \in list, \forall OL \in list. \ sort1(IL, OL) \leftrightarrow perm(IL, OL) \ \& \ ordered(OL) \quad (1)$$

This specification of **sort1** can itself be readily 'compiled' into a prototype sorting program **sort2**. For instance, in Prolog we could write:

---

2 We will adopt the convention of using typewriter font for programs and maths font for specifications.

```
sort2(IL,OL) :- perm(IL,OL), ordered(OL).
```

The program sort2 would be very inefficient — it would generate each permutation in turn and then test to see if it was ordered — but it would work. Now the proof of theorem (1) can be re-interpreted as a process of transformation of a very inefficient sorting program, sort2 into a more efficient program sort1 meeting the same specification.

By a specification of a program we will mean a formula of the form:

$$\forall \overrightarrow{Args} \in \overrightarrow{types}. \; head(\overrightarrow{Args}) \leftrightarrow body(\overrightarrow{Args})$$

where:

- *head* is the name of the program being specified;
- $\overrightarrow{Args}$ is some sequence of distinct variables and $\overrightarrow{types}$ is a pairwise corresponding sequence of their types; and
- $body(\overrightarrow{Args})$ is an arbitrary first-order logic formula, with no defined functions, whose free variables are in $\overrightarrow{Args}$.

In the interests of readability we will sometimes reduce clutter by omitting the $\forall \overrightarrow{Args} \in \overrightarrow{types}$ part.

The condition excluding defined functions from the body means that only undefined or constructor functions are allowed. That is, we can use functions to form data-structures, *e.g.* $s(0)$ or $[Hd|Tl]$, but not to define relations between objects, *e.g.* $x + y$.

1.3 Partial Evaluation of Logic Programs

This process of verifying/transforming logic programs can be regarded in two ways:

- as the proof of a mathematical theorem (*e.g.* (1)) using the definitions of the programs and specifications as axioms; or
- the symbolic and partial execution of the logic program.

Because of the second interpretation this process it is often called *partial evaluation*. We will see that to reason with recursive logic programs we will have to add *mathematical induction* to partial evaluation. Because we use induction we must work with a typed logic, *cf.* theorem (1) above in which $\forall IL \in list$ means for all objects, IL, of type *list*.

1.4 Abstract Interpretation of Logic Programs

It is also possible to reason with abstractions of logic programs, *i.e.* with programs in which some details of the program are generalised. For instance, we can reason with an abstraction in which the parameters to each procedure are replaced with labels representing their types or with their input/output mode,

[Bruynooghe & De Schreye, 1986]. The purpose of such reasoning is to deduce the implied type or mode of some parameters from others. For instance, suppose we know that the parameters of our sort2 program are both of type list. Its definition can be abstracted to:

```
sort2(list,list) :- perm(list,list), ordered(list).
```

From which we can deduce that the parameters of **perm** and **ordered** are also all lists (or some more general type).

The method of reasoning with abstract programs is the same as that used for concrete programs, *e.g.* partial evaluation, but the reasoning tends to be simpler because of the loss of detail caused by abstraction.

2 What Shall we Reason About?

In this section we discuss the tradeoffs between reasoning with impure programs, pure programs or specifications. There are advantages and disadvantages with each approach.

2.1 Incomplete Information

A disadvantage of reasoning directly with either pure or impure logic programs is that their declarative meaning does not quite correspond with their intended meaning. To see why not consider, for instance, the Prolog test for list membership, member/2.

```
member(El,[El|Tl]).
member(El,[Hd|Tl]) :- member(El,Tl).
```

Note that this program says nothing about the case when the list is empty. If you try to find out about this case by calling the goal:

```
?- member(X,[]).
```

then you will find that it fails. The standard interpretation of this failure is to say that member(X,[]) is false for all X. But this is to go beyond the declarative meaning of the member/2 program. It is to adopt the *closed world assumption*: anything that is not true is false.

One way to formalise this assumption is to say that the meaning of a logic program is not the immediate declarative meaning of its clauses, but the *Clark completion*. To form the Clark completion of the set of clauses defining a predicate we replace them all with a single equivalence. This equivalence states that a program head is true if *and only if* its body is true. For instance, the Clark completion of the member/2 program is[3]:

$$member(El, L) \leftrightarrow (\exists Tl.\ L = [El|Tl]) \vee$$
$$(\exists Hd, Tl.\ L = [Hd|Tl]\ \&\ member(El, Tl))$$

---

3 Maths font is used here because we intend to use typed versions of Clark completions as the specifications of the programs they complete.

Note the following general properties of Clark completions.

- All the clauses for **member/2** are replaced with one equivalence. Each of the old clauses corresponds to one disjunct in the body of this new equivalence.
- The head of the new equivalence consists of the predicate with distinct variables as its arguments. The head arguments in the old clauses are represented by equalities in each disjunct between these old arguments and the new variables, *e.g.* $L = [El|Tl]$.
- Each variable that appears in the body of the equivalence but not the head is existentially quantified in the body, *e.g.* Hd and Tl.

When reasoning about logic programs it is sometimes necessary to use this missing information. It is, therefore, more convenient to reason directly with the Clark completions than with the programs themselves.

A further piece of missing information is the types of the expressions in the programs. Consider the goal clause:

```
?- member(X,42).
```

We normally want to regard this call, not as false, but as ill-formed. Furthermore, if we want to reason using mathematical induction it will be vital to circumscribe the types of argument that a program can take. To do this we must add type declarations to our language. For instance, we might declare the type of *member* as:

$$member \in Type \times list(Type)$$

This is to be read as saying that *member* can take objects of any type as its first argument but must take lists of objects of this type as its second argument.

The types of the variables in an equivalence can be declared as $Obj \in Type$ statements within the quantifier declarations. For instance, we can enrich the Clark completion of **member/2** as follows.

$$\forall El \in Type, L \in list(Type).$$
$$member(El, L) \leftrightarrow (\exists Tl \in list(Type).\ L = [El|Tl]) \lor$$
$$(\exists Hd \in Type, Tl \in list(Type).\ L = [Hd|Tl]\ \&\ member(El, Tl))$$

Note that this extended version of the Clark completion of **member/2** fits the logical form of a program specification as defined in §1.2. In fact, we can use an extended Clark completion as the specification of the program it completes. We can reason directly with extended Clark completions, compiling them into their corresponding programs when the reasoning is complete. In this way we can reason with complete information and only 'forget' it when it is no longer required. This provides an argument for reasoning at the specification level.

Notation Henceforth, we will use the term *completion* to mean extended Clark completion.

2.2 What does 'Equivalent' Mean?

The simplest kind of reasoning with programs is to transform one program into an equivalent one. Unfortunately, there are many rival senses of 'equivalent'. A transformation that is legal under one sense of 'equivalent' may be illegal under another.

Impure Logic Programs Consider, for instance, the Prolog clauses:

```
p1(X) :- q(X), r(X).
p2(X) :- r(X), q(X).
q(a).
r(b) :- r(X).
r(a).
```

The programs p1 and p2 differ only in the order of the two literals in their body. Logically these literals are conjoined. Thus logically the two programs are equivalent. Unfortunately, they have very different operational behaviour. The goal p1(X) will succeed with X=a and stop, but the goal p2(X) will loop for ever with no success. Thus p1 and p2 are logically equivalent but operationally different.

The operational differences are due to the Prolog search strategy. It always evaluates literals left to right and applies clauses top to bottom. If it evaluated clauses bottom to top then p2(X) would also succeed with X=a, but would then backtrack and loop for ever. If it evaluated literals right to left then it would be p2(X) that would succeed with X=a and p1(X) that would loop for ever.

The most popular form of program transformation is partial evaluation. For instance, we can transform the program p1 by partially evaluating the goal p1(X). We resolve the literal q(X) with the unit clause q(a) and then the literal r(a) with the unit clause r(a). This transforms the clause p1(X) :- q(X), r(X). into the logically and operationally equivalent clause p1(a). We can apply the same process to the program p2 to transform clause p2(X) :- r(X), q(X). into p2(a). However, in this case we get a clause that is logically but not operationally equivalent. For instance, p2(X) will not now loop for ever.

If we want to preserve the operational behaviour of Prolog programs then we must restrict partial evaluation to use the same search strategy as Prolog. This will permit the above partial evaluation of p1 but prevent the one of p2. To partially evaluate the clause for p2 we must resolve the literals in its body in left/right order. The literal r(X) can be resolved with either of the two clauses for r. This produces the partially evaluated clauses:

```
p2(b) :- r(X), q(b).
p2(a) :- q(a).
```

If we continue to partially evaluate the first clause then we will get into a loop, so we should stop now. The second clause can be further partially evaluated to p2(a).

Such a restriction throws away a major reason for using logic programs: their logical semantics. It also makes the reasoning machinery dependent on the details of the Prolog interpreter. Small 'enhancements' of the interpreter might render the reasoning machinery obsolete.

Pure Logic Programs One reaction to this problem is to define conditions that a pure logic programming language must fulfil and then assume these conditions are met when designing reasoning machinery. One such condition is to insist on a *fair* search strategy, *i.e.* that each node in the search space is considered at some point during search. Prolog's search strategy is not fair; it can easily get trapped down an infinite branch of the search space and never return, *cf.* p2(X) above. This means that we must reject Prolog as an implementation of logic programming and confine ourselves to more restrictive implementations.

Unfortunately, even if we restrict ourselves to pure logic programs, there are still rival notions of equivalence. For instance, [Maher, 1987] considers ten different definitions of equivalence and shows that only two pairs of these give the same notion. This leaves eight distinct notions of equivalence. Consider, for instance, the following pure logic programs[4] from [Maher, 1987].

$$even1(0)$$
$$even1(s(s(N))) \leftarrow even1(N)$$
$$odd1(s(0))$$
$$odd1(s(s(N))) \leftarrow odd1(N)$$

$$even2(0)$$
$$even2(s(N)) \leftarrow odd2(N) \qquad (2)$$
$$odd2(s(N)) \leftarrow even2(N)$$

At first sight the definitions of *even1/odd1* and *even2/odd2* look like alternative, but equivalent programs for testing for even and odd numbers. [The natural numbers are represented in unary notation, where, for instance, 3 is represented by $s(s(s(0)))$.]

However, these programs are not logically equivalent, *i.e.* it is not the case that:

$$even1(N) \leftrightarrow even2(N)$$
$$odd1(N) \leftrightarrow odd2(N)$$

To see this consider the clause:

$$even2(s(0)) \leftarrow odd2(0) \qquad (3)$$

---

4 We will use maths font for pure logic programs.

This clause is a logical consequence of the definitions of $even2/odd2$ — it is an instance of clause (2) above. If the two programs were equivalent then the implication:

$$even1(s(0)) \leftarrow odd1(0) \tag{4}$$

would also be a logical consequence of the definitions — but it is not. To see this consider the non-standard model in which $even1(N)$ is true for all even numbers N, but $odd1(N)$ is true for *all* numbers, both odd and even. In this model all the clauses of the program are true, but clause (4) is false.

However, there is a notion of equivalence in which these two definitions are equivalent: their completions are logically equivalent[5].

$$even1(N) \leftrightarrow N = 0 \vee \exists M \in nat. \ N = s(s(M)) \ \& \ even1(M)$$
$$odd1(N) \leftrightarrow N = s(0) \vee \exists M \in nat. \ N = s(s(M)) \ \& \ odd1(M)$$

$$even2(N) \leftrightarrow N = 0 \vee \exists M \in nat. \ N = s(M) \ \& \ odd2(M)$$
$$odd2(N) \leftrightarrow \exists M \in nat. \ N = s(M) \ \& \ even2(M)$$

Both clauses (3) and (4) are logical consequences of this definition because $odd1(0) \leftrightarrow odd2(0) \leftrightarrow false$.

Thus, whether we regard these two programs as equivalent depends on whether we take logical equivalence of programs, logical equivalence of completions or one of the other six rival notions, as the definition of equivalence.

2.3 Summary

We can summarise these different tradeoffs as follows.

Impure Programs

Pros It is possible to reason directly with existing programs.

Cons There are a wide variety of different transformation schemes depending on what properties of the program it is desired to preserve. A different semantics is required to define each kind of preservation and to justify the corresponding transformations. These transformations are very complex and restricted. The semantics and the transformations are sensitive to small changes in the definition of the programming language. The logical and intended meaning of the program do not coincide. This means that some information required to reason with the programs is not represented directly.

---

5 Our use of 'logically equivalent' is non-standard in that we use more than just rules of logic in proofs of equivalence; we also use mathematical induction.

Pure Programs

Pros The kinds of transformation are general across a wide range of programming languages.

Cons There are still several different notions of program equivalence and, hence, several different transformation schemes. These transformations do not preserve the operational behaviour of impure programs, *e.g.* Prolog programs. The mismatch between the logical and intended meaning of the program causes problems during reasoning.

Specifications

Pros There is only one notion of equivalence. The logical and intended meaning of the specification coincide.

Cons Specifications of programs must be available for transformation. A specification can be 'compiled' into alternative programs which are not operationally equivalent.

Conclusion It seems to us that the benefits of logic programs can best be realised by a combination of reasoning with specifications and with impure programs. The basic algorithm is best determined by reasoning at the specification level, where the notion of equivalence is unambiguous and the reasoning invariant under changes in programming language. However, various implementational aspects can only be dealt with at the programming language level, so some tuning transformations at this level must also be catered for. The rest of these tutorial notes will assume this viewpoint.

3 Logic Specifications and Programs

Adopting this viewpoint makes it vital that we can translate freely between specifications and the programs they specify. In this section we describe two algorithms: one for lifting programs into specifications and one for compiling specifications into programs.

3.1 From Programs to Specifications

In this section we describe an algorithm for translating a logic program into a specification of itself. We will call this the *Clark completion algorithm*. It works by constructing the program's completion and using this as the specification.

Definition 1 (The Clark Completion Algorithm) *The algorithm consists of four stages.*

1. *Rewrite the program into logical notation, e.g. for Prolog clauses turn :-
   into ←s, commas into &s, semi-colons into ∨s, negation as failure into ¬s,
   provide definitions for system predicates, etc.*
2. *Make the head arguments into distinct variables. This is done by replacing
   each clause of the form:*

$$p(s_1, \ldots, s_n) \leftarrow body$$

with the clause:

$$p(X_1, \ldots, X_n) \leftarrow X_1 = s_1 \, \& \, \ldots \, \& \, X_n = s_n \, \& \, body$$

*where the X_i are new variables. The same X_is are used for each clause for
p.*
*Note that if s_i is already a variable distinct from s_j for $j < i$ then rather than
add $X_i = s_i$ to the body we can replace s_i by X_i throughout. In practice, we
will adopt this option when possible, since it will lead to simpler completions.*
3. *Existentially quantify each variable which occurs in the body but not in the
   head. Replace each clause of the form:*

$$p(X_1, \ldots, X_n) \leftarrow body(Y_1, \ldots, Y_m)$$

*where $X_i \neq Y_j$, for all i and j, and Y_j occurs in $body(Y_1, \ldots, Y_m)$, with the
clause:*

$$p(X_1, \ldots, X_n) \leftarrow \exists Y_1 \in t_1, \ldots, \exists Y_m \in t_m. \, body(Y_1, \ldots, Y_m)$$

*where t_i is the type of Y_i. We postpone the problem of discovering these types
to §9.*
4. *Combine the clauses for each predicate into a single equivalence. Suppose
   there are k clauses defining p, each of the form:*

$$p(X_1, \ldots, X_n) \leftarrow body_i$$

We replace these k clauses with:

$$p(X_1, \ldots, X_n) \leftrightarrow body_1 \vee \ldots \vee body_k$$

This algorithm is adapted from [Lloyd, 1987][§14], in which further details
may be found.

Example: Constructing the completion of *subset* Here is a simple example of the algorithm. Consider the following program for **subset/2**.

```
subset([],J).
subset([Hd|Tl],J) :- member(Hd,J), subset(Tl,J).
```

A goal of the form subset(I,J) succeeds if I is a subset of J, where sets are represented as lists.

The first stage of the algorithm is to rewrite the program into logical notation.

$$subset([\,], J)$$
$$subset([Hd|Tl], J) \leftarrow member(Hd, J) \,\&\, subset(Tl, J)$$

The second stage is to make the head arguments into distinct variables.

$$subset(I, J) \leftarrow I = [\,]$$
$$subset(I, J) \leftarrow I = [Hd|Tl] \,\&\, member(Hd, J) \,\&\, subset(Tl, J)$$

Note that in both clauses we exercised our option not to replace J.

The third stage is to existentially quantify each variable which occurs in the body but not in the head.

$$subset(I, J) \leftarrow I = [\,]$$
$$subset(I, J) \leftarrow \exists Hd \in Type, \exists Tl \in list(Type).$$
$$I = [Hd|Tl] \,\&\, member(Hd, J) \,\&\, subset(Tl, J)$$

The fourth stage is to combine these two clauses into a single equivalence.

$$subset(I, J) \leftrightarrow I = [\,] \vee$$
$$\exists Hd \in Type, Tl \in list(Type).$$
$$I = [Hd|Tl] \,\&\, member(Hd, J) \,\&\, subset(Tl, J)$$

3.2 From Specifications to Programs

In this section we describe an algorithm for compiling a specification into the logic program it directly specifies.

One way to make such a compilation algorithm would be to reverse the Clark completion algorithm. Unfortunately, this would not be a general-purpose compilation algorithm. This is because the class of specifications is much larger than the class of completions. For instance, the equivalence:

$$subset(I, J) \leftrightarrow (\forall El \in Type.\ member(El, I) \rightarrow member(El, J)) \qquad (5)$$

is a specification, but is not a completion. Our algorithm must cover these non-completions too.

Our algorithm will compile any specification. We will call it the *Lloyd-Topor compilation algorithm.*

Definition 2 (The Lloyd-Topor Compilation Algorithm) *The key idea of the algorithm is to put the specification into clausal form. It consists of three stages.*

1. *Turn the main* ↔ *into a* ←.
2. *Put the specification into clausal form using the Lloyd-Topor rules. They are too complicated to give in full here. We have illustrated the general idea by giving some examples in figure 3.2. The complete set is given in [Lloyd, 1987][p113].*
3. *Turn the logical symbols into program symbols, i.e. invert stage 1 of the Clark completion algorithm.*

Name	Input Clause	Output Clause(s)
∃	$head \leftarrow \ldots \ \& \ \exists \vec{X} \in \vec{t}. \ body \ \& \ \ldots$	$head \leftarrow \ldots \ \& \ body \ \& \ \ldots$
∨	$head \leftarrow \ldots \ \& \ (a \vee b) \ \& \ \ldots$	$head \leftarrow \ldots \ \& \ a \ \& \ \ldots$ $head \leftarrow \ldots \ \& \ b \ \& \ \ldots$
¬ →	$head \leftarrow \ldots \ \& \ \neg(a \rightarrow c) \ \& \ \ldots$	$head \leftarrow \ldots \ \& \ a \ \& \ \neg c \ \& \ \ldots$
∀	$head \leftarrow \ldots \ \& \ \forall \vec{X} \in \vec{t}. \ body \ \& \ \ldots$	$head \leftarrow \ldots \ \& \ \neg \exists \vec{X} \in \vec{t}. \ \neg body \ \& \ \ldots$
¬∃	$head \leftarrow \ldots \ \& \ \neg \exists \vec{X} \in \vec{t}. \ body \ \& \ \ldots$	$head \leftarrow \ldots \ \& \ \neg q(Y_1, \ldots, Y_k) \ \& \ \ldots$ $q(Y_1, \ldots, Y_k) \leftarrow body$

In the last rule, q is a new predicate symbol and the Y_i are the free variables in $\exists \vec{X} \in \vec{t}. \ body$.

These rules are applied as rewrite rules to the specification until no more apply. The specification is then in clausal form.

Fig. 1. Selected Lloyd-Topor Transformation Rules

We can now see the need for the restriction, given in §1.2, to exclude defined functions from specifications. There is no provision in the Lloyd-Topor algorithm to turn these defined functions into predicate definitions. Consider, for instance:

$$plus(X, Y, Z) \leftrightarrow (X = 0 \ \& \ Z = Y) \vee$$
$$\exists X' \in nat. \ X = s(X') \ \& \ Z = s(X' + Y)$$

Lloyd-Topor compiles this into the Prolog program:

```
plus(X,Y,Z) :- X=0, Z=Y.
plus(X,Y,Z) :- X=s(X'), Z=s(X'+Y).
```

But this program will fail for non-zero X because X'+Y is undefined. There is no provision in Prolog to define it as a function nor in Lloyd-Topor to compile it into a predicate. This problem could be solved by a suitable modification of the Lloyd-Topor algorithm.

The Lloyd-Topor algorithm is a near inverse of the Clark completion algorithm.

- Stage 4 of Clark completion is inverted by Stage 1 and rule ∨ of Lloyd-Topor.
- Stage 3 of Clark completion is inverted by rule ∃ of Lloyd-Topor.
- Stage 1 of Clark completion is inverted by stage 3 of Lloyd-Topor.

We could also invert stage 2 of Clark completion by introducing an additional rule into stage 2 of Lloyd-Topor which removed $X_i = s_i$ literals from the body be replacing X_i by s_i in the head. Unfortunately, it would be impossible to prevent this rule from removing X_i literals that were present in the original program. Similarly, rule ∨ can remove disjunctions that were present in the original program, if the programming language allows them.

Note that Clark completion is not an inverse of Lloyd-Topor, since the original specification may not be a completion.

Example: Translating a *subset* specification into a program We will illustrate the Lloyd-Topor algorithm on the specification of *subset*, (5) above. The first stage is to turn the ↔ into a ←.

$$subset(I, J) \leftarrow (\forall El \in Type.\ member(El, I) \rightarrow member(El, J))$$

The second stage is to use the Lloyd-Topor compilation to put this into clausal form. The rules from figure 3.2 apply as follows:

$$subset(I, J) \leftarrow \neg \exists El \in Type.\ \neg(member(El, I) \rightarrow member(El, J))$$

$$subset(I, J) \leftarrow \neg not\_subset(I, J)$$
$$not\_subset(I, J) \leftarrow \neg(member(El, I) \rightarrow member(El, J))$$

$$subset(I, J) \leftarrow \neg not\_subset(I, J)$$
$$not\_subset(I, J) \leftarrow member(El, I)\ \&\ \neg member(El, J)$$

The third stage is to rewrite the clauses into program notation.

```
subset(I,J) :- not not_subset(I,J).
not_subset(I,J) :- member(El,I), not member(El,J).
```

4 Equivalence of Specifications

The problem we will consider in this section is how to prove that two logic program specifications are equivalent. That is, suppose $Spec_1$ and $Spec_2$ are two logic program specifications. We will consider how to prove:

$$\forall \overrightarrow{Args} \in \overrightarrow{Types}.\ Spec_1 \leftrightarrow Spec_2 \tag{6}$$

In the interests of readability we will sometimes omit the $\forall \overrightarrow{Args} \in \overrightarrow{Types}$ part.

We saw in §3 that each of these two specifications compiles directly into a logic program. Some of these logic programs are more practical than others. Suppose $Prog_1$ is the logic program corresponding to $Spec_1$ and $Prog_2$ to $Spec_2$.

If $Prog_2$ is an impractical program and $Prog_1$ is a practical program then we can view the proof of equivalence (6) as verification of $Prog_1$ in terms of $Spec_2$. If both $Prog_1$ and $Prog_2$ are practical programs then we can view this proof as the transformation of $Prog_2$ into $Prog_1$. Typically, $Prog_1$ will be more efficient than $Prog_2$.

Thus, when reasoning at the level of specifications, the processes of verification and transformation coalesce; They differ in degree not kind. Of course, the real challenge in transformation is to be given an inefficient program, $Prog_2$, and to construct a more efficient program, $Prog_1$. This is essentially the same problem as synthesising a practical program, $Prog_1$, which meets a given specification, $Spec_2$. We will tackle this joint problem in §5 below.

Example: Equivalence of two *subset* definitions Here is a simple example. Consider the verification theorem:

$$\forall I \in list(Type), J \in list(Type). \tag{7}$$
$$subset(I, J) \leftrightarrow (\forall El \in Type. \; member(El, I) \rightarrow member(El, J))$$

where *subset* and *member* are defined by the specifications:

$$subset(I, J) \leftrightarrow I = [\,] \vee$$
$$\exists Hd \in Type, Tl \in list(Type).$$
$$I = [Hd|Tl] \; \& \; member(Hd, J) \; \& \; subset(Tl, J)$$
$$member(El, L) \leftrightarrow (\exists Hd \in Type, Tl \in list(Type). \; L = [El|Tl]) \vee$$
$$(\exists Hd \in Type, Tl \in list(Type). \; L = [Hd|Tl] \; \& \; member(El, Tl))$$

The quantification of the outer universal variables has been omitted to reduce clutter.

Using the Lloyd-Topor compilation, the left and right hand sides of equivalence (7) compile into the two Prolog programs:

```
subset_1([],J).
subset_1([Hd|Tl],J) :- member(Hd,J), subset_1(Tl,J).

subset_2(I,J) :- not not_subset(I,J).
not_subset(I,J) :- member(El,I), not member(El,J).
```

The details of these compilations were given in §3. Program subset_1/ will run in any input mode, but subset_2 will flounder in all input modes except subset_2(+,+). The proof of equivalence (7) can thus be regarded as an exercise in verification of subset_1.

To prove (7) we must use induction on the recursive structure of lists with I as the induction variable. This generates a base case and a step case.

The base case is obtained from (7) by substituting the empty list for I.

$$subset([\,], J) \leftrightarrow (\forall El \in Type. \; member(El, [\,]) \rightarrow member(El, J))$$

Applying the definitions of *subset* and *member* this reduces to the problem of proving:

$$true \leftrightarrow (\forall El \in Type.\ false \rightarrow member(El, J))$$

which further reduces to *true*, using the rules of predicate logic.

The step case consists of an induction conclusion which can be proved with the aid of an induction hypothesis. The induction hypothesis and conclusion are obtained from (7) by substituting tl and $[hd|tl]$, respectively, for I.

$$subset(tl, J) \leftrightarrow (\forall El \in Type.\ member(El, tl) \rightarrow member(El, J))$$
$$\vdash subset([hd|tl], j) \leftrightarrow (\forall El \in Type.\ member(El, [hd|tl]) \rightarrow member(El, j))$$

The turnstile symbol $\vdash$ indicates that the induction hypothesis on the left can be assumed when proving the induction conclusion on the right. Note that the universal variable J in the induction hypothesis can be instantiated, if necessary, whereas it should not be instantiated in the induction conclusion. We have ensured this by representing J by a free variable (upper case) on the left and a constant j (lower case) on the right.

Using the definitions of *subset* and *member* the induction conclusion can be reduced to:

$$member(hd, j)\ \&\ subset(tl, j)$$
$$\leftrightarrow (\forall El \in Type.\ (El = hd \lor member(El, tl)) \rightarrow member(El, j))$$

which, using the rules of predicate logic, can be rewritten as follows:

$$member(hd, j)\ \&\ subset(tl, j)\ \leftrightarrow\ (\forall El \in Type.\ (El = hd \rightarrow member(El, j))\ \&$$
$$(member(El, tl) \rightarrow member(El, j)))$$
$$member(hd, j)\ \&\ subset(tl, j)\ \leftrightarrow\ ((\forall El \in Type.\ El = hd \rightarrow member(El, j))\ \&$$
$$(\forall El \in Type.\ member(El, tl) \rightarrow member(El, j)))$$
$$member(hd, j)\ \leftrightarrow\ ((\forall El \in Type.\ El = hd \rightarrow member(El, j))$$
$$member(hd, j)\ \leftrightarrow\ member(hd, j)$$
$$true$$

This completes the proof of the step case. The whole of equivalence (7) is now proved.

5 Synthesis of Specifications

The problem we will consider in this section is how, given an initial specification, we can synthesise an equivalent new one. That is, suppose we are given a specification $Spec_2$, how can we construct a specification $Spec_1$ such that:

$$\forall \overrightarrow{Args} \in \overrightarrow{Types}.\ Spec_1 \leftrightarrow Spec_2$$

As discussed in §4 above, this will enable us to synthesise a practical program, $Prog_1$, from a specification $Spec_2$ and/or to transform an inefficient program, $Prog_2$, into an efficient program, $Prog_1$.

The essential idea underlying our solution to this problem will be to proceed with the proof of the equivalence theorem as if $Spec_1$ were known, and to pick up clues as to its definition as we proceed. This is best illustrated with an example.

Example: Synthesis of *subset* We will repeat the proof of equivalence (7), given in §4 above, but with *subset* left undefined. This will leave us with some unprovable subgoals, which we can use to form the definition of *subset*.

We repeat below the equivalence to be proved.

$$subset(I, J) \leftrightarrow (\forall El \in Type. \; member(El, I) \rightarrow member(El, J))$$

Many of the steps in the previous proof of this equivalence can be repeated, but the proof cannot be completed due to the absence of the definition of *subset*. The residue of subgoals is:

$$subset([\,], J) \leftrightarrow true$$
$$subset([hd|tl], J) \leftrightarrow member(hd, J) \; \& \; subset(tl, J)$$

This residue can be used to suggest a definition of *subset*. We regard the residue as a logic program and take its completion, which we then adopt as the definition. In this case the definition this suggests is:

$$subset(I, J) \leftrightarrow I = [\,] \lor$$
$$\exists Hd \in Type, Tl \in list(Type).$$
$$I = [Hd|Tl] \; \& \; member(Hd, J) \; \& \; subset(Tl, J)$$

as required. The residue of subgoals can be readily proved from this definition, so the proof is completed.

6 Automated Theorem Proving

The equivalences between specifications are usually straightforward to prove. However, it does require some facility with mathematical ideas to find these proofs. It is desirable to automate this proof discovery process as much as possible to reduce the burden on the program developer. In this section we discuss how this can be done.

The main technical problem in automated theorem proving is guiding the search for a proof. It is easy to represent the theorem and the rules for manipulating it. It is particularly easy in a logic programming language since the necessary data-structures are provided directly. It is also easy to write a program to apply these rules exhaustively until the required proof is found. Unfortunately, for all but trivial theorems, this process rapidly becomes bogged down in an explosion of partially generated proofs. This phenomenon is called the *combinatorial explosion.*

To defeat the combinatorial explosion we need to use heuristic methods to guide the proof building process along the most promising paths. To illustrate such heuristic methods, consider the problem of rewriting the induction conclusion so that the induction hypothesis may be applied to it. The form of the step case of an inductive proof is:

$$P(tl) \vdash P(\boxed{[hd|\underline{tl}]}^{\uparrow})$$

Note that the induction conclusion on the right differs from the induction hypothesis on the left by inclusion of the induction term $[hd|\ldots]$. We have emphasised this by drawing a box around the induction term and underlining the induction variable, tl inside it. We call this boxed sub-expression a *wave-front*. The arrow, $\uparrow$, represents the direction of movement of the wave-front: upwards (or outwards depending on your point of view) through the induction conclusion.

The presence of this wave-front prevents us from using the induction hypothesis to prove the induction conclusion. To enable the induction hypothesis to be used we need to move the wave-front to the outside of the induction conclusion, *i.e.* we need to rewrite the induction conclusion into the form:

$$P(tl) \vdash \boxed{Q(hd, tl, \underline{P(tl)})}^{\uparrow}$$

so that the induction hypothesis can be used, giving:

$$P(tl) \vdash Q(hd, tl, true)$$

We call this rewriting process *rippling*. It consists of applying rewrite rules which move the wave-fronts outwards but leave the rest of the induction conclusion unchanged. Rewrite rules of this form are called *wave-rules*. They have the form:

$$f(\boxed{g(\underline{X})}^{\uparrow}) \Rightarrow \boxed{h(\underline{f(X)})}^{\uparrow}$$

Some examples are given in figure 6. For more information about rippling see [Bundy *et al*, 1991].

Example: Rippling in the *subset* proof To illustrate rippling consider the proof of equivalence (7) from §4. We start by putting a wave-front around each occurrence of the induction term and then we ripple these outwards.

$$subset(\boxed{[hd|\underline{tl}]}^{\uparrow}, j) \leftrightarrow$$
$$(\forall El \in Type.\ member(El, \boxed{[hd|\underline{tl}]}^{\uparrow}) \rightarrow member(El, j))$$

$$subset(\boxed{[hd|\underline{tl}]}^{\uparrow}, j) \leftrightarrow$$
$$(\forall El \in Type.\ \boxed{El = hd \lor \underline{member(El, tl)}}^{\uparrow} \rightarrow member(El, j))$$

$$subset(\boxed{[hd|\underline{tl}]}^{\uparrow}, j) \leftrightarrow$$
$$(\forall El \in Type.\ \boxed{El = hd \rightarrow member(El, j)\ \&\ \underline{member(El, tl) \rightarrow member(El, j)}}^{\uparrow})$$

$$\boxed{member(hd, j)\ \&\ \underline{subset(tl, j)}}^{\uparrow} \leftrightarrow$$
$$\boxed{\forall El \in Type.\ El = hd \rightarrow member(El, j)\ \&\ \underline{\forall El \in Type.\ member(El, tl) \rightarrow member(El, j)}}^{\uparrow}$$

$$\boxed{member(hd, j) \leftrightarrow \forall El \in Type.\ El = hd \rightarrow member(El, j)\ \&}$$
$$\boxed{subset(tl, j) \leftrightarrow \forall El \in Type.\ member(El, tl) \rightarrow member(El, j)}^{\uparrow}$$

The wave-rules that enable this rewriting are given in figure 6.

Rippling is now complete and a copy of the induction hypothesis is embedded within the induction conclusion. The induction hypothesis can now be used to replace this copy with *true*. This leaves the subgoal:

$$member(hd, J) \leftrightarrow \forall El \in Type.\ El = hd \rightarrow member(El, J) \qquad \&\ true$$

which is readily proved. This completes the step case.

There are many other search control problems in inductive inference, *i.e.* choosing an induction rule, guiding the base case, deciding on case splits, constructing witnesses of existential variables, generalising the induction formula, conjecturing suitable lemmas. Rippling is often the key to solving these problems. For instance, an induction rule can be chosen by a look-ahead mechanism to see which choice will most facilitate subsequent rippling. Unfortunately, there is insufficient space to explore these issues further here. The interested reader is referred to [Bundy *et al*, 1990].

7 Termination of Logic Programs

The problem we will consider in this section is how to prove that a logic program terminates. To solve this problem we must reason about programs rather than specifications. This is because whether a program terminates depends on both the code and the interpreter, that is, we must consider how the search strategy evaluates the code. There is a wider variation of interpreters in logic programming than in other kinds of programming. This factor makes termination a more complicated problem for logic programs than it is for other kinds of programs.

$$subset_1(\boxed{[Hd|\underline{Tl}]}^\uparrow, J) \Rightarrow \boxed{member(Hd, J) \& \underline{subset_1(Tl, J)}}^\uparrow$$

$$member(El, \boxed{[Hd|\underline{Tl}]}^\uparrow) \Rightarrow \boxed{El = Hd \lor \underline{member(El, Tl)}}^\uparrow$$

$$\boxed{A \lor \underline{B}}^\uparrow \to C \Rightarrow \boxed{(A \to C) \& \underline{(B \to C)}}^\uparrow$$

$$\forall X \in T. \boxed{A \& \underline{B}}^\uparrow \Rightarrow \boxed{\forall X \in T. A \& \underline{\forall X \in T. B}}^\uparrow$$

$$\boxed{(A_1 \& \underline{B_1})}^\uparrow \leftrightarrow \boxed{(A_2 \& \underline{B_2})}^\uparrow \Rightarrow \boxed{(A_1 \leftrightarrow A_2) \& \underline{(B_1 \leftrightarrow B_2)}}^\uparrow$$

If the left-hand side of a wave-rule matches a sub-expression of a goal then this sub-expression can be replaced by the instantiated right-hand side of the wave-rule. In this match the wave-fronts in the rule and the goal must be aligned. Note that this causes the wave-front to move outwards through the goal leaving the rest of the goal unchanged.

Wave-rules can be derived from the step parts of recursive definitions, e.g. the subset and member rules above. They can also come from distributive laws, e.g. the other three rules above, from associative laws, substitution laws and many other sources. Our proofs run backwards from the theorem to the axioms. Thus, in wave-rules based on implications, the direction of rewriting is opposite to the direction of implication, cf. the last rule above.

Fig. 2. Wave-Rules for the *subset* Example

Another complication is caused by the relational nature of logic programs, *i.e.* the fact that they work in different input modes and that they can return alternative outputs on backtracking. Termination is no longer a simple concept. A program may terminate in some input modes, but not in others. Hence, whether a program terminates depends on the goal. A program may return one or more outputs, but then fail to terminate on backtracking. We must define different kinds of non-termination.

Universal termination: the search space is finite.
Existential termination: if the search space is infinite, then it has success branches at finite depth.

Note that existential termination includes universal termination as a special case.

Whether a search space is finite depends, among other things, on the order in which subgoals are solved. For instance, consider the clause defining $p2$ below:

$$p2(X) \leftarrow r(X) \& q(X).$$
$$q(a).$$
$$r(b) \leftarrow r(X).$$
$$r(a).$$

If $r(X)$ is solved before $q(X)$ then the search space will be infinite, but if $q(X)$ is solved first then the search space is finite.

In the discussion below we will describe a simple termination technique that establishes universal termination of definite programs[6] Our technique is to associate a well-founded measure with each procedure call and to show that this strictly decreases as computation proceeds. A well-founded measure is one that cannot decrease for ever. This ensures that the computation cannot proceed for ever. An example of a well-founded measure is the natural numbers[7] ordered by $>$. This is well-founded because there is no infinite sequence of the form: $n_1 > n_2 > n_3 > \ldots$. Eventually, one of the n_i would be 0, and there is no natural number smaller than that. This natural number, well-founded measure will suffice for our purposes. To prove termination we will associate a natural number with each procedure call and show that the number associated with the current procedure call strictly decreases as the computation proceeds.

Example: The Termination of *subset* A simple example will illustrate this technique.

Consider the pure logic *subset* program in input mode $subset(+, +)$.

$$subset([\,], J). \tag{8}$$
$$subset([Hd|Tl], J) \leftarrow member(Hd, J) \,\&\, subset(Tl, J). \tag{9}$$

$$member(El, [El|Tl]). \tag{10}$$
$$member(El, [Hd|Tl]) \leftarrow member(El, Tl). \tag{11}$$

We will define a measure on procedure calls of the form $subset(I, J)$ and $member$ (El, J), where both I and J are ground.

Definition 3 (List length norm) *The* list length norm *is a function from a list, L, to a natural number, $len(L)$, defined recursively as:*

$$len([\,]) = 0 \quad len([Hd|Tl]) = len(Tl) + 1$$

Definition 4 (procedure call measure) *The* procedure call measure *is a function, $|\ldots|$, from a literal, $subset(I, J)$ or $member(El, J)$, where I and J are ground, to a natural number defined as:*

$$|subset(I, J)| = len(I) + len(J)$$
$$|member(El, J)| = len(J)$$

We now show that the procedure call measure is strictly decreased as the computation proceeds. Each step of the computation is a resolution with one of the clauses (8) — (11). Each resolution replaces a call of the head literal with

---

6 *I.e.* those without negation as failure.

7 *I.e.* the non-negative integers: 0,1,2,...

calls to each of the body literals. We show that the measure of the head literal is strictly greater than the measure of each of the body literals. In the case of the clauses (8) and (10) this is trivial, since their bodies contain no literals. In the case of clauses (9) and (11) the following calculations establish our claim.

$$
\begin{aligned}
|subset([Hd|Tl], J)| &= len([Hd|Tl]) + len(J) = len(Tl) + 1 + len(J) \\
&> \quad len(J) \quad = |member(Hd, J)| \\
|subset([Hd|Tl], J)| &= len([Hd|Tl]) + len(J) = len(Tl) + 1 + len(J) \\
&> \quad len(Tl) + len(J) \quad = |subset(Tl, J)| \\
|member(El, [Hd|Tl])| &= \quad len([Hd|Tl]) \quad = len(Tl) + 1 \\
&> \quad len(Tl) \quad = |member(El, Tl)|
\end{aligned}
$$

Note that this argument breaks down if any of the lists in the procedure call arguments are non-ground. Indeed, a procedure call $member(a, L)$, for instance, will not terminate in the universal sense. It will return an infinite number of results of the form:

$$
[a|Tl], \quad [Hd1, a|Tl], \quad [Hd1, Hd2, a|Tl], \quad \ldots
$$

Similar remarks hold for $subset(I, J)$ where either I or J is non-ground. These are examples of existential termination.

An introduction to more elaborate techniques for proving termination can be found in [Hogger, 1990][Theme 59] and [De Schreye, 1992].

8   Tuning Impure Programs

Suppose we have specified a logic program, transformed this into an equivalent specification and then compiled this new specification into a Prolog program using the Lloyd-Topor algorithm. We have seen that the completion of this Prolog program is logically equivalent to the original specification. Unfortunately, due to the impure aspects of Prolog, this program may not behave operationally as desired. For instance,

- It may not terminate in its intended input mode.
- Some of its uses of negation as failure may flounder.
- Its side effects may occur in the wrong order.

For a complete list of the possible problems see [Deville, 1990][p240].

Sometimes these problems can be fixed by tuning the program.

Examples: Reordering clauses and literals We can illustrate this with some examples.

Consider the following clauses for r/1:

```
r(b) :- r(X).
r(a).
```

These clauses do not existentially terminate in input mode r(-). This problem can be readily fixed by reordering the two clauses to:

```
r(a).
r(b) :- r(X).
```

This gives a logically equivalent,but operationally different program. The program for r/1 now existentially terminates.

Now consider the following clauses:

```
p2(X) :- r(X), q(X).
q(a).
r(a).
r(b) :- r(X).
```

The program for p2/1 existentially terminates in input mode p2(-). We can transform this into a program that universally terminates by reordering the literals in the p2 clause to give:

```
p2(X) :- q(X), r(X).
```

As we have seen, in §2.2, this clause can be partially evaluated to the logically and operationally equivalent clause:

```
p2(a).
```

9 Abstract Interpretation of Logic Programs

The techniques we have discussed above can be used not only to reason with programs in their original form, but also with abstractions of these programs. For instance, we can infer mode and type information about programs, or we can discover whether variables are independent or aliasing during execution. This information can be used, for instance, to optimise programs. Abstract reasoning can be automated and incorporated into optimising compilers.

To apply abstract reasoning to a logic program we must define the abstraction as follows.

Abstract the domain: We must describe the objects that can appear as arguments to the program. For instance, for reasoning with modes we will want objects to represent a free variable and an instantiated variable. We might also want objects for various degrees of instantiation, *e.g.* totally ground *vs* partially instantiated. For reasoning about types we will choose abstract objects to represent the various types of concrete objects, *e.g.* numbers, strings, lists, *etc.*

In addition, to these basic abstract objects we must usually specify how to form the least upper bound of any two objects. This is because finding the least upper bound is often used to combine different instantiations of the same variable. This is usually done by putting the objects in a lattice with a top element, *e.g.* any mode, any type, *etc.*

Abstract the programs: We must describe how the program operates on this abstract domain. For instance, we must define the built-in predicates over the abstract domain. We must adapt the interpreter, as necessary. For instance, if a variable is instantiated in different ways on different branches of the search space these values might be combined into one by taking the least upper bound. We must describe how to unify abstract objects.

Non-termination is far more common with abstract programs, because distinct concrete procedure calls are often identical when abstracted. Therefore, it is usually necessary to modify the interpreter to trap looping and ensure that an output is returned.

Example: Inferring the Types of Predicates We will illustrate these ideas by the use of a simple example: the inference of the types of a whole program, given some partial information about types.

Our abstract domain will consist of the simple lattice of types given in figure 3.

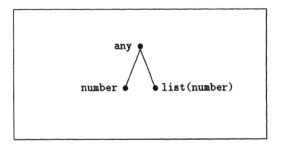

Fig. 3. Simple Lattice of Types

Suppose we want to infer the types of **subset** and **member** as defined by the clauses:

```
subset([],J).
subset([Hd|Tl],J) :- member(Hd,J), subset(Tl,J).

member(El,[El|Tl]).
member(El,[Hd|Tl]) :- member(El,Tl).
```

given that the type of the first argument of **subset** is known to be *list(number)*, say.

To infer the remaining types we partially evaluate the procedure call:

```
?- subset(list(number),X).
```

If we resolve this goal against the first clause for **subset** then we must unify `list(number)` with `[]` and X J. We should define abstract unification so that these unifications both succeed, but this resolution does not tell us more than we already knew.

More interesting is the resolution against the second clause for **subset**. We should define abstract unification so that unifying `list(number)` and `[Hd|Tl]` succeeds, instantiating Hd to number and Tl to `list(number)`. The new procedure calls are:

```
?- member(number,X), subset(list(number),X).
```

We can now repeat this process for the **member(number,X)** procedure call. This gives us two instantiations of X, both of which are `list(number)`. The new procedure call generated by the second clause is:

```
?- member(number,list(number)).
```

but this is subsumed so by the earlier call, so can tell us nothing new. Our loop detection mechanism should note this and terminate this call.

The remaining procedure call is:

```
?- subset(list(number),list(number)).
```

which is also subsumed by an earlier call and should also be terminated.

The abstract interpretation is now complete. We have inferred the types of our predicates to be:

```
subset(list(number),list(number))
member(number,list(number))
```

10   Conclusion

In these notes we have outlined various techniques for reasoning about logic programs. These techniques can be combined in various ways to form a methodology for logic program development. Ideally, for the reasons summarised in §2.3, this methodology should take logic program specifications as its central representation. The original description of the desired program should take the form of a specification, as defined in §1.2.

This specification can be used to synthesise a more efficient specification using the techniques of §5. It may seem odd to discuss the 'efficiency' of a specification. One way to measure this is to compile the specification into a logic program using the Lloyd-Topor algorithm (see §3.2) and then to measure the efficiency of this program. However, there are some aspects of efficiency that are independent of the particular target programming language, *e.g.* the complexity associated with any forms of recursion used in the specification. These aspects can be measured more directly.

Having synthesised an acceptable specification, this can be compiled into a program using the Lloyd-Topor algorithm. The termination, mode and similar properties of this program can be analysed using the techniques of §7 and 9. If necessary, the program can then be tuned using the techniques of §8.

If a program, rather than a specification, is available, then this can be lifted into a specification using the Clark completion algorithm, §3.1. The above methodology is then applicable.

All these techniques can be automated to a greater or lesser extent. Automation makes possible machine aids to program development that remove some of the tedium and error from the proof, compilation and analysis steps.

Recommended Reading

In these notes it has only been possible to give a rough outline of the range and complexity of the techniques available. If you want to find out more, here are some suggestions for further reading.

Elementary and very readable introductions to the ideas outlined in these notes can be found in the two books by Chris Hogger, [Hogger, 1984, Hogger, 1990]. John Lloyd's book, [Lloyd, 1987], is the standard reference for the theoretical background on logic programming. A more detailed account of verification and synthesis of logic programs can be found in Yves Deville's book [Deville, 1990]. Deville adopts the same position that we have on reasoning with specifications, wherever possible, rather than with programs. A discussion of the different notions of logic program equivalence can be found in the paper by Michael Maher, [Maher, 1987]. An introductory survey of termination proving techniques can be found in the tutorial notes by Danny De Schreye and Kristof Verschaetse, [De Schreye, 1992], and a similar survey of work on abstraction can be found in the tutorial notes by Maurice Bruynooghe and Danny De Schreye, [Bruynooghe & De Schreye, 1986].

References

[Bruynooghe & De Schreye, 1986] Bruynooghe, M. and De Schreye, D. (1986). Tutorial notes for: abstract interpretation in logic programming. Technical report, Department of Computer Science, Katholieke Universiteit Leuven, Belgium, Tutorial given at ICLP-86, Seattle. Notes available from the authors. Department of Computer Science, Katholieke Universiteit Leuven, Celestijnenlaaan 200A, 3001 Heverlee, Belgium.

[Bundy et al, 1990] Bundy, A., Smaill, A. and Hesketh, J. (1990). Turning eureka steps into calculations in automatic program synthesis. In Clarke, S.L.H., (ed.), *Proceedings of UK IT 90*, pages 221–6. Also available from Edinburgh as DAI Research Paper 448.

[Bundy et al, 1991] Bundy, A., Stevens, A., van Harmelen, F., Ireland, A. and Smaill, A. (1991). Rippling: A heuristic for guiding inductive proofs. Research Paper 567, Dept.

of Artificial Intelligence, Edinburgh, To appear in Artificial Intelligence.

[De Schreye, 1992] De Schreye, K. D.and Verschaetse. (1992). Termination of logic programs: Tutorial notes. Technical Report CW-report 148, Department of Computer Science, Katholieke Universiteit Leuven, Belgium, To appear in the proceedings of Meta-92.

[Deville, 1990] Deville, Y. (1990). *Logic programming: systematic program development*. Addison-Wesley Pub. Co.

[Hogger, 1984] Hogger, C.J. (1984). *Introduction to logic programming*. Academic Press.

[Hogger, 1990] Hogger, C.J. (1990). *Essentials of Logic Programming*. Oxford University Press.

[Lloyd, 1987] Lloyd, J.W. (1987). *Foundations of Logic Programming*. Symbolic Computation. Springer-Verlag, Second, extended edition.

[Maher, 1987] Maher, M.J. (1987). Equivalences of logic programs. In Minker, J., (ed.), *Foundations of Deductive Databases and Logic Programming*. Morgan Kaufmann.

Software Formal Specification by Logic Programming:
The example of Standard PROLOG*

AbdelAli Ed-Dbali[1,2] and Pierre Deransart[2]

[1] Université d'Orléans
LIFO - B.P. 6759
45067 ORLEANS Cedex 2 FRANCE
E-mail : AbdelAli.Ed-Dbali@univ-orleans.fr
[2] INRIA - Projet ChLoÉ
Domaine de Voluceau - B.P. 105
78153 LE CHESNAY Cedex
E-mail : Pierre.Deransart@inria.fr

Abstract. The aim of this presentation is to show how logic programming can be used to make formal specifications. This approach is illustrated by the project of standard PROLOG, as discussed in the ISO working group, which has been completely described by a formal specification.

1   Introduction

The aim of this presentation is to show how logic programming can be used to make formal specifications. This approach will be illustrated by the project of standard PROLOG, as discussed in the ISO working group SC22/WG17 [8], which has been completely described by a formal specification.

The specification method has been introduced by P. Deransart and G. Ferrand in 1987 [4]. A specification has a program part and a comment part.

The program part is based on stratified normal programs (Horn clauses with negation in their bodies).

A methodology has been developed to write the comment part. Without any comment a formal specification is too difficult to understand. Each predicate definition is thus commented in order to facilitate its reading. The link between the comment part and the program part is explained by the proof method.

The proof method is based on the structural induction and a tractable refinement of it [2]. It permits to prove some properties without knowing the semantics of the specification. It has been implemented in a semi automatic system which permits to handel the specification and the comments together to verify their consistency [7].

Another important point concerning this formal specification method is the ease in deducing an executable specification from the formal one. As most of

---

* This work is partly supported by the GRECO-programmation and the MRT.

the relations in the program part are new, it is essential to be able to "run" the specification in order to test them.

We show how this specification language has been used to specify the project of Standard PROLOG in the ISO working group SC22/WG17 (The full Standard PROLOG specification can be found in [8]), and present some tools which have been built in order to help to validate the project: test of stratification, automatic generation of the error cases from templates, systematic verification of the cited examples with an executable specification [5], validation tools.

This presentation is organized as follow: there are mainly two sections. In section 2 we introduce the specification language: stratified normal program, the proof method used to define the comments and the way to design a specification. In section 3 we present the Standard PROLOG ISO project and the application of the specification method to specify it.

2 The specification method

2.1 Program part of the specification

Syntax: The program part uses exclusively *normal clauses*. A normal clause is a clause of the form: $A \leftarrow L_1, ..., L_m$ $(m \geq 0)$ where A is an atom and $L_1, ..., L_m$ are literals (positive or negative). The atoms are terms of the form $p(t_1, t_2, ..., t_n)$ where p is a predicate symbol defining a relation in the formal semantics.

A formal specification consists of a collection of packets. Each packet consists of a collection of normal clauses defining a relation denoted by a predicate name. All the predicate names are unique. We recommend not to use the same predicate name for relations with different arities (different number of arguments). Each packet has an associated comment whose syntax is defined in the subsection 2.2. The order of the packets, clauses in the packets or literals in the body of the clauses is irrelevant.

The order of presentation is only aimed at facilitating the reading of the specification. There is no methodology to define the best order of presentation. In practice this order may correspond to the presentation of the clauses in the version of the specification executable on some "standard" interpreter.

Example 1. Here is an example of specification:

- $includ(L_1, L_2)$ then *if* L_1, L_2 are lists *then* $L_1 \subseteq L_2$
    $includ(L_1, L_2) \leftarrow not \ $ $ninclud(L_1, L_2)$
- $ninclud(L_1, L_2)$ then *if* L_1, L_2 are lists *then* $L_1 \not\subseteq L_2$
    $ninclud(L_1, L_2) \leftarrow elem(E, L_1), not \ elem(E, L_2)$
- $elem(E, L)$ then *if* L is a list *then* $E \in L$
    $elem(E, [E|L]) \leftarrow$
    $elem(E, [H|L]) \leftarrow elem(E, L)$

Notice that the language used in the comments is not constrained, so it may contain informal statments. The reader of a specification is supposed to understand completely the meaning of the comments.

Semantics: We consider the semantics of the program part of a specification.

Our aim is to take advantage of the high level of expressiveness given by the negation, without requiring any specific knowledge on its semantics.

A program P has a purely declarative semantics which is formalized by a set of ground atoms, called $DEN(P)$ (for *denotation of P*). A program is stratified [1]. In this case all the approaches of the semantics of the negation based on the choice of a privileged model coincide [4]. But for our purpose the most important is that this semantics is very well suited to methodology, especially proof methods. The simple definition we give defines uniquely the semantics as a set of ground atoms.

Our view of the declarative semantics is a constructive one, based on a notion of generalised proof-tree. Given a normal program P, a generalised proof-tree is a proof-tree built with ground clause instances such that the positive leaves are ground instance of facts and the negative leaves *not a* are such that *a* is not the root of any ground generalised proof-tree. The semantics of a program is the set of the ground generalised proof-tree roots.

Notice that the definition of the semantics limits the class of program which can be used: not any general program has such a semantics. But all stratified programs have.

A normal program is stratified if the predicates used in the negative literals of a packet defining the predicate p never use p in their definition directly or indirectly.

The program of the previous example is stratified. The following program is not:

$$even(zero) \leftarrow$$
$$even(succ(X)) \leftarrow not\ even(X)$$

2.2 Comment part of the specification

The link between the comment part and the program part is based on the proof method for partial correctness and completeness.

Proof method: Let us assume that all comments are formal.

A *formal comment associated to a specification* is given by two families of assertions (expressed in some logical language) s.t. each predicate of the specification has exactly one associated assertion of each family. The families correspond respectively to the assertions of partial correctness and completeness.

Let us denote by S and C the formulas defining respectively the properties of partial correctness and completeness and by $DEN(P)$ the semantics of a program P represented by some formula. One says that a program P is *partially correct* and *complete* w.r.t. the properties S, C iff $C \Rightarrow DEN(P) \Rightarrow S$. The properties S, C can be viewed as an approximation of the semantics expressed in an other language.

In [3] a complete proof method is described. It uses structural induction in a bottom up manner for the partial correctness and in a top down manner for the completeness.

For partial correctness the idea is to replace the atoms in the clauses by their specification S and to prove the resulting new formulas. For the completeness one has to prove the opposit implication using C instead of S, after having broken the completeness assertion into as many assertions as there are clauses in the packet defining a predicate. Some additional decreasing criterion must also be found.

The question arises which assertion should be associated to a negative literal. For partial correctness (resp. completeness) one uses *not C* (*not S*).

Here are the comments which could be associated to the previous program in order to perform the proof:

Predicate	S	C
$includ(L_1, L_2)$	L_1, L_2 lists $\Rightarrow L_1 \subseteq L_2$	L_1, L_2 lists and $L_1 \subseteq L_2$
$ninclud(L_1, L_2)$	L_1, L_2 lists $\Rightarrow L_1 \not\subseteq L_2$	L_1, L_2 lists and $L_1 \not\subseteq L_2$
$elem(E, L)$	L list $\Rightarrow E \in L$	L list and $E \in L$

It is not difficult to apply the method and verify that the assertions S hold at every ground generalised proof-tree root. See [4] or [3] for a complete treatment of this example.

To complete the formal verification the stratification must be checked in the full program.

Building the comments: Notice that in the previous example the form of the assertions which describe S and C are respectively $A \Rightarrow B$ and $A \wedge B$. So *not S* and *not C* are described respectively by $A \wedge not\ B$ and $A \Rightarrow not\ B$. If one may prove that all the generalised proof-tree of interest satisfy A, the remaining properties to prove are such that S and C become the same, hence the proof is simplified.

We claim that simpler is the proof, simpler and more readable is the specification.

Hence following and generalising this idea we recommend to write the comments in the form of an implication which is an assertion of partial correctness $(A \Rightarrow B)$ from which the assertion of completeness $(A \wedge B)$ can be easely deduced. The assertions of completeness define the subset of the generalised proof-tree roots of interest; but the assertions of partial correctness are more useful to read the clauses.

To sum up, the methodology is the following: all the comments must have the form:

$$p(X_1, \ldots, X_n) \text{ then } if\ A\ then\ B$$
$$\text{or}$$
$$p(X_1, \ldots, X_n) \text{ iff } B$$

The first form is used if $A \Rightarrow B$ is the assertion of partial correctness (with such comment the corresponding assertion of completeness is intended to define exactly the subset of proof-tree roots of interest), the second if B specifies exactly the denotation of the defined relation.

One assumes that A can be easely checked, thus the form of the proof for the remaining assertions B becomes simpler.

In the case of the previous example, instead of writing all the assertions used in the proof, one gives the comments as presented in the section 2.1:

2.3 Building the full specification

A specification will be presented with many layers such that all the relations defined in some layer use in their comments relations defined in the same or in previous layers only.

There is no formal methodology defining the way to organize the layers. Notice that this definition permits to have one layer only. So the choice of the number of layers is a matter of organization which better facilitates the reading.

Here is a short example illustrating a hierarchical organization of a specification:

First layer: very simple relations. The comments are totally informal.

is-an-integer(N) iff N is an integer built by the constant *zero* and the function *succ*.
    is-an-integer(*zero*) $\leftarrow$
    is-an-integer(*succ*(I)) $\leftarrow$ *is-an-integer*(I).

is-a-list(L) iff L is a list built by any term of the language as elements and 'nil' and '.' as function symbols.
    is-a-list(*nil*) $\leftarrow$
    is-a-list($X.L$) $\leftarrow$ *is-a-list*(L).

plus(X, Y, Z) iff Z is the sum of X and Y (axioms of the addition).
    plus(*zero*, X, X) $\leftarrow$
    plus(*succ*(X), Y, *succ*(Z)) $\leftarrow$ *plus*(X, Y, Z).

Second layer: more complex relation. The comments use some formally defined predicates.

member(X, L) then *if is-a-list*(L) *then* X is one of the elements of L.
    member($X, X.L$) $\leftarrow$
    member($X, Y.L$) $\leftarrow$ *member*(X, L).

is-a-list-of-integers(L) iff *is-a-list*(L) $\wedge (\forall X, member(X, L) \Rightarrow is\text{-}an\text{-}integer(X))$.
    is-a-list-of-integers(*nil*) $\Leftarrow$
    is-a-list-of-integers($N.L$) $\Leftarrow$ *is-an-integer*(N), *is-a-list-of-integers*(L).

Third layer: the most complex relations.

list-sum(L, S) then *if is-a-list-of-integers(L) then is-an-integer(S)* and S is the sum of all the elements of L.

list-sum(nil, zero) ←

list-sum(N.L, S) ← *list-sum(L, M), plus(M, N, S)*.

3 The example of standard PROLOG

3.1 History: The formal specification in the draft proposal of standard PROLOG. (ISO, JTC1/SC22/WG17)

The ISO working group WG17 has been working during several years on a project of draft for a standard PROLOG. Very early the need of a formal description has been recognised in order to clarify the concepts. It is the first attempt in the history to produce a complete description of a logic programming dialect and the group has been faced to new descriptional problems. In particular the need to standardize a particular SLD resolution scheme or the unification without occur-check as it is usually implemented. This has raised difficult and original problems that the efforts in developing a formal specification helped to solve.

A draft proposal for standard PROLOG has been issued in March 1992 with a formal specification which has been developed at INRIA-Rocquencourt by P. Deransart, A. Ed-Dbali, S. Renault and G. Richard [8]. The purpose has been to design the specification not only to help the design of the standard but also for a possible inclusion in the standard.

To include a formal specification appears to be the best way to achieve correctness and completeness objectives for the description of the standardized language. Although standard PROLOG will contain relatively few primitives (around 65 built-in predicates) it is a real challenge to ensure the consistency and the readability of the full description.

We expect to achieve these objectives thanks to the presented methodology which permited to develop validation tools.

3.2 General presentation

The underlying control model of the formal specification is the SLD-tree [6]. It is wellknown and has been deeply studied since the early years of logic programming and clarified. It is the underlying model of many approaches of logic programming (dialects with delaying primitives, parallelism, constraints) and is thus open to many extensions. The specification describes a meta-interpreter of standard programs. On the other side it avoids overspecification in the sense that not the whole process of SLD-tree construction as described by the formal specification is standardized, but only some aspects. By defining the SLD-tree in a fully declarative manner we avoid to refer to some particular algorithm. Thus it is possible to explain in a very clear way which characteristics of the SLD-tree are standard.

After an informal introduction of the underlying control model, the formal specification of standard PROLOG is organized in two layers only:

1. The data structure part which contain all relations (called D- or L- relations) defining data structures (search-tree, goal, program, ...). The D-predicates are defined formally. The L- ones are an interface between the formal specification and other concepts provided elsewhere in the standard description. These predicates are not defined formally. Their semantics is implicitly given by a possibly infinite set of ground facts (for example the elementary arithmetical operations are defined by the corresponding table).

2. The kernel of the specification (which mainly contains the relation $semantics(P, G, E, T)$ where P is a standard program, G a goal, E an environment and T the search-tree build from P, G and E) and the definition of all the control constructs and built-in predicates of standard PROLOG.

The size of the full description (program and comments) reaches 40 pages in the ISO standard description format (10 for an informal introduction and 30 for the specification).

The program part contains about 400 clauses defining 200 predicates.

3.3 Validation tools

Many validation tools have been developed in order to guarantee as far as possible the correctness of the whole standard PROLOG definition.

Runnable specification: Among all possible validation tools, a runnable specification is the main tool from which many validation tools may be grafted around. The formal specification is not executable as it is and, in general, it is very difficult to prove the adequacy between the formal specification and an executable one. However, in our approach, this verification is considerably facilitated by the fact that the clauses of the specification are a subset of the clauses of the runnable specification.

Testing the standard: The second validation tool which depends directly on the executable specification is the systematic verification of the cited examples in the standard PROLOG draft. All the cited examples are collected into a file. One uses the executable specification to verify that the examples are correct.

For example, in the case of the program of Example 1, one could have:

```
1- Goal: elem(a, [a,b])        Result: success
2- Goal: elem(X, [a,b])        Result: X <- a or X <- b
3- Goal: includ([c], [b,a])    Result: failure
```

where $X < - a$ is the representation of the substitution of X by a.

Now we have to assume that the program part of Example 1 contains some errors (assuming here that the executable specification and the program part are the same), in particular the second clause of *elem* is missing. The validation tool will respond:

```
1- Yes : Intended result = Computed result
2- No  : More substitutions in the intended result.
                        Computed result = X <- a
3- Yes : Intended result = Computed result
```

Each time the standard (examples or formal specification) are modified, the tool is rerun and as soon as some "No" answer occurs some correction must be made.

Stratification verification: An other validation tool is the verification of the stratification.

The stratification testing algorithm builds, in a first step, a dependency graph between the predicates of the program part. There are two categories of edges:

- *positive edges* denoted $(p,q)^+$ if a clause
  $p(x_1, ..., x_n) \leftarrow ..., q(y_1, ..., y_m), ...$ is in the packet defining p.
- *negative edges* denoted $(p,q)^-$ if a clause
  $p(x_1, ..., x_n) \leftarrow ..., not \ q(y_1, ..., y_m), ...$ is in the packet defining p.

In a second step, the algorithm tries to find a negative cycle in the graph. A negative cycle is a cycle which contains, at least, one negative edge. If such a cycle exists then the program is not stratified, otherwise the program is.

In this last case, the algorithm builds the strict partial order, denoted $<$, between predicates in the program. This order is defined as follow: $p < q$ iff there is a path from p to q which contains exaclty one negative edge. The stratification levels are obtained from this order.

Example 2. (continuation of Example 1)
The corresponding graph is:
$\{(includ, ninclud)^-, (ninclud, elem)^+, (ninclud, elem)^-, (elem, elem)^+\}$
and the order is: $elem < ninclud < includ$.

The output given by the stratification algorithm will be:

```
The program is stratified. There are 3 stratification levels:
     level 0 : elem
     level 1 : ninclud
     level 2 : includ
```

The formal specification of standard PROLOG has 4 stratification levels.

Automatic error cases generation from templates: An independent tool is the automatic generation of built-in predicate error cases. In the body of standard PROLOG draft, a built-in predicate informal description contains an informal collection of error cases and also a declaration of the type of its arguments by means of *templates*. The relation between the types and the categories of errors is defined informally inside the body.

A template for a predicate p has the form:

$template(\ p(\mu_1\tau_1, ..., \mu_n\tau_n)\ )$ where $\tau_i (1 \le i \le n)$ is the type of the ith argument and μ_i its *mode*. The mode is specified by one of the following characters: '+', '?' and '-'.

- '+': means that the argument must have the given type. An *instantiation-error* is raised if the argument is a variable.
- '?': means that the argument must be a variable or have the given type. In this case there is no instantiation error.
- '-': The argument is not constrained (i.e. it may be any term).

In the formal specification of standard PROLOG, a built-in predicate is described by two parts: The specification of its behaviour when it is called correctly and all its error cases defined by clauses of the *in_error(G, E)* predicate, where G is a built-in predicate goal and E the corresponding error category.

The purpose of the error cases generation tool is to establish the consistency between the error definitions in the formal specification (*in-error* clauses) and the error cases in the body. It helps also to identify all the relevant error cases.

To achieve this, we formalize, at first, the type definitions and their corresponding error categories. A type τ and its corresponding error categories are defined as follow:

$$\tau = \{X \text{ in } \tau' \mid Condition(X) \Rightarrow error\ category\}$$

where X is a variable, τ' is an other type such that $\tau \subset \tau'$ and *error category* is the identification of the error raised if the relation *Condition(X)* is true when X is instantiated by a given term. The type τ inherits all error categories of τ'.

There is a type called *term* which is the superset of all the other types and has no corresponding category of error.

Example 3. Let $open(F, M)$ be a built-in predicate whose semantics is to open a file of the name F for input or output as indicated by the input-output mode M. The template specification of this predicate is:

$template(\ open(+\ identifier, +\ io\text{-}mode)\ )$ where *identifier* is an unstructured object used to denote a file name and *io-mode* one of the two characters 'i' (for input) and 'o' (for output).

As shown above, the specification mode '+' in this template means that the first argument of *open* must be instantiated by an element of the type *identifier* and the second one must be an *io-mode*, otherwise an instantiation error (if one argument is a variable) or other error categories will be raised.

In the following, we describe formally what is an *identifier* and an *io-mode* and what category of error will be raised if the argument does not have the expected type. In the present state of the standard, there are 17 different error categories. Among them, we can cite:

- *instantiation-error*: raised if the argument is insufficiently instantiated.
- *type-error*: raised if the argument has an incorrect type.
- *range-error*: raised if the argument has a correct type but the value is outside the permitted range.

For the cited example, the specification of the type *identifier* and its error categories is:

```
identifier = {I in term | not D-is-an-identifier(I) => type-error}
```

where *D-is-an-identifier* is a data structure relation defined in the formal specification. This definition must be read as follow: A term *I* is an *identifier* iff it is in *term* and *D-is-an-identifier*(I) is true (positive sentence form). A term *I* produces a *type-error* iff it is in *term* and *D-is-an-identifier*(I) is false (negative sentence form).

The corresponding condition to *instantiation-error* is *D-is-a-var* for any type.

The specification of the type *io-mode* and its error categories is:

```
io-mode    = {X in char | not D-member(X,[i,o]) => range-error}
```

where *D-member* is a data structure relation.

Finally *char* is an other type defined as follow:

```
char       = {C in term | not D-is-a-char(C) => type-error}
```

Here are some erroneous calls of the predicate *open* followed by what errors the system raises:

```
   Goals:                    Errors:
1: open(F, i).               instantiation-error(1)
2: open(12, o).              type-error(1)
3: open(file1, a).           range-error(2)
4: open(file2, 5).           type-error(2)
```

The argument of an error category is the number of the erroneous argument in the given goal. Remark that in the last example the error is *type-error* and not *range-error* because the second argument has an incorrect type (5 ∉ *char*). In the 3rd example, the argument has a correct type but is out of range (a ∈ *char* but not *D-member(a,[i,o])*).

The error cases generation tool takes as input the template specification and types definition and produces all clauses for the *in-error* predicate of the formal specification. For the given example, the tool's output is:

```
in-error( open(F, M), instantiation-error(1) ) <- D-is-a-var(F).
in-error( open(F, M), instantiation-error(2) ) <- D-is-a-var(M).
in-error( open(F, M), type-error(1) ) <- not D-is-a-var(F),
                                    not D-is-an-identifier(F).
in-error( open(F, M), type-error(2) ) <- not D-is-a-var(M),
                                    not D-is-a-char(M).
in-error( open(F, M), range-error(2) ) <- D-is-a-char(M).
                                    not D-member(M,[i,o])
```

The comparison between the generated error cases and those given in the body of standard is made manually.

Validation by proofs: The last validation tool helps (partially) to prove consistency between the comment part and the program part. It uses a proof system based on the annotation method [2] and developed by S. Renault [7].

In the presentation of the formal specification, the comments are written in a natural language style. This avoids to overload the reader with more formalism. So part of the comments are encoded into formal assertions of partial correctness and completeness and splitted (by hand) into a collection of shorter assertions. Then a system tries to organize a proof by building all the assertions to be proven in every clauses. Finally it tries to prove as much assertions as possible and shows to the user the assertions it cannot prove automatically (in practice around 30% of the assertions).

There is no way to make a complete formal proof of the formal specification, because most of the relations are not completely defined by the part of the comments which can be expressed formally. However the proof activity permits to detect many errors in the clauses as in the comments as well and to improve them.

4 Conclusion

The specification method presented here is an interesting compromise between all the requirements that a quality specification should fulfill: abstraction, simplicity, extensibility, formally verifiable and testable.

To summarize, it is now clear that the effort of writting formal specifications helped in many ways to design standard PROLOG (it helped in particular to clarify the unification definition, the data-base update view, the error treatment and to complete the informal descriptions in many places). It is also a way to make it clear and to shorten the whole description thanks to the conciseness of the formal specification. In fact its size remains relatively short and practice showed that few days are needed to understand it.

Finally we cannot guarantee to produce a standard absolutely free of errors, but the use of (partially) automatized tools greatly helps to reach such goal. Although the size of the formal specification (and the whole standard) is relatively short, practice has proved that the use of automatized tools remains necessary.

References

1. Apt, K. R., Blair, H., Walker, A.: Toward a theory of declarative Knowledge. LITP, RR **86-10** (Fed. 1986)
2. Deransart, P.: Proof Methods of Declarative Properties of Definite Programs. RR **1248**, INRIA-Rocquencourt, (June 1990) 64p (to appear in TCS).
3. Deransart, P., Ferrand, G.: Proof Method of Partial Correctness and Partial Completeness for Normal Logic Programs. RR **92-4**, LIFO, University of Orléans, (April 1992) 12p. To appear in ICLP'92 proceedings (Washington DC, Nov 1992).
4. Deransart, P., Ferrand, G.: An Operational Formal Definition of PROLOG: A Specification Method and its Application. New Generation Computing **10** (1992) 121–171.

5. Ed-Dbali, A., Deransart, P., Scowen, R. (ed.): PROLOG, A guide to the executable specification. ISO/IEC JTC1 SC22 WG17 **N71**, (Jan 1991)
6. Lloyd, J. W.: Foundations of logic programming (First ed.). Springer-Verlag, Berlin, (1984)
7. Renault, S.: Logic Programs Validation (in French). DEA Repport, University of Paris 7, (September 1991) 65p.
8. Scowen, R. (ed.): PROLOG Part 1 - General Core. ISO/IEC JTC1 SC22 WG17 **N92** (March 1992)

Executable specification: Address request to: AbdelAli.Ed-Dbali@univ-orleans.fr or P. Deransart, INRIA-Rocquencourt, BP 105, F-78153 Le Chesnay

Standardization documents: Address request to: Roger Scowen (Convenor) DITC/93, NPL, Teddington, Middlessex, UK, TW11 0LW. email: rss@seg.npl.co.uk

The Art of Computer Un-Programming: Reverse Engineering in Prolog

Peter T. Breuer

Oxford University Computing Laboratory, 11 Keble Road, Oxford OX1 3QD, UK.
Email: Peter.Breuer@comlab.oxford.ac.uk

Abstract. A suite of Prolog tools for reverse-engineering and validating COBOL programs has been developed as part of the ESPRIT REDO project [8]. These tools produce functional abstractions, object-oriented designs and documentation from raw source code, with the aim of improving comprehensibility and maintainability, and this article discusses the tools and aspects of their programming.

1 Reverse-Engineering and Validating COBOL

The COBOL language [13] hinders program comprehension in several ways:

- by its use of formats instead of types (which means that each assignment between variables usually involves an implicit conversion of data from one format to another);
- by the mixed use of PERFORM statements and the ordinary *fall-through* execution of paragraphs;
- by the inclusion of unstructured constructs such as GO TO's and ALTER GO TO's (which establish computed aliases for labels in the code); and
- by excessive proliferation of specialised verbs and variants of verbs, the exact semantics of which in turn requires very specialised knowledge on the part of the programmer.

These features of COBOL make recognising reusable program components in monolithic programs difficult, and therefore mean that most of the extant COBOL code is difficult to re-engineer for integration within a modern software suite. Many software houses today find themselves with products inherited from an earlier era that are comprehensible only to the original development team, who may well have left. This is part of the software crisis that has been predicted for many a year now (other parts concern the proliferation of programming languages, dialects and implementations), but whose consequences continue to hamper the advancement of practised software science. When a modification or an improvement to functionality is required, old code has to be either jetisoned or laboriously researched, and it is in response to the research requirement that the ESPRIT II project REDO [1] [8] was launched in 1989.

---

1 "REengineering, DOcumentation and validation of systems", ESPRIT project no. 2487.

This project has sought to develop methods of maintaining and improving old FORTRAN and COBOL applications, and has met with some success. The project partners as a whole early on came to the view that reverse engineering was the key technology to investigate, and in this article such methods and tools for reverse engineering are described as have been developed at Oxford as part of the REDO project.

These tools take in (primarily) COBOL code and produce *specifications* via an interactive process. The specifications are not exact transcriptions of the code, but deliberate *abstractions* away from the programming detail, produced by engineers engaged in documenting, decomposing and analysing the functionality of an application code. The Oxford tools, in distinction to contributions from other partners in the project and, so far as we know, attempts at reverse engineering worldwide,

- produce specifications that are written in a recognised formal specification language (Z [14]), and
- are arranged, structured and presented in object-oriented packages of varyng levels of abstraction,

These outputs satisfy the growing requirement of many contractors for

- formal specifications and
- a recognised top-down development method

as part of the contract in order to guarantee future maintainability and keep the servicing costs down.

The objects and classes which come out of the analysis also provide reusable software components worthy of inclusion in a development software library, which contractors may also insist upon in an effort to protect themselves from excessive development costs in future contracts.

The Oxford tools are unique in attempting to capture the *functionality* of the code, not merely the formal arrangement of its control-flow, the data-flow, or the types of data structures, although all these views are available from the toolset too.

The features of COBOL pointed out above, and the correctness requirements on the formal specifications which have to be produced, make software tools and a properly understood *formal method* for going about reverse engineering a necessity. It is here that Prolog has proved most useful, because we have been able to give methods phrased at the level of code-to-code relationships (between one formal language and the same or a different language) and transcribe them into Prolog quite quickly. Moreover, the Quintus Prolog [16] environment has supplied a satisfactory front end to the prototype toolset, and the compiled code has operated efficiently, and, more importantly, convincingly, in actual trials on medium sized applications. In additon, we have been able to integrate theorem provers and tools which validate the code against the derived specifications within the whole ensemble. AI techniques generate heuristics which aid the user in guessing suitable specifications in those cases which are not automatic, And we are also able to interpret some of the derived specifications themselves within Prolog in order to obtain executable simulations which are useful in visualization.

Below, the *process* followed by the toolset is described.

On loading, the application source code is parsed and automatically transformed into an internal intermediate language, UNIFORM [15], which is somewhat simpler than COBOL. This is in order to facilitate later operations. All the tools work from this representation of the application, which is maintained in a persistent database [5] and is available to the user during certain operations. An external database is used for reasons of compatibility with the rest of the project developments, but a Prolog database would have been satisfactory otherwise. An example of UNIFORM is shown in Figure 3(b), corresponding to the COBOL code in Figure 3(a). It may be observed that the UNIFORM code differs in kind only in having proper scoping constructs (**begin, end**). This permits the use of local variables and private functions in UNIFORM codes.

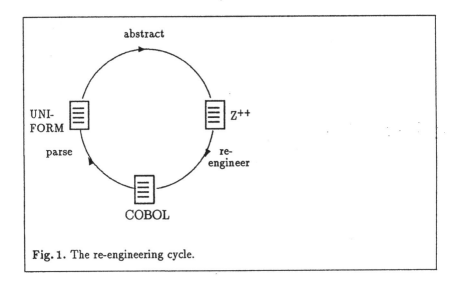

Fig. 1. The re-engineering cycle.

The parser library in the Quintus Prolog [16] environment has been most useful in enabling us to construct a parser where no single-token lookahead parser would have sufficed. The parser is superior in speed to a version constructed using the standard UNIX *lex* and *yacc* tools [7] by other partners in the project, though it takes a little programming discipline to write the parser in a portable way.

The semantics of UNIFORM has been completely formally specified, so the transformation apparently succeeds in giving an unambiguous formal semantics to COBOL (!) but it should be borne in mind that formal specifications need not be fully deterministic, and in fact many different implementations will satisfy the specification used. The tools work on the logic of the common abstraction.

Outline object designs are then suggested to the user by heuristics which identify significant subsections of the code, using the data-flow graph for guidance (this is derived from the UNIFORM intermediate code). These *objects* group together data which appears to be dependent or inter-dependent and list as their outline *methods* those sections of code which update or access it. At this stage, classes have only one object instance each; an abstraction step later establishes proper hierarchies.

The functionality of these methods is next abstracted up into a simple first-order functional language [2], via an interactive process with an automatic normalization feature. An *entirely* machine-led abstraction is possible at this point, and still results in a very readable, although lower level, specification. The normalization method used often re-identifies differing implementations of the same standard concept [4] and does therefore achieve a significant degree of abstraction on its own.

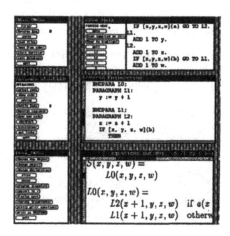

Fig. 2. Reverse-engineering session.

Many transformations and operations on the initial abstraction are available on the menus programmed into the toolset at this point (see Figure 2), and these help the user to produce an improved (structured) and functionally correct higher level specification of the application.

Some of the operations on the menus are just the inverses of the individual steps taken during automatic normalisation, but some are much more general and require considerable user guidance, demanding selection of a piece of program or statement, and some typed input: selected equational specifications may be substituted in each other, for example, selected equations may be dropped, or replaced by a generalisation (marked as un-validated until the proof procedure has been discharged), and so on.

Validation of the code and/or specifications is an integral part of the transformational stage, because the abstractions may sometimes require proof if they are significantly different in form from the literal specification obtained by examining the code.

If the specification contains a conjunction, for example, then one may drop one conjunct to obtain a proper abstraction, and this requires no proof. This is supported by a single operation on the menu of abstraction techniques. But if the specification contains a recursive feature, then one may wish to assert a non-recursive abstraction, which will always require proof. Most theorem-proving tools for validation use the method of Gries [6] to insert extra 'checkpoint' assertions into the COBOL code itself, and whilst the Oxford tools support this methodology, they are also able to prove statements about the specifications themselves. They are particularly adept at proving statements about the first order functional language, which is distinguished by having only four or five distinct constructs and therefore only very few atomic proof procedures.

A standard repertoire of refinement operations then allow the code to be re-engineered, replacing abstract specifications in Z by Uniform code which implements the abstractions, and which can be pretty-printed as COBOL.

The functional abstractions, written in the first order functional language (which can be interpreted as Z with some slight reprinting) then replaces the source code within the prototype object descriptions that were obtained from analysis of the data flow. The functional abstractions form specifications of the objects methods.

Restructuring of the abstracted objects then takes place. The whole ensemble – object prototypes with formal abstractions instead of code – form constitutes a high-level Z^{++} specification of the application [11], and specifications, methods and data can be migrated from one prototype to another in order to reduce duplication and isolate the essential functional components.

Further documentation is then produced from this level, if required. This documentation includes flow diagrams, transcriptions into english of the specifications, and COBOL pseudo-code representations of selected parts of the reverse engineered module.

The components of the complete system are:

- Verification condition generator.
- Parser and translator for COBOL.
- Abstraction generator and normaliser.
- Design recognizer.
- Regenerator of code from abstractions.

Operations have also been included to generate pseudocode from or directly animate the Z and Z^{++} [11] specifications.

```
      DATA DIVISION.
      WORKING-STORAGE SECTION.
       02 X PIC 99.

      PROCEDURE DIVISION.
      pd SECTION.
      p1.
      MOVE 0 TO X.
      PERFORM p2 THRU p4.
      p2.
      PERFORM p3.
      p3.
      ADD 1 TO X.
      IF X = 5 MOVE 0 TO X.
      p4.
      ADD 3 TO X.
```

Fig. 3. (a) COBOL program.

```
      X : N[2]
      begin
       pd section
         paragraph p1:
           X := 0;
           perform thru p2, p3, p4
         endpara p1;
         paragraph p2:
           perform p3
         endpara p2;
         paragraph p3:
           X := X + 1;
           if  X = 5 then X := 0
           end if
         endpara p3;
         paragraph p4:
           X := X + 3
         endpara p4
      end
```

Fig. 3. (b) UNIFORM translation.

$$pd(X) = p1(X)$$
$$p1(X) = p2(5)$$
$$p2(X) = p3(0) \quad \text{if } (X+1) = 5$$
$$\qquad = p3(X+1) \text{ otherwise.}$$
$$p3(X) = p4(0) \quad \text{if } (X+1) = 5$$
$$\qquad = p4(X+1) \text{ otherwise.}$$
$$p4(X) = X + 3$$

Fig. 3. (c) First order functional abstraction.

2 Techniques

The toolset has been written in Quintus Prolog with Quintus ProWindows [16]. Prolog was chosen because it could express the many rewrite transforms conveniently and succinctly. The basic relation is

$$\text{deduce}(P, A \xrightarrow{*} B)$$

which expresses the idea that under the hypotheses P, there is a chain of elementary transforms $(\_ \rightarrow_i \_)$ such that

$$A = X_0 \xrightarrow[1]{} X_1 \xrightarrow[2]{} \cdots \xrightarrow[n-1]{} X_{n-1} \xrightarrow[n]{} X_n = B$$

The hypothesis P will usually be a statement like $A \xrightarrow{*} A$. An elementary transform

$$X \rightarrow X'$$

is encoded as the clause:

$$\text{deduce}(P, X \xrightarrow{*} B) :- \text{deduce}(P, X' \xrightarrow{*} B).$$

and the facts

$$\text{deduce}(A \xrightarrow{*} B, A \xrightarrow{*} B).$$

are included in the Prolog database.

Prior to performing the normalization, the functional expressions have to be produced from the the initial code. We assert the specification relationship between (procedural) programs and specifications:

```
program_spec(Prog,Vars,Spec) : - globals_of_program(Prog,Globals),
                                  decls_of_program(Prog,Decls),
                                  decls_spec(Decls,union(Vars,Globals),Spec).
```

which claims that (in the presence of the previously declared variables Vars) Prog is related to Spec if its declared functions and subroutines Decls together are related to Spec, having declared in addition the global variables of the program, Globals. It requires only five such high-level predicates [1] to define the transformation from procedural code to literal specifications, and then the normalizingtransforms may be aplied automatically in order to smooth away such inficities as '$(x, (\text{if}(y = 1)\text{then}(2)\text{else}(2)))$'. The prog_spec predicate is one of these five, whose types are given below:

```
prog_spec :: predicate(program,environment,equations)
decl_eqn  :: predicate(declaration,environment,equation)
code_spec :: predicate(code,environment,equations,environment)
expr_expr :: predicate(expression,environment,expression)
stmt_expr :: predicate(statement,environment,equations,environment)
```

in terms of the semantic domains

```
equations      ≡ equation*
program        ≡ environment × code
code           ≡ statement*
environment    ≡ variable_name*
equation       ≡ lhs × rhs
lhs            ≡ pattern* + function_name × pattern*
pattern        ≡ variable_name + 1 + pattern × pattern
rhs            ≡ specification
specification ≡ expression × equations
```

The idea is that a program should give rise to a set of equations, each corresponding to a function declared (and defined) in the program. The right hand side of an equation is a (non-deterministic) expression with subequations which define its locals – a specification.

Knowledge-based techniques were also employed: in implementing Gries' heuristics [6] for generating candidate loop invariants, as part of the verification process, and in the heuristics which guess at program purpose during the reverse-engineering process. Prolog turns out to be particularly useful in the part of the toolset which allows the engineer to formally prove contentions about the code or specifications, because a *non-ground* theorem may be proved. As the proof progresses, the theorem takes shape, being determined both by the proof method attempted and the code or specification it is applied to.

The tools all have an object-oriented design, and this is itself visible to the user. The current software fragment is always an object instance, for example, with attributes: *Text*, *Data-flow Graph*, and *Functional Abstraction*, presented in the window, with methods: *Abstract*, *Generate Design*, etc., represented by buttons embedded in the frame.

Building object-class descriptions from code involves extensive user interaction in order properly to incorporate the maintainers knowledge, but the abstraction steps and translation between languages are automatic (albeit controllable). During development, the tools accessed a Prolog database modeled on the main Eclipse object-oriented database [5] used within the rest of the REDO project, but this has now been replaced with an interface to the main database,

The toolset consists of over 10,000 lines of Prolog code, and has been used in the reverse-engineering and validation of large data-processing applications [2]. The code-generating tool subset has been applied to large systems, for example, a radar track-former [10], whose specification comprised 20 pages of Z schemas.

3   The Examples

As remarked above, COBOL has several features which serve to obscure the meaning of a piece of code. The use of unstructured control statements (GOTO's and PERFORM THRU's), the use of formats instead of types, the mixing of PERFORM calls on paragraphs with ordinary *fall-through* semantics, and the proliferation of specialised verbs all serve to make COBOL programs difficult to understand.

298

class *pd_class*

owns
$X: \mathbb{N}$;

operations
$pd: () \rightarrow ()$;
$pd \overset{def}{=} p1$
$p1: () \rightarrow ()$;
$p1 \overset{def}{=} X := 5; p2$
$p2: () \rightarrow ()$;
$p2 \overset{def}{=} X := 0; p3 \qquad$ if $X + 1 = 5$
$\overset{def}{=} X := X + 1; p3$ otherwise.
$p3: () \rightarrow ()$;
$p3 \overset{def}{=} X := 0; p4 \qquad$ if $X + 1 = 5$
$\overset{def}{=} X := X + 1; p4$ otherwise.
$p4: () \rightarrow ()$;
$p4 \overset{def}{=} X := X + 3$

algebra
$pd = X := 10$
$\dots$

end class

Fig. 4. A low-level abstraction object for the example COBOL program.

The initial translation into UNIFORM which is carried out within the toolset serves to remove these complications. It clarifies the semantics, and also converts implicit COBOL functionality into explicit form, where it can. Further translation into a first-order functional programming language then removes any possibility of unpleasant semantic surprises and the result can, moreover, be directly manipulated by the theorem-proving and re-engineering components of the toolset.

Abstraction: the functional language [4] contains only three constructs: expressions, functional application, and conditional choice, in contrast to the more than 200 keywords in COBOL '74. An example of the gain in comprehensibility and the ability to reason from the representation is given in Figure 3, where a COBOL program (a), its UNIFORM translation (b) and the *automatically* calculated normalized representation (c) in the functional programming language are shown together. From (c) the toolset allows the user to prove easily that $pd(X) = 10$ always holds, which is quite remarkable. An examination of the code (a) reveals only a hopelessly twisted structure even to the expert eye. It is clear that this code has been modified several times, probably whilst attempting to preserve the existing semantics, and the engineers who modified it would probably be surprised to learn that it inevitably sets the value of x to 10.

Much of the confusion in the code comes about because paragraph p2 first calls paragraph p3 like a subroutine, then drops through into it again when the call returns. Since paragraph p1 calls paragraphs p1 to p4 in sequence (before dropping into them on the final return), the situation is even more obscure. Add to that the fact that COBOL `perform` commands do not have the precise semantics of a subroutine call – a second call of the same paragraph *overwrites* the return address of the first call – and it can be guaranteed that no COBOL programmer can predict the functionality of this short piece of code with certainty.

An (imperatively coded) abstraction class corresponding to the example is shown in Figure 4, and a list of some of the normalization rules for the underlying first order functional language is shown in Figure 5. This class structure (data x and methods pd, p1 etc.) was the first alternative offered by the toolset, and just corresponds to a monolithic program, in which all the global variables are the data for the object, and all the paragraphs are the methods. The code for the whole program then reduces to a declaration and a method call:

```
pd_class main;
main.pd;
```

Design Recognition: in practice, COBOL data-processing applications are often large and complex, and involve system calls to CICS or TOTAL. In order to make functional abstraction practical, and to provide an improvement in application comprehensibility, these applications must be separated into meaningful subsystems and operations on these subsystems. CICS is an example of such a subsystem, and an outline specification of it can be given in an object-oriented style, enabling a concise representation of its facilities and call formats [9]. Similarly, the components of an application can often be expressed in a natural way using object-oriented design, even when the original design (if any) of the system was certainly not in this style.

Component construction starts with the data-flow analysis of the code; related variables are grouped together, and sections of code which correspond to operations on this group of variables are identified, using techniques of *slicing* and *phasing* [12, 2]. The details of the functionality of the identified operations are stored in the database, but will be hidden in the *top-level* design visible to the user.

4 Concluding Remarks

Development of the tools followed an object-oriented methodology, which was 'semi-formal' and consistent with the rapid-prototyping methodology supported by Prolog. The relative independence of sections of the system, whilst they all operated on a standard format of data, enabled development to proceed within the projects agreed timetable. The software has now been used in other institutions, both within and outside the project consortium, for practical reverse-engineering tasks.

References

1. Breuer P. The First Step Backwards, *REDO project document 2487-TN-PRG-1031*, Oxford University Computing Laboratory, 11 Keble Road, Oxford OX1 3QD, UK.
2. Breuer P., Lano K., From Code to Specifications: Reverse Engineering Techniques, *Software Maintenance – Research and Practice*, Sept. 1991.
3. Breuer P., Lano K., Reverse Engineering COBOL *Software Maintenance – Research and Practice*, 1992.
4. Breuer P., et al. *Understanding Programs through Formal Methods* PRG-TR-15-91, Oxford University Computing Laboratory, 1991.
5. Cartmell J., Alderson A., *The Eclipse Two-Tier Database*, Chapter 5, ECLIPSE Manual, IPSYS Software Ltd, Marlborough Court, Pickford St., Macclesfield, Cheshire, 1990.
6. Gries D., *The Science of Programming*, Springer-Verlag, 1981.
7. S.C. Johnson and M.E. Lesk. Language development tools. *The Bell System Technical Journal* 57(6) part 2, pp. 2155–2175, July/August 1978.
8. Katsoulakos P., Reverse Engineering, Documentation and Validation: The REDO Project, *ESPRIT 89 Conference*, Brussels, 1990.
9. Lano K., *An Outline Specification of the CICS Application Programmers Interface* REDO Document TN-2487-PRG-1025, Oxford University Programming Research Group, 1989.
10. Lano K., *The Transformation of Specifications into Code*, REDO Document TN-2487-PRG-1023, Oxford University Programming Research Group, 1990.
11. Lano K., Z^{++}, An Object-Oriented Extension to Z, *Proc. 5th Annual Z User Meeting, December 1990*, Springer-Verlag Workshops in Computer Science, 1991, to appear.

$$\frac{\text{if e1 then (if e2 then d1 else d2) else d3}}{\text{if e1\&e2 then d1 else (if e1 then d2 else d3)}}$$

$$\frac{((a,b),c)}{(a,(b,c))} \qquad \frac{((\ ),a)}{(a)} \qquad \frac{((a))}{(a)}$$

$$\frac{x_1}{\text{head } x} \qquad \frac{x_n}{(\text{tail } x)_{n-1}}[n > 1] \qquad \frac{\perp_n}{\perp} \qquad \frac{n_1}{n}$$

$$\frac{\text{head }(a,b)}{a} \qquad \frac{\text{tail }(a,b)}{(b)} \qquad \frac{\text{head }(\ )}{\perp} \qquad \frac{\text{tail }(\ )}{\perp}$$

$$\frac{(f \circ g) \circ h}{f \circ (g \circ h)} \qquad \frac{(f \circ g)(a)}{f(g(a))} \qquad \frac{\text{id} \circ f}{f} \qquad \frac{f \circ \text{id}}{f} \qquad \frac{\text{id}(a)}{a}$$

$$\frac{(\text{if e then c1 else c2})_n}{\text{if e then c1}_n \text{ else c2}_n} \qquad \frac{f(\text{if e then c1 else c2})}{\text{if e then } f(c1) \text{ else } f(c2)}$$

$$\frac{\text{if e then \{c1 where v=d\} else c2}}{\text{\{if e then c1 else c2\} where v=d}}[v \notin fv(c2)]$$

$$\frac{\text{if e then c1 else \{c2 where v=d\}}}{\text{\{if e then c1 else c2\} where v=d}}[v \notin fv(c1)]$$

$$\frac{\text{e where v=d}}{e[v/d]} \qquad \frac{(\text{if e then a1 else a2) op b}}{\text{if e then (a1 op b) else (a2 op b)}}$$

$$\frac{\text{a op (if e then b1 else b2)}}{\text{if e then (a op b1) else (a op b2)}}$$

$$\frac{(\text{if e then a1 else a2)\&b}}{\text{if e then (a1\&b) else (a2\&b)}}$$

$$\frac{(\text{if e then a1 else a2) or b}}{\text{if e then (a1 or b) else (a2 or b)}}$$

$$\frac{\text{a\&(if e then b1 else b2)}}{\text{if e then (a\&b1) else (a\&b2)}}$$

$$\frac{\text{a or (if e then b1 else b2)}}{\text{if e then (a or b1) else (a or b2)}}$$

$$\frac{\text{if (if e then b1 else b2) then c1 else c2}}{\text{if e then(if b1 then c1 else c2)else(if b2 then c1 else c2)}}$$

$$\frac{\neg(\text{if e then b1 else b2})}{\text{if e then }(\neg b1)\text{ else }(\neg b2)} \qquad \frac{\text{if }(\neg e)\text{ then b1 else b2}}{\text{if e then b2 else b1}}$$

$$\frac{\text{if (e1 or e2) then b1 else b2}}{\text{if e1 then b1 else (if e2 then b1 else b2)}}$$

$$\frac{\text{if (e1\&e2) then b1 else b2}}{\text{if e1 then (if e2 then b1 else b2) else b2}}$$

$$\frac{\text{e1\&e2}}{\text{if e1 then e2 else False}} \qquad \frac{\text{e1 or e2}}{\text{if e1 then True else e2}}$$

$$\frac{\neg e}{\text{if e then False else True}} \qquad \frac{\text{if True then b else c}}{b}$$

$$\frac{\text{if False then b else c}}{c} \qquad \frac{\text{if p then True else False}}{p}$$

$$\frac{\text{if p then b else c}}{\text{if p then b[p/True] else c[p/False]}}[p \in fv(b) \cup fv(c)]$$

Fig. 5. Some Normalization rules for Presentations in a 1st Order Functional Language

12. Linger R., Hausler P., Pleszlioch M., Heruer A., Using Functional Abstraction to Understand Program Behavior, *IEEE Software*, Jan. 1990.

13. Parkin A., COBOL *for Students*, Edward Arnold, London, 1984.

14. Spivey M., *The Z Notation : A Reference Manual*, Prentice Hall, 1989.

15. Stanley-Smith C., Cahill A., *UNIFORM: A Language Geared To System Description and Transformation*, University of Limerick, 1990.

16. *Quintus Prolog Version 2.5 Manual*, Artificial Intelligence International Ltd., Watford, U.K., 1990.

Opium - An Advanced Debugging System

Mireille Ducassé

European Computer-Industry Research Centre
Arabellastr. 17, D-8000 Munich 81, Germany
email: mireille@ecrc.de

1 Introduction

The data used by program analysis in general is often restricted to the source code of the analysed programs. However, there is a complementary source of information, namely *traces of program executions*. Usual tracers, which extract this trace information, do not allow for general trace analysis. Opium, our debugger for Prolog, sets up a framework where program sources and traces of program executions can be jointly analysed.

As the debugging process is heuristic and not all the debugging strategies have been identified so far, Opium is programmable. In particular, its trace query language gives more flexibility and more power than the hard coded command sets of usual tracers. This trace query language is based on Prolog. Opium is therefore both a helpful tool for Prolog and a nice application of Prolog.

The most innovative extensions of Opium compute abstract views of Prolog executions to help users understand the behaviours of programs. In particular they help them understand how error symptoms have been produced.

In the following we briefly recall some general information about Opium. A debugging session is then commented in detail. The commands used in the session are described in appendix. A detailed description of Opium can be found in [7].

2 Introducing Opium

This section briefly recalls the principles of Opium. A trace query mechanism enables users to specify precisely which trace lines they want to see or verify. A trace database supports backward tracing. Abstract views of executions are the basis for a trace browser.

2.1 A trace query language based on Prolog

We model an execution into a trace which is a *stream* of events. Execution events have a uniform representation, and can be analysed by programs. At a conceptual level, only two tracing primitives are necessary to retrieve trace information on the fly: one to retrieve the information related to the current event, another one to retrieve information related to the next event. A trace can be considered as a history of execution events, and the two primitives can be matched on the notions of today (*current event*) and tomorrow (*next event*). These two primitives, plus Prolog, make a powerful trace query language. In particular, general conditional breakpoints can be set on the fly. For example, the following goal can be read as *"repeatedly get the next event until the CONDITION on Event is true, then display the Event which passed the checking"*.

```
?- next(Event), CONDITION(Event), DISPLAY(Event).
```

The use of the `current` primitive is illustrated in section 3.

The previous model makes a simple, powerful and precise trace query language. Unfortunately, systematically processing all the information at every event is impracticable for large executions, especially when the data of the program are large. Opium incorporates some optimizations which improve the response time while keeping close to the model. In particular, one primitive optimizes queries on control flow information which is cheap to process. It allows the encoding of more than the usual tracing commands with no loss in performance. It can parse several million execution events with reasonable response times.

2.2 A trace database

To enable backward tracing we have chosen to implement a trace database. Some tracers provide their users with a `retry` command which can ask to re-execute part of the current execution. This is not accurate here because we want to be able to roll back the execution and this is extremely difficult to achieve by re-execution.

Our trace query mechanism is based on the fact that the information related to the current event is stored in a record. It is then straightforward to have a simple database by storing the records of all the events in their chronological order. The "pre-filtering" optimizations described in the previous section can then be applied straightforwardly to backward tracing.

If an execution is made up of millions of events, it is of course not reasonable to store the exhaustive trace. This would result in a database of several megabytes. Considering that this database is not permanent, the costs in time and space are currently prohibitive. It is, nevertheless, no problem to store several *thousands* of events and thus to analyse portions of executions a posteriori. We have introduced dedicated primitives so that users and extensions can control the amount of trace which is stored. It should be noted, however, that much can be achieved without storing anything.

2.3 Abstract views of Prolog executions

Line oriented tracers usually present histories of execution events. Each event represents a low-level step in the execution process. The relationship between the executed program and the execution history is not straightforward to derive. For example, goals are natural notions in Prolog but the events related to a given goal are scattered throughout the history. Hence, to build up a picture of a goal behaviour the user usually has to sift through a large piece of the trace history to extract the goal related events. Abstract views filter out irrelevant details; they may restructure the remaining information; they may compact it so that the amount of information given at each step has a reasonable size; and they eventually print the resulting information.

There are many levels of abstractions at which an execution can be examined in Opium. For example, control flow information can be selected to illustrate a goal behaviour or a recursion behaviour. The data of the program can be complex and large, and different levels of abstraction can be used to display data values. The data flow of the program can be abstracted with respect to some variables. Trace information can be selected to explain how a particular bug symptom has been produced (see [6, 11]). If programs are interpreted by a Prolog application (for example an interpreter written in Prolog), tracing these programs can be done at the level of the application and not at Prolog level. It is worth mentioning that these abstract views, presented in [5], include all the abstract views detected by the cognitive study described in [1].

3 A debugging session with Opium

We show a session illustrating Opium facilities. More commented examples can be found in [9] and Opium user manual [8].

As a debugging session mainly depends on the understanding of the programmer, there is no deterministic way to debug. We only show possible paths. In the following examples, the full names of the commands are used to ease the understanding. However most of them have abbreviations which can be used to save typing. Whenever queries repeat part of previous ones the corresponding abbreviations are used. The description of the Opium commands can be found in appendix in alphabetical order.

In the following the trace lines contain the following information:

ExecTime InvocationNumber [ExecDepth] Port PredicateName (Arguments)

The analysed program is the Nqueens program taken from [17, p210, program 14.2], in which a bug has been added to the attack/3 predicate. "The N queens problem requires the placement of N pieces on a N-by-N-rectangular board so that no two pieces are on the same line: horizontal, vertical or diagonal". The toplevel goal of the traced execution is nqueens(4, Qs) which should give the solutions for a 4x4 board. We can use the continue command, which traces breadth-first, to see whether the goal succeeds or fails.

```
    1 1[1] call nqueens(4, Qs)
[opium]: continue.
    2746 1[1] fail nqueens(4, Qs)
```

The computation fails without producing any solution. In Opium there is an extension which enables users to track the leaf failures in a structured way. It is an example of abstract views:

```
[opium]: leaf_failure_tracking(1).
    1 1[1] call nqueens(4, Qs)
      56 23[2] call safe([1, 2, 3, 4])
     106 23[2] fail safe([1, 2, 3, 4])
                    subgoal failures abstracted (24 altogether).
    2643 697[2] call safe([4, 3, 2, 1])
    2689 697[2] fail safe([4, 3, 2, 1])
    2746 1[1] fail nqueens(4, Qs)
```

We can see that all the 24 invocations of **safe/1** fail whereas two should succeed.
Note that it was important to abstract away these failures. Otherwise there would
have been 50 lines traced in one go. It would have been too much information for
a user to grasp in one glance, it would not therefore be an abstract view. In order
to examine the failures of **safe/1** we can use the trace query language previously
mentioned. We can set a breakpoint on **safe/1** using the **spy** command.

```
[opium]: spy safe/1.
```

We go back to the first line of the traced execution using goto, then the **leap**
command retrieves and prints lines related to spied points.

```
[opium]: goto(1).
    1 1[1] call nqueens(4, Qs)
[opium]: leap.
      56 23[2] call safe([1, 2, 3, 4])
[opium]: leap.
      57 23[2] unify safe([1, 2, 3, 4])
```

At this point we can realize that a simple breakpoint is not precise enough. What
we really wanted to see were the *failures* of **safe/1** and not all the lines related to it.
We can refine the query by adding a condition after the leap. We then use **leap_np**
which leaps to the next breakpoint but does not print the line. A line is only printed
when it is a failure.

```
[opium]: leap_np, curr_port(fail), print_line.
     104 25[4] fail safe([3, 4])
```

This query still is not precise enough, we are only interested in the toplevel invocations of **safe/1** (ie at depth 2). We can add this new condition to the query. For
the repeated commands we use their abbreviations. Note that an experienced user
would have specified this query in one step but we wanted to show how progressive
the use of the query language can be.

```
[opium]: l_np, c_port(fail), curr_depth(2), p.
     106 23[2] fail safe([1, 2, 3, 4])
```

In order to determine more easily whether the failure of **safe/1** is correct or not,
we can use a program which displays the 4x4 board. We first retrieve the value of
the arguments using the **curr_arg** primitive and display them in an adapted way
using the **show_queens** predicate. This predicate is an example of tracing facilities
dedicated to a particular application.

[opium]: curr_arg([X]), show_queens(X).

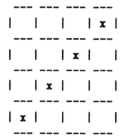

```
--- --- --- ---
|   |   |   | x |
--- --- --- ---
|   |   | x |   |
--- --- --- ---
|   | x |   |   |
--- --- --- ---
| x |   |   |   |
--- --- --- ---
```

X = [1, 2, 3, 4]

All the four positions are aligned, this failure is valid. We can now retrieve the
next failure of **safe/1** at toplevel, displaying the board as previously, by simply
concatenating the two previous queries.

[opium]: l_np,c_port(fail),c_depth(2),p, c_arg([X]), show_queens(X).
    328 67[2] fail safe([1, 3, 2, 4])

```
--- --- --- ---
|   |   |   | x |
--- --- --- ---
|   | x |   |   |
--- --- --- ---
|   |   | x |   |
--- --- --- ---
| x |   |   |   |
--- --- --- ---
```

X = [1, 3, 2, 4]   More? (;)

This failure is also valid. Using the usual toplevel backtracking facility of Prolog we
can ask Opium to find the next appropriate line by simply typing ";".

    450 95[2] fail safe([1, 3, 4, 2])

```
--- --- --- ---
|   |   | x |   |
--- --- --- ---
|   | x |   |   |
--- --- --- ---
|   |   |   | x |
--- --- --- ---
| x |   |   |   |
--- --- --- ---
```

X = [1, 3, 4, 2]   More? (;)

```
553 127[2] fail safe([1, 4, 2, 3])
```

```
--- --- --- ---
|   | x |   |   |
--- --- --- ---
|   |   |   | x |
--- --- --- ---
|   |   | x |   |
--- --- --- ---
| x |   |   |   |
--- --- --- ---
```

```
X = [1, 4, 2, 3]      More? (;)
```

The previous failures are also valid. It can become tedious to examine the 24 failures.
If we are testing the program we know that **safe/1** should succeed for [2,4,1,3]
and [3,1,4,2]. We can directly ask Opium to check whether **safe/1** would not fail
for these values. In such a case printing the full line is not even needed, showing the
board is enough.

```
[opium]: l_np, c_port(fail), c_depth(2), c_arg([X]),
         (X == [2, 4, 1, 3] ; X == [3, 1, 4, 2]),
         show_queens(X).
```

```
--- --- --- ---
|   | x |   |   |
--- --- --- ---
|   |   |   | x |
--- --- --- ---
| x |   |   |   |
--- --- --- ---
|   |   | x |   |
--- --- --- ---
```

```
X == [2, 4, 1, 3]      More? (;)
```

The **safe/1** predicate indeed fails for this, actually safe, position. We can use the
leaf failure tracking to understand why. We need the invocation number of this goal
which can be retrieved using **curr_call**.

```
[opium]: curr_call(C), lft(C).
```

```
1282 320[3] call not attack(2, [4, 1, 3])
1335 320[3] fail not attack(2, [4, 1, 3])
```

This is the end of the failing path.

The execution fails, because **not attack(2, [4, 1, 3])** fails, i.e. because **attack(2,
[4, 1, 3])** succeeds. The leaf failure tracking stops because the symptom changes.
Indeed, the problem is no longer that a predicate fails whereas it should succeed, but
the opposite. The **attack(2, [4, 1, 3])** goal should not succeed as can be seen
from the previously displayed board. Here we can either list the code of **attack**:

```
[opium]: listing(attack).
    attack(X, Xs) :-
            attack(X, 1, Xs).

    attack(X, N, [Y|_Ys]) :-
            X is Y + N.
    attack(X, N, [Y|_Ys]) :-
            X is Y - N.
    attack(X, N, [_Y|Ys]) :-
            N1 is N - 1,
            attack(X, N1, Ys).
```

and see that the recursive call of attack/3 is badly constructed, it should be N1 is N + 1 and not N1 is N - 1. Or we can imagine that the programmer has not understood enough yet and wants to trace further the execution of attack/2. We can go to the call of attack/2 (therefore skipping the details of execution of not), and there zoom into its execution.

```
[opium]: f_get_np(_,_,_, call, attack/2), c_arg([2, [4,1,3]]), zoom.
    1286  322[5]  call attack(2, [4, 1, 3])
        1288  323[6]  call attack(2, 1, [4, 1, 3])
        1318  323[6]  exit attack(2, 1, [4, 1, 3])
```

The latter goal should not succeed. As we know that the predicate has several clauses we can ask which one has been used to produce the solution:

```
[opium]: unif_clause.
    attack(X, N, [_Y|Ys]) :-
            N1 is N - 1,
            attack(X, N1, Ys).
```

Here again the error could be detected, but it is more obvious if we zoom into the execution of attack(2, 1, [4, 1, 3]).

```
[opium]: zoom.
    1296  323[6]  next attack(2, 1, [4, 1, 3])
        1298  326[7]  call N1 is 1 - 1
        1299  326[7]  exit 0 is 1 - 1
```

It seems valid that one minus one is equal to zero. We ask to trace further at this level using the continue command.

```
[opium]: continue.
        1300  327[7]  call attack(2, 0, [1, 3])
        1317  327[7]  exit attack(2, 0, [1, 3]
```

Here it should strike the programmer that the second argument should never be 0 and that the recursive clause of attack/3 is thus badly constructed.

4 Related work

Existing practicable programmable tracers try to model the *debugging process* and offer a complicated set of constructs to perform basic debugging actions (see for example [12, 3]). There is no guarantee that these basic actions actually cover the whole debugging process. In Opium we model the *debugging data*, i.e. the execution trace, and program the debugging processing with a general programming language. Some systems also model the debugging data, but they require that an actual trace database containing the whole execution is systematically created (see for example [15, 13]). This is too costly for large executions. In Opium execution traces can be parsed on the fly. The trace database is only used when it is really needed, and under the control of the programmer. Hence, Opium offers the performance of practicable tracers with the power of the "database-oriented" debuggers.

Algorithmic debugging systems (see [16] and its followers) provide algorithms which compare concrete behaviour of a program against its specification. They use the programmer as an oracle when there is no formal specification. The method is algorithmic only insofar as the programmer makes a reliable oracle, indeed if one answer is wrong the result is irrelevant. Some of the graphical debuggers have recognized the need for more flexibility and offer several levels of abstraction [4, 2, 10]. However, so far they provide only a limited and hard coded variety. In Opium any abstract view of the execution can be programmed.

5 Discussion

Opium as such is already helpful for Prolog programmers. It can furthermore be used to support further research on automated debugging. The trace query principles could be reused for other sequential languages. The general approach is also worth noting: we are using Prolog as a nice interface language and we have optimized the few key points. Opium is fully implemented and the prototype is currently being distributed to academic sites together with Sepia, ECRC's Prolog system [14].

Acknowledgements Anna-Maria Emde engineered Opium into its pre-release stage and contributed many ideas. Micha Meier and Joachim Schimpf have helped to port Opium to Sepia. Marc Bourgois gave fruitful comments on an early draft of this paper. Michael Ratcliffe helped with the English.

References

1. D. Bergantz and J. Hassell. Information relationships in Prolog programs: how do programmers comprehend functionality? *International Journal of Man-Machine Studies*, 35:313–328, 1991.
2. M. Brayshaw and M. Eisenstadt. Adding data and procedure abstraction to the Transparent Prolog Machine (TPM). In *Proceedings of 5th Int. Conference and Symposium on Logic Programming*, pages 532–547, Seattle, August 1988.
3. B. Bruegge and P. Hibbard. Generalized path expressions: A high-level debugging mechanism. *The Journal of Systems and Software*, 3:265–276, 1983.

4. A.D. Dewar and J.G. Cleary. Graphical display of complex information within a Prolog debugger. *International Journal of Man-Machine Studies*, 25(5):503–521, November 1986.

5. M. Ducassé. Abstract views of Prolog executions in Opium. In V. Saraswat and K. Ueda, editors, *Proceedings of the International Logic Programming Symposium*, pages 18–32, San Diego, October 1991. ALP, MIT Press.

6. M. Ducassé. Analysis of failing Prolog executions. In *Actes des Journées Francophones sur la Programmation Logique*, Mai 1992. University of Lille, France.

7. M. Ducassé. *An extendable trace analyser to support automated debugging*. PhD thesis, University of Rennes, France, June 1992. Numéro d'ordre 758. *European Doctorate*. In English.

8. M. Ducassé and A.-M. Emde. A high-level debugging environment for Prolog. Opium user's manual. Technical Report TR-LP-60, ECRC, May 1991.

9. M. Ducassé and A.-M. Emde. Opium: a debugging environment for Prolog development and debugging research. *ACM Software Engineering Notes*, 16(1):54–59, 1991.

10. M. Eisenstadt and M. Brayshaw. The Transparent Prolog Machine(TPM): an execution model and graphical debugger for logic programming. *Journal of Logic Programming*, 5(4):277–342, 1988.

11. A.-M. Emde and M. Ducassé. Automated debugging of non-terminating Prolog programs. In *Proceedings of the ICLP'90 Pre-conference Workshop on Logic Programming Environments*, June 1990.

12. M.S. Johnson. Dispel: A run-time debugging language. *Computer languages*, 6:79–94, 1981.

13. C.H. LeDoux and D.S. Parker. Saving traces for Ada debugging. In *Proceedings of the ADA International Conference*, pages 97–108, 1985.

14. M. Meier et al. SEPIA - an extendible Prolog system. In *Proceedings of the IFIP '89*, 1989.

15. M.L. Powell and M.A. Linton. A database model of debugging. In M.S. Johnson, editor, *ACM SIGSOFT/SIGPLAN Software Engineering Symposium on high-level debugging*, pages 67–70. ACM, March 1983.

16. E.Y. Shapiro. *Algorithmic Program Debugging*. MIT Press, Cambridge, MA, 1983.

17. L. Sterling and E. Shapiro. *The Art of Prolog*. MIT Press, Cambridge, Massachusetts, 1986.

A Command descriptions

Here follow the descriptions of the commands which have been used in the previous debugging session. The abbreviations of the commands are in brackets.

continue {c}

Command which is used to trace one level of an execution subtree. It traces forwards to the next line whose depth is equal to or less than the zooming depth. The zooming depth is reset to the current depth as soon as a trace line with a lower depth is met, that is when the subtree is left.

curr_arg(ArgList) {c_arg}

Primitive which gets or checks the value of the "argument" slot of the current trace line.

curr_call(Call) {c_call}

Primitive which gets or checks the value of the "call" slot of the current trace line.

curr_depth(Depth) {*c_depth*}
Primitive which gets or checks the value of the "depth" slot of the current trace line.

curr_port(Port) {*c_port*}
Primitive which gets or checks the value of the "port" slot of the current trace line.

f_get(Chrono, Call, Depth, Port, Pred) {*fg*}
Command which searches forward, *in an optimized way* the *first* line corresponding to the specified slot values, and prints it. For Chrono, Call, and Depth you can specify an exact value, a list of values, or a semi-interval, for example "< 4". Port and Pred may be lists of values, or negated lists, for example "¯[next, fail]".

f_get_np(Chrono, Call, Depth, Port, Pred) {*fg_np*}
Primitive which does the same as command f_get except printing a trace line.

goto(C) {*g*}
Command which moves the trace pointer to the line with chronological number C.

leaf_failure_tracking(GoalNumber) {*lft*}
Command which helps to locate suspicious leaf failures, on an automated top-down zooming way. It works on directly failing goals (ie goals which fail and have never been proved). The algorithm proceeds as follows. lft retrieves all the directly failing subgoals of the reference goal. If there is only one, lft is recursively applied to this subgoal. If there are no failing subgoals or if one of them is a "not", the tracking process is finished and the directly failing subgoal(s) are displayed. If there are several failing subgoals lft displays them, and the user should decide how to go on.

leap {*l*}
Command which prints the next line related to a spypoint and whose port is one of the "traced_ports". To set a spy point use "spy". To see the very next spypoint line use "f_leap".

leap_np {*L_np*}
Primitive which does the same as command leap except printing a trace line.

listing(Pred) {*ls*}
Command which lists the source code of a predicate defined in the traced session. A predicate P/A is listed if it is visible in the current toplevel module. It is also possible to give the module using M:P/A. If the arity is omitted then all predicates with name P are listed. The predicate may be static or dynamic.

print_line {*p*}
Command which prints the current trace line according to the value of the display parameters.

spy(Pred)
Command which adds a spypoint flag on Pred. The spypoints can be traced with the basic commands "f_leap" and "b_leap". See also "leap" and "leap_back" of the step_by_step scenario.

unif_clause
Command which prints the clause unified in the current trace line.

zoom {*z*}
Command which is used to enter the next level of the execution subtree below the current goal. Especially useful at a "quit" line. Shows the most recent "entry" line corresponding to the current goal, increases the zooming depth, and shows the next trace line one level deeper if the current goal has a subtree.

This article was processed using the LaTeX macro package with LLNCS style

Automatic Theorem Proving
within the Portable AI Lab

Fabio Baj and Michael Rosner
IDSIA, Corso Elvezia 36, CH-6900 Lugano

Abstract

The Portable AI Lab[1] is a joint research project concerned with the design and implementation of an integrated environment to support teaching of Artificial Intelligence at University level. The system is made of several modules implementing basic AI techniques in a uniform way. This paper focuses primarily on the module dealing with Automated Theorem Proving (ATP).

Keywords
artificial intelligence, logic, integrated environment.

1 Introduction

Logic provides a set of tools for the analysis, expression and solution of problems that might typically be posed from within the artificial intelligence (AI) paradigm. Although opinions differ as to whether these tools are appropriate or optimal for the study of intelligence, there is little doubt that AI problems are a good testing ground for the logicist approach in general, and automated theorem proving devices in particular.

The work described in this paper stems from the assumption that typical AI problems involve the intersection of several subdomains – one of which happens to be automated theorem proving. The AI practitioner who has acquired a significant level of expertise will understand not only the theory and practice of this subdomain, but also the extent to which the techniques within it apply to a given class of problems, as well as the way in which, in solutions to a broader class of problems, they might relate to techniques in other subdomains.

The system described in the next section is the result of incorporating this assumption into the design of a piece of software called the Portable AI Lab that is primarily intended to support AI teaching activities.

---

[1] The system is being developed under Swiss National Research Programme PNR 23 on AI and Robotics by IDSIA, Lugano in collaboration with the Institut für Informatik, University of Zürich and the Laboratoire d'Intelligence Artificielle at the École Polytechnique Fédérale, Lausanne.

2 The Portable AI Lab

The Portable AI Lab is a computing environment containing a collection of state-of-the-art AI tools, examples, and documentation. It is aimed at those involved in AI courses (i.e. in both teaching and learning) at university level or equivalent.

The design of Portable AI Lab is motivated by the conviction that the acquisition of expertise in AI depends on extensive practical experience with a broad range of AI problems. It is important that the student come to appreciate the typically *interdisciplinary* nature of such problems: they often involve more than one domain. The system has enough built-in functionality to enable its users to get such experience without having to build all the supporting tools from scratch.

This functionality is provided though a number of modules covering different AI subdomains domains and associated techniques. The system currently consists of prototype versions of modules concerned with automated theorem proving, natural language processing, rule-based inference, truth maintenance, learning, knowledge acquisition, genetic algorithms, neural networks, and planning. Each module comprises

- a fully implemented computational theory over the domain in question;

- a set of autonomous demonstrations that illustrate the operation of the module;

- a representative set of example problems to which the computational theory can be applied;

- extensive documentation that includes a literature guide, a user's guide, and a programmer's guide.

The whole system is implemented in Common Lisp + Common Windows in a a uniform style. Together with a *pool* facility for sharing data, this is designed to encourage the user to design and implement interfaces between different modules. The system thus intended to support teaching activities by providing enough material to encourage the exploration of domains and the relationships between them.

A full discussion of the system is beyond the scope of both the article and the proceedings. Below therefore, we concentrate on the goals, structure and content of the module devoted to automatic theorem proving (ATP). The reader should see [18] for details of how the system might be used to investigate relationships between logic and language. Allemang et. al [1] illustrates some other interdisciplinary problems that are within the scope of the system.

3 The Automated Theorem Proving Module

The importance of having a theorem prover as a module of the system is strictly related to the importance of logic in AI. The "Logic approach to AI" (a term coined by Lifschitz [11] according to which study of the deductive relations between propositions will yield or provide inspiration for a theory of mind) has no shortage of either protagonists (eg McCarthy [13], Hayes [7], Israel [10], Moore [16], Genesereth and Nilsson [4] and others) or antagonists (eg McDermott [14], Minsky [15] and Wilks [21]).

In this article we will not be concerned with the arguments advanced from either camp, since the decision to include a module devoted to automatic theorem proving (ATP) in Portable AI Lab is based on what we hope is a modest claim: that didactically speaking, an operational ATP is, for several reasons, a tool of immense value to students and teachers of AI. A few such reasons are as follows:

- An important part of AI expertise consists of the ability to represent knowledge and problems in a formal language. First order logic is a good example of a such a language, and is the lingua franca of ATPs.

- Unlike many experimental notations, the model-theoretic semantics of first order logic is well understood and immutable. These are useful properties when learning about formalisation (even if the models themselves are inadequate for many knowledge representation tasks).

- An operational ATP is useful for evaluating a formalisation by providing a set of behaviours that can often suggest the way in which it should be modified in order to achieve better results.

- ATPs are interesting as problem solving devices in their own right. They are generally useful for coping with all those problems that can be expressed in first order logic: mathematics, natural language analysis, program validation and synthesis, circuit design and verification, planning, and so on.

- The use of logic as a formalisation tool is often confused with logic programming, and logic programming is often confused with Prolog. An ATP can, when suitably designed, be used to bring out these distinctions and reveal the relationships between logic, control and programming.

- The integration of ATPs with other AI modules provides a way to investigate the relationships between deductive reasoning and other aspects of intelligent behavior.

We expect the module to be of interest to users with different degrees of experience. For novices without specific knowledge of either symbolic logic or automated theorem proving, we would expect the autonomous demonstrations to provide an initial route to the workings of the prover. Subsequently the system might be invoked manually on some of the example problems provided.

A student more familiar with the basic concepts may investigate such particular subfields as term rewriting, paramodulation, demodulation and so on. Normally he will start using the tool on example problems, interacting with the prover by means of the graphic interface, experimenting with different proof strategies, selecting combinations of inference rules, output format etc. Later on he may want to to modify examples and their solutions.

The advanced user will have acquired some expertise in the use of an ATP system, will understand how the basic algorithms work, and might be interested in reusing pieces of source code, or in connecting the ATP module with the others provided by the Portable AI Lab. For instance he may want to export Prolog proof trees to the Explanation Based Generalization module, or to import rules from the Inductive Learning module, or to connect the theorem prover with the Natural Language module.

3.1 Overview

The module, called LENprover, provides a significant range of well-known techniques for theorem proving including resolution [17], paramodulation [22] term rewriting [9] [8], equational reasoning [3], semantic attachment [5] [20], question answering [6], and Prolog [19]. The system is a refutation based theorem prover for first order logic with equality working on the clausal representation of formulas as follows:

- The input language is full first order logic, with quantifiers and connectives. The user may define infix or postfix functors. The conventional Prolog notation for clauses is also recognized. The system reads a theorem contained in a text file, making some preprocessing steps (for instance tranforming first order formulas into clauses).

- Starting from the initial database of clauses, the main loop chooses a pair of clauses according to some criterion and applies an inference rule to them (from the ones selected by the user) to deduce a new clause which is added to the database. When a new clause is generated, some operations are performed to simplify the clause database (subsumption, simplification, demodulation, and semantic simplification), according to the options selected by the user.

- The general theorem proving algorithm used by the ATP module is summarised in figure 1. This generative process terminates when a contradiction (or an answering clause) has been found or no more deductions are possible.

The behaviour of this process is affected by a number of user-settable parameters as indicated in figure 2.

The following sections describe the main features provided by the ATP tool.

3.2 Question Answering

The use of a theorem prover as a question answering system was introduced by C. Green in 1969 [6]. The ATP module provides this classical approach to question answering. A simple theorem prover is itself a question answerer, but

```
Prove(F)
  CL := Clauses_of(F)
  Repeat
    choose clauses c1 and c2 from CL
    new-c  := an-inference-rule(c1,c2)
    new-c1 := simplify new-c with CL
    CL     := simplify CL with  new-c1
  until Empty-Clause is found OR
        no more deductions are possible
  IF Empty-Clause is found THEN
     return ''contradiction''
  ELSE return ''consistent''
```

Figure 1: The theorem-proving algorithm

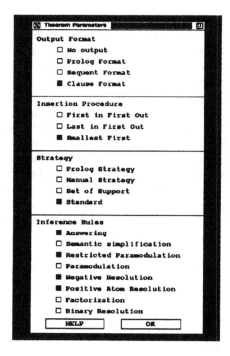

Figure 2: User settable parameters in the ATP module

it is limited to stating whether a given fact is true or false. If we want to answer questions like "which individual has property P?" some extensions to the prover are needed. The usual way is to express such a question as a theorem of the form $\exists x P(x)$, and to trace the substitutions occurring to the variable x along the proof. These substitutions are recorded with a special marker (the *answering literal* [2]), and alternative answers can be computed. This mechanism is also used to provide answers when the prover is in Prolog mode. Here are examples of how the user may ask questions to the prover:

```
all x ( fact(10, x) | $ans(x))
all x (fact(10, x) => ?(x))
:- fact(10, X).
```

3.3   Semantic Attachment

Since automated theorem proving consists in the manipulation of linguistic expressions according to some rules, no reference to the meaning of symbols is made during computations. Nevertheless these linguistic constructs generally denote objects, functions and relations in a particular domain, and there are cases in which the intended interpretation of symbols is a standard and computable one [5]. For instance, if we are using the predicate symbol $>$ with the intended meaning of *being greater than*, then we might expect the prover to consider as true an expression like $5 > 3$, simply *evaluating* it, instead of *demonstrating* it or adding to the axioms set facts like $(5 > 3)$ for all numbers. The evaluation of the expression $(5 > 3)$, can be done with a call to a computer program that compares numbers within the *finite* domain of the program. Let us call this program **greaterp**: when the linguistic expression has to be evaluated we associate with the symbols 5 and 3 the computer objects 5 and 3, and with the symbol $>$ the computer program **greaterp**, then we run **greaterp** on inputs 5 and 3. For example, consider the two formulas:

```
all x all y all z   ((x + y) > (z - x)) => P((z + y)) | Q(x,y,z).
~Q(1,2,3).
```

Normally the ATP module will deduce

```
((1 + 2) > (3 - 1)) => P((2 + 3))
```

but if the user selects the option of semantic attachment the prover will derive

```
P(5).
```

Notice that we need to precisely define the domain of the attached functions. For instance the prover must call the program **plus** attached to $+$ only when both the arguments of $+$ are numerals; expressions like $(x + 3)$ must not be evaluated. Semantic evaluation of formulas is a process of interpretation with respect to a partial model [20] which in some cases can be the underlying Lisp interpreter. The possibility of semantic simplification of complex linguistic expressions is of great importance in improving the efficiency of theorem provers. It is also a powerful way to interface declarative and procedural knowledge in complex AI systems like PAIL. For instance we can combine the symbolic reasoning features of a planner for the blocks world with a computable description of its physical characteristics, like stability, gravity.

3.4 Prolog

Since a Prolog interpreter can be seen as a theorem prover which uses only a particular inference rule (SLD resolution [12]), a particular search strategy, and works on a restricted class of formulas (Horn clauses), it is interesting to convey to the user how a general theorem prover can be adapted to operate exactly as a Prolog system. Of course, it is equally important to understand which class of problems can be succesfully confronted with Horn clauses rather than with full first order logic.

The module provides the standard *cut* operation for reducing the search space of a query, and for implementing *negation as failure.* Implementing the cut does not affect in any way the general structure of the theorem proving algorithm (it can be seen as an evaluable predicate whose semantics is to modify the goal stack): it was included primarily to significantly extend the class of Prolog programs that the module might handle. The traditional Prolog notation for lists is recognised, but not predicates such as assert, retract, clause ... which lie outside the domain of pure theorem proving activity.

Figure 3 shows the module at work on a Prolog theorem.

3.5 Syntax

Amongst the problems faced by the student of computational logic is the variety of syntactic conventions for representing clauses. The system therefore allows the selection of different output formats for clauses for establishing the equivalences between

Prolog syntax: a(X,Y) :- b(X), c(Y).

Clausal format: a(x,y) | b(x) | c(y).

Sequent format: b(x) & c(x) => a(x,y)

as three equivalent ways of expressing the same clause.

3.5.1 Inference rule and Strategy

The SLD resolution inference rule [12] is simply one of the available inference rules, just as the linear input depth first search strategy is only one of the available strategies: their combination results in a Prolog system, but the user can

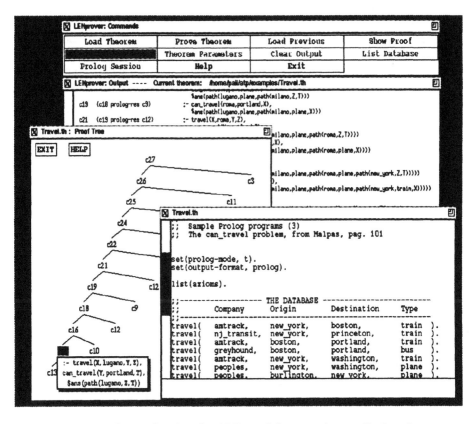

Figure 3: A screendump showing the ATP module at work on a Prolog theorem

play with every other combination. Running the prover in Prolog mode a user can see in detail the process of goal matching, and the graphic display of proof trees gives a clear idea of the Prolog search strategy as shown in figure 3. But the general theorem proving mechanism is open to more complex experiments: let us consider for example the classical left recursive Prolog program:

```
ancestor(X,Y) :-  ancestor(Z,Y),
                   parent(X,Z).
ancestor(X,Y) :-  parent(X,Y)
```

Running this program with the Prolog strategy the user will observe an infinite growth of the database of deductions. What he can do at this point is select a different search strategy (for instance with a breadth first component): in this way the prover will find the solution in few steps.

From this kind of experiment a student can learn that sometimes a logically correct description may be useless if the proof method used is not appropriate: in particular, he cannot write Prolog programs by simply describing *what* to perform: he do has to consider *how* the program will be executed. On the other hand the ATP module also provides a set of examples that are almost intractable with a general theorem proving system, but which became straightforward if the Prolog strategy is selected.

3.5.2   Answering mechanism and semantic evaluation in Prolog mode

This section shows how two general features of the ATP module, question answering and semantic attachment, match with Prolog. Standard Prolog systems implement basic arithmetic with the special predicate **is**, which explicitly indicates a request of semantic evaluation. This mechanism fits quite naturally with the one provided by LENprover. Furthermore the user is not committed to use the **is** predicate to do arithmetic: the following definitions of factorial are both legal in LENprover, and provide the same answers:

```
fact(0,1).
fact(X,Y) :- Z is (X - 1),
             fact(Z, FZ),
             W is (X * FZ).

fact(0,1).
fact(X, (X * FX1)) :- fact((X - 1), FX1).
```

Notice that a normal Prolog interpreter signals an error if the arguments of arithmetic functors in an **is** statement are not numbers. Compatibility with programs using the **is** predicate can be attained simply adding the clause

```
(X is Y) :- (X = Y).
```

Here are some possible ways to specify the *greatest common divisor* function to the ATP module: they are equivalent in the sense that they all provide the same answers.

```
gcd(X,0,X)    :- !.
gcd(X,Y,Res) :-  (X>Y),!,
                 gcd(Y,(X - Y),Res).
gcd(Y,X,Res) :-  gcd(Y,(X - Y),Res).

:- gcd(218,478,X).
------------------------------------------------
all x all y(
   gcd(0,y,y) & gcd(x,0,x) &
   (gcd(y,(x - y),xr) & ~(y > x) => gcd(y,x,xr)) &
   (gcd(y,(x - y),xr) & (x > y)  => gcd(x,y,xr))).

gcd(218,478,x) => ?(x).
------------------------------------------------
(gcd(x,y) = ($IF($EQ(x,0), y,
$IF($EQ(y,0), x,
                      $IF((x<y), gcd(x,(y-x)),
                          gcd(y,(x-y)))))).
?(gcd(218,478)).
------------------------------------------------
```

The first definition works with a Prolog strategy, the second can with a generic theorem proving strategy (plus semantic attachment). The last formulation shows some features of functional programming: since this prover is able to handle the equality predicate, equations can be used to define computable functions. The evaluable predicate $IF allows the definition of conditional expressions.

3.6   Further documentation and availability of the system

In addition to the usual documentation provided for Portable AI Lab modules, LENprover includes an extended set of commented working examples which also provide additional information about these concepts in the form of comment lines intended to provide a useful guide to the user who wants to write his own theorems. An additional tutorial document, entitled **Logic with LENprover** introduces symbolic logic and theorem proving through the use of the ATP tool.

Portable AI Lab is distributed free of charge in source form. Readers who are interested in obtaining a copy are asked to contact one of the authors.

References

[1] D. Allemang, R. Aiken, N. Almassy, T. Wehrle, and T. Rothenfluh. Teaching machine learning principles with the portable ai lab. In *Proceedings of CALISCE, EPFL, Lausanne, Switzerland*, 1991.

[2] Chung-Liang Chang and Richard Char-Thung Lee. *Symbolic logic and Mechanical Theorem Proving*. Academic Press, 1971.

[3] Knuth D. E. and Bendix P. Simple word problems in universal algebras. In *Proceedings of the conference: Computational problems in Abstract Algebras*, pages 263–298. Pergamon Press, 1970.

[4] Michael R. Genesereth and Nils J. Nilsson. *Logic Foundations of Artificial Intelligence*. Morgan Kaufmann, Los Altos, CA, 1987.

[5] Michael R. Genesereth and Nils J. Nilsson. *Logic Foundations of Artificial Intelligence*. Morgan Kaufmann, Los Altos, CA, 1987.

[6] C. Green. Theorem-Proving by Resolution as a Basis in Question-Answering Systems. In B. Meltzer and D. Michie, editors, *Machine intelligence 4*, pages 183–205, Edinburgh, UK, 1969. Edinburgh University Press.

[7] Patrick J. Hayes. In defence of logic. In *Proceedings of the International Joint Conference in Artificial Intelligence, Cambridge Massachussetts*, pages 559–565, 1977.

[8] Jeh Hsiang. Refutational theorem proving using term rewriting systems. *Artificial Intelligence*, 3(25):225–230, 1985.

[9] Jeh Hsiang. Rewrite method for theorem proving in first order logic with equality. *Journal of Symbolic Computation*, 1(3):133–151, 1987.

[10] D. Israel. The role of logic in knowledge representation. *IEEE Computation*, 16(10):37–42, 1985.

[11] V. Lifschitz. The logic approach to AI. *Stanford Computer Science Video Journal*, 1987.

[12] J. W. Lloyd. *Foundations of Logic Programming, Second , Extended Edition*. Springer-Verlag, New York, 1987.

[13] J. McCarthy. Programs with common sense. In R. J. Brachman and H. J. Levesque, editors, *Readings in Knowledge Representation*. Kaufmann, Los Altos, CA, 1958.

[14] D. McDermott. A critique of pure reason. *Computational Intelligence*, 1987.

[15] Marvin Minsky. *The society of mind*. Simon and Schuster, New York, 1986.

[16] R.C. Moore. The role of logic in artificial intelligence. In I. Benson, editor, *Intelligent machinery: theory and practice*. Cambridge University Press, Cambridge, UK, 1986.

[17] J. A. Robinson. A machine-oriented logic based on the resolution principle. *Journal of the ACM*, 1(12):23–41, 1965.

[18] M. Rosner and F. Baj. Portable ai lab for teaching artificial intelligence. In F. Lovis, editor, *Proceedings of the IFIP Working Conference on Informatics at university Level*. Elsevier, forthcoming.

[19] Clocksin W. and Mellish C. *Programming in Prolog*. Springer Verlag, 1981.

[20] R Weyhrauch. Prolegomena to a theory of mechanized formal reasoning. *Artificial Intelligence*, 1(13):133–170, 1980.

[21] Y. Wilks. Form and content in semantics. In M. Rosner and R. Johnson, editors, *Computational Linguistics and Formal Semantics*. Cambridge University Press, Cambridge, 1992.

[22] L. Wos and G. A. Robinson. Paramodulation and set of support. In *Symposium on Automatic Demonstration*, pages 276–310. Springer-Verlag, 1970.

Lecture Notes in Artificial Intelligence (LNAI)

Lecture Notes in Computer Science